THE ROLE OF ACTIVE ELEMENTS IN THE OXIDATION BEHAVIOUR OF HIGH TEMPERATURE METALS AND ALLOYS

Proceedings of the European Colloquium organised by:

Commission of the European Communities
Directorate General: Science, Research and Development

and held at the Institute of Advanced Materials, Joint Research Centre,
Petten Establishment, Petten, (N.H.), The Netherlands, on 12-13 December 1988.

THE ROLE OF ACTIVE ELEMENTS IN THE OXIDATION BEHAVIOUR OF HIGH TEMPERATURE METALS AND ALLOYS

Edited by

E. LANG

CEC Joint Research Centre, Institute of Advanced Materials,
Petten, The Netherlands

ELSEVIER APPLIED SCIENCE
LONDON and NEW YORK

ELSEVIER SCIENCE PUBLISHERS LTD
Crown House, Linton Road, Barking, Essex IG11 8JU, England

Sole Distributor in the USA and Canada
ELSEVIER SCIENCE PUBLISHING CO., INC.
655 Avenue of the Americas, New York, NY 10010, USA

WITH 19 TABLES AND 209 ILLUSTRATIONS

Softcover reprint of the hardcover 1st edition 1989

British Library Cataloguing in Publication Data

The role of active elements in the oxidation behaviour
of high temperature metals and alloys.
1. Metals. High temperature oxidation
I. Lang, E.
546'.3

ISBN-13: 978-94-010-7009-6 e-ISBN-13: 978-94-009-1147-5
DOI: 10.1007/978-94-009-1147-5

Library of Congress CIP data applied for

Publication arrangements by Commission of the European Communities, Directorate-General Telecommunications, Information Industries and Innovation, Scientific and Technical Communication Unit, Luxembourg

EUR 12327

Preface

Metals and alloys rely for their application at high temperature on the formation and retention of oxide scales, which act as a barrier between the metallic substrate and the reactive species in the environment such as O, S, N, C, Cl, etc. This protection concept requires that the oxide grows slowly, develops a dense, uniform layer, is well adherent, has sufficient ductility to accommodate plastic deformation of the substrate and is resistant to thermal cycling.

For many years it has been known that small concentrations of certain "active elements" such as Y and the rare earths, as well as carbon and sulphur, can exercise a significant influence upon the oxidation/corrosion behaviour of high temperature metals and alloys. An increasing number of experimental studies on this topic have been published recently. However no generally accepted understanding with regard to the detailed mechanisms and the way in which alloy composition and structure, temperature and environmental conditions, etc., are interacting has yet been achieved, although many - often controversial - theories and ideas have been presented.

It therefore seemed to be an appropriate time to bring together a group of experts to review and evaluate the current state-of-the-art and to discuss various aspects of this important topic.

OBJECTIVES of this Colloquium were:

- To provide a forum for the presentation and discussion of an up-to-date and comprehensive evaluation of recent progress and unsolved problems regarding the role of active elements in the oxidation and corrosion behaviour of HT-metals and alloys.
- To bring together experts and other scientists with interests in this topic in order to facilitate and encourage mutual exchange of knowledge and experiences.
- To aid in the identification of research needs and the orientation of future studies.

The $1^1/_2$ days Colloquium was based upon a number of keynote papers by invited speakers on the different aspects of the subject. Further short communications relevant to the respective topics were presented and poster contributions on original and unpublished material were displayed. This structure is retained in the proceedings.

The Editor.

Programme Committee:

- M.J. Bennett : UKAEA Harwell, U.K.
- A.M. Huntz : Université de Paris-Sud, F.
- E. Lang : C.E.C. JRC Petten, Establishment, Petten, NL.
- M. Merz : C.E.C. JRC Petten, Establishment, Petten, NL.

LIST OF CONTENTS

Opening of Colloquium

THE MAGIC OF ACTIVE ELEMENTS

by E.D. HONDROS

Institute of Advanced Materials
Commission of the European Communities

It is a pleasure to welcome you to the Institute of Advanced Materials of the Joint Research Centre and to this Colloquium on a well targetted and fascinating subject area.

It is indeed an intriguing theme with ramifications throughout the materials sciences and technologies. In fact, as early as 1781, Bergman a Swedish metallurgist observed closely the dramatic effects of small additions of carbon on the properties of iron. These observations, now fundamental knowledge in modern ferrous metallurgy, rather perplexed him. He described the phenomenon in terms of astonishment, as "the mystery of mysteries".

I mention this historical anecdote in order to illustrate a class of phenomena encountered in metallurgical practice in which the presence of a minor amount of an elemental constituent can induce changes in properties - physical, mechanical, chemical - quite out of proportion to the amount added. Today we know that various residual elements, or deliberate additives, or dopants may procure such effects, changing significantly the materials properties. This is seen most spectacularly in connection with electronic properties of semiconductors. To those whose knowledge of metallurgy was in its infancy, such phenomena might well have represented truly magical effects, of the type sought after by alchemists of the middle ages.

This phenomenon is being looked at today in the context of alloys which must perform at high temperatures in agressive operational conditions of service and which depend on their performance on the formation and the retention of a highly

adherent oxide scale. The "active element effect", which is the subject of our business today, is concerned with how certain additives, in particular yttrium, help to retain the scale that is developed and prevent early spalling or breakaway oxidation.

To digress for a moment into other fields, the effects of trace elements have now been well documented and studied in a number of branches of modern materials sciences. The effects can be both negative or positive, that is, deleterious or beneficial. In fact, some of the best known phenomena involving residual trace elements or impurities produce clearly unwanted effects, for example, in the classical phenomenon of temper embrittlement which had catastrophic results in the manufacturing of large guns during the First World War (perhaps from another point of view a beneficial effect). As a result of the growing understanding of microchemical processes at grain boundaries and surfaces, we now have a good understanding of the deleterious effects of naturally occuring residuals in metallurgical phenomena such as temper embrittlement or creep embrittlement.

Let us mention some of the positive effects of minor constituents. Here I do not have to mention the most famous example of all, namely carbon in iron. It is known for instance that boron in steel can procure a distinct toughening effect as can for example, the addition of rare earth elements. The latter, it is suspected, may procure the effect through a sort of gettering action. For example, additions of lanthanum and cerium will scavenge phosphorus or sulphur in embrittled steel. More recently, it has been discovered that boron has a beneficial effect on intermetallics, producing an improvement in toughness through the segregation of the boron at the grain boundary where the adhesion of the grains is increased. It has been said that this recent observation of the boron ductilising effect may have triggered off the present day re-examination of the whole field of inter-metallics.

The effect of active elements such as yttrium on the stability and the integrity of oxide scales represents another important example of the positive effects of additives. This meeting aims to examine how this effect is procured. What is the mechanism of this remarkable phenomenon?

There have been many speculations and publications in the past two decades and a variety of mechanisms has been proposed. These will be examined in detail during this meeting. They may be divided into two broad classes: that in which it is supposed that the effect comes via the metal/oxide interface, and secondly, that where it is supposed that the mechanism is due to the yttrium acting on the bulk scale itself. Interfacial effects may include the elimination of porosity between the metal and the oxide scale. Other ideas relate to the possibility of formation of intrusions at the interface, the "oxide pegging" effect which may produce a better mechanical bond at the interface. Another possibility is that the yttrium may reside at the metal/oxide interface and improve the intrinsic adhesion between metal and oxide. Indeed, some time ago some colleagues and I observed experimentally that yttrium does concentrate at metal/oxide interfaces in certain super-alloys. Considering the second type of possibility, the active element may affect the plastic properties of the scale, or it may affect the growth mechanism by provoking short circuit diffusion especially along the grain boundaries. In a more recent hypothesis, it is supposed that the active element or trace element acts as a sort of getter for the presence of sulphur impurities, which reduce the intrinsically high bond strength of the metal/oxide interface. Indeed, one may have sympathy for this idea, through analogies with other metallurgical phenomena already alluded to, in which the role of the additive is to scavenge the noxious impurities. The deleterious effects of sulphur on bulk properties of steels are well known and over the years this effect has been controlled by metallurgists through a calculated balance with manganese which "fixes" chemically the sulphur.

All these points will be taken up and discussed in detail by the specialists among us and we are looking forward to considering the latest information.

Taking this opportunity, I would like to mention a few words in connection with the Joint Research Centre. It is timely to announce that at the present time the various Establishments of the Joint Research Centre, which have been in existence for some 20 to 30 years, are being restructured (in common with many other similar institutions). The Joint Research Centre, which is one of the instruments used by the European Community to implement its policies in the industrial, social, scientific and technological sphere, is now being transformed into a number of autonomous Institutes. Organisationally, these Institutes operate differently from the cost-shared action type of activity with which you are probably familiar, in which the Community shares in the cost of research projects. The Institutes will cover a number of important technological orientations, in particular Environment, Nuclear Safety, Safety Engineering and in recognition of the importance of materials for the industrial competivity of Europe, there will be one large Institute devoted to Materials. Here, the materials activities in two Centres, Petten in the Netherlands and Ispra in Italy, will be combined in one Institute where programmes will be highly coordinated in a broad range of materials subject areas. The Institute will be based on a total coherence and coordination in the various facilities and specialisms in the regional Centres. In turn, these various resources and activities will be made available to the industry of Europe either on a research contract basis or in the form of closely linked or shared programmes. In this way, we are optimistic in being able to make our contribution to the materials challenge in Europe.

I have taken the opportunity to inform you at this early stage of the planned structure of our organisation. Among other materials themes, the subject matter of the present meeting will be important for the future work of our Institute. I would like to add how we would in the future like to reach out to you, not only in this present Conference, which is a very enjoyable form of dialogue between scientists, but by other means such as the exchange of staff as well as the joining of allied projects in cooperative networks. Cooperation is the keynote of the future: carrying out projects cooperatively will enhance the totality and effectiveness of the European effort.

With these words I reiterate: welcome to this Institute in the fervent expectation that our deliberations on this very interesting theme will be fruitful.

Session 1

INTRODUCTION

PRINCIPLES OF GROWTH AND ADHESION OF OXIDE SCALES

F. H. Stott

Corrosion and Protection Centre
University of Manchester Institute of Science and Technology
P.O. Box 88, Manchester M60 1QD, UK.

ABSTRACT

Some of the important principles which govern the establishment and growth of the protective scales, Cr_2O_3 and Al_2O_3, on high-temperature alloys are considered and discussed. Emphasis is placed on the transient oxidation stages and how these influence the development of the healing layer of the protective oxide. The importance of alloy composition, alloy interdiffusion coefficients and oxygen solubility and diffusivity in the alloy is indicated and correlated with the relative ease of establishment of the protective layers on iron-, nickel- and cobalt-base alloys. It is shown how alloy grain boundaries can assist in the development of Cr_2O_3 healing layers by providing enhanced transport paths for chromium to the surface, particularly for nickel-base alloys where the alloy interdiffusion coefficient is relatively low.

Following establishment of the steady-state scale, its subsequent development is influenced by the transport paths for the reactants, the microstructure of the oxide and the properties of the alloy substrate. The importance of short-circuit diffusion in the growth of Cr_2O_3 and Al_2O_3 scales is discussed and the possible processes by which poorly-adherent, convoluted configurations can develop are considered and correlated with the observed morphologies.

A major cause of failure of high-temperature alloys is cracking and spallation of the Cr_2O_3 or Al_2O_3 scales, particularly under thermal-cycling conditions. The main factors which influence the scale-adhesion characteristics are considered and discussed while possible scale-failure mechanisms are described.

1. INTRODUCTION

Many alloys for use at high temperatures rely on the establishment of either Cr_2O_3 or Al_2O_3 healing layers for protection against oxidation. These oxides are thermodynamically very stable with respect to the metal and have high melting points while transport processes through the scales are generally slow. Hence, they are reasonably effective in this respect. However, both types of scale have major disadvantages in practice which limit their usefulness during long exposure periods. For applications up to about 850°C to 900°C, Cr_2O_3-forming alloys can be very successful. However, at higher temperatures, Cr_2O_3 reacts further with oxygen to form the volatile species, CrO_3. Loss of protective scale by conversion to the higher oxide thus limits the long-term application of such alloys to temperatures below about 900°C. Aluminium does not form such volatile oxide species while transport processes in Al_2O_3 scales are significantly slower even than those in Cr_2O_3 scales. Thus, for applications in oxygen-rich environments at higher temperatures up to 1200°C, metallic components are often designed to develop an Al_2O_3 surface layer.

Unfortunately, Al_2O_3 scales and, to a lesser extent, Cr_2O_3 scales are often very susceptible to mechanical failure. This may involve development of through-thickness cracks normal to the scale-alloy interface, or delamination cracks parallel to the interface. The most severe mode of failure is complete separation of the scale from the metal, known as spallation, which occurs particularly on thermal cycling. This results in loss of the protective scale; healing of the scale requires regrowth of the protective oxide and re-establishment of the complete layer. In some alloys, where the chromium or aluminium concentration is relatively low, repeated spallation depletes the alloy in this element until, eventually, the protective oxide cannot reform. This can lead to rapid breakaway oxidation as less-protective base metal oxides develop. In any case, even if the protective oxide reforms, repeated loss of scale may be life-limiting. In more aggressive environments, such as those of high sulphur or chlorine potential, where the oxide scale may give some protection, spallation can allow access of the gas to the bare alloy surface, with the possibility of catastrophic attack.

The present paper reviews the mechanisms of establishment and growth of Cr_2O_3 and Al_2O_3 scales on high-temperature alloys, with emphasis on the influence of alloy composition and alloy microstructure. Consideration is given to stress-generation and stress-relief processes during oxidation and how these can lead to scale failure and enhanced oxidation.

2. ESTABLISHMENT AND GROWTH OF Cr_2O_3 SCALES

Most research on the development of Cr_2O_3 scales has been concerned with binary M-Cr alloys, where M is Ni, Fe or Co, since these form the basis of most commercial materials. The principles governing the process are generally similar for the three systems and will be discussed here mainly in terms of the Ni-Cr system.

The factors determining the establishment of a healing Cr_2O_3 surface layer on the binary alloys are well documented.[1,2] As Cr_2O_3 and the base metal oxide are stable in 1 atm oxygen or air, the overall development of the scale is influenced by both thermodynamics and reaction kinetics. Chromium forms the more stable oxide while the base metal oxide is less stoichiometric than Cr_2O_3 and grows at a considerably faster rate. Thus, although thermodynamic considerations predict the formation of Cr_2O_3, they do not determine the composition of the initial transient oxide or even of the final steady-state scale. The bulk alloy chromium content determines mainly whether Cr_2O_3 forms as internal-oxide precipitates in the alloy matrix[3,4] or as a complete surface layer. The latter requires 10 to 20 wt% Cr, depending on various factors such as the nature of the base metal.[1] It is favoured by a high alloy interdiffusion coefficient which promotes establishment of a Cr_2O_3 layer by ensuring replenishment of chromium to the surface as the oxide is formed, and low oxygen solubility and diffusivity in the alloy which also enable chromium to diffuse to the surface without being immobilized as internal-oxide precipitates. Other pertinent factors include the growth rate of the base metal oxide and the ease of establishment of the spinel oxide, MCr_2O_4. The latter is associated with the compositional range of the oxide. The Fe–Cr spinel oxide has a wide range and can thus form relatively easily. The Ni–Cr spinel oxide, $NiCr_2O_4$, has a fixed composition and is established only with difficulty.

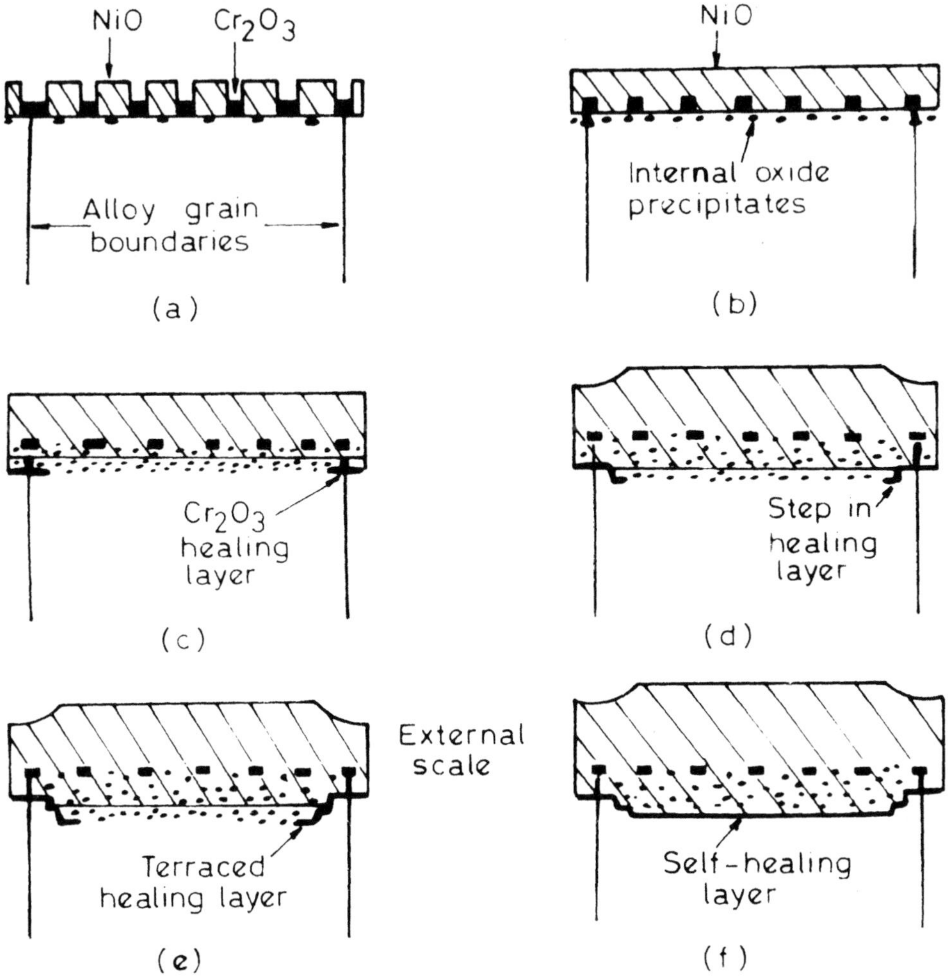

Fig. 1. Schematic diagram showing transient oxidation and progressive development of a healing Cr_2O_3 layer on Ni–20%Cr at 1000 °C.[5]

Overall, a surface layer of Cr_2O_3 is developed the most readily on Fe–Cr alloys, due to a relatively high alloy interdiffusion coefficient and low oxygen solubility and diffusivity in the alloy. The formation of the spinel oxide prevents the very fast-growing FeO from dominating the early, transient oxidation stages. Conversely, it is most difficult to establish such a Cr_2O_3 layer on Co–Cr alloys, due to a less favourable combination of properties, while the Ni–Cr system is intermediate in this respect.

Recent studies have examined the development of the steady-state Cr_2O_3 healing layer on Ni–Cr alloys containing 10 to 20 wt% Cr.[5,6] When the alloy is exposed to 1 atm oxygen at 1000°C, nuclei of all the stable phases, NiO, Cr_2O_3 and, possibly, $NiCr_2O_4$ form on the surface (Fig. 1). In the diagram, $NiCr_2O_4$ nuclei are omitted for clarity. The fast-growing NiO overgrows the other nuclei and a layer of essentially this oxide is developed. As Cr_2O_3 is more stable than NiO, precipitates of the former are able to nucleate at or near the scale/alloy interface by reaction with oxygen which is at the dissociation pressure of the base metal oxide at this location. Eventually, the Cr_2O_3 particles are of sufficiently high population density to exceed the critical volume fraction[7] and coalesce to form a continuous healing layer. The final steady-state scale is an outer NiO layer, an inner NiO layer containing incorporated Cr_2O_3 particles (which react with the NiO to form $NiCr_2O_4$ particles) and a basal layer of Cr_2O_3 (Fig. 2(a)).

However, in alloys containing 10 to 20 wt% Cr, there is insufficient chromium for the healing Cr_2O_3 to form immediately and it is only established following diffusion of chromium from the bulk alloy substrate to the surface to replace that taken up as oxide. It has been shown[5,6] that the healing layer forms initially at the intersection of the alloy grain boundaries with the alloy/scale interface, following enhanced transport of chromium to the surface along these boundaries. Subsequently, the layer penetrates in a stepwise manner across the grains (Fig. 1). At other surface locations, the volume fraction of internal-oxide particles remains below the critical value and the internal-oxidation front progresses into the alloy substrate. The resulting depletion of chromium allows lateral diffusion of chromium from the alloy grain boundaries, to extend the Cr_2O_3 layer laterally, parallel to the surface, following growth and coalescence of the precipitates.

This lateral progress of the healing layer has to incorporate the advancing internal-oxidation front and, hence, a step is formed in the layer. However, the rate of advancement of the front slows down as it penetrates further from the oxygen supply. This enables the Cr_2O_3 layer to progress again in a lateral direction, to form a 'terrace' (Fig. 1). Only two such 'terraces' are shown in the schematic figure, although, in practice, a significant number (typically 10 μm wide by 3 μm deep) form. At the same time, for Ni–20% Cr, the bulk alloy is able to supply sufficient chromium across the width of the internal-oxidation front until the volume fraction of particles can exceed the critical value and a complete self-healing layer is able to develop rapidly (Figs. 1, 2(b) and 2(c)). In alloys containing lower alloy chromium concentrations, this takes much longer to occur and a series of steps can form before the layer is complete, (Fig. 2(d)). Indeed, for Ni–10%Cr, the process takes many hours and the final layer is comprised of a terraced morphology across the full grain width.[5,6]

In the corresponding Fe–Cr alloys, where the alloy interdiffusion coefficient is much higher, grain boundaries have a lesser role in establishment of the healing Cr_2O_3 layer. Here, the transport of chromium in the substrate is high enough to develop rapidly a complete layer, even for alloy concentrations of only 10 to 12 wt%Cr. This results in the presence of significantly less base-metal oxide in the outer scale than for the corresponding Ni–Cr alloys. As discussed earlier, the wide compositional range of the Fe–Cr spinel oxide ensures that this outer scale is the spinel oxide rather than the less protective FeO.

After completion of the Cr_2O layer, the oxidation rate is determined by its effectiveness as a diffusion barrier and its resistance to damage in service. It is generally accepted that Cr_2O_3 scales on chromium and binary M–Cr alloys grow

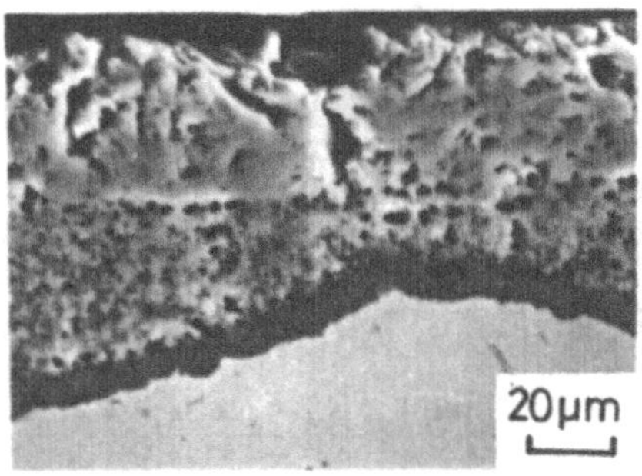

a) Ni–15%Cr, for 24h, cross section of scale.

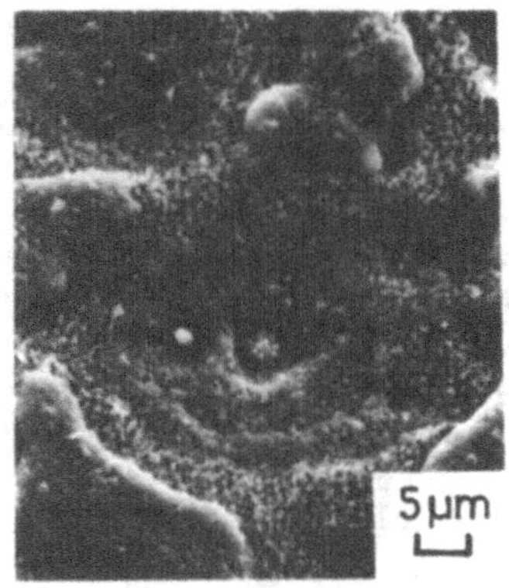

b) Ni–20%Cr, for 20 min, underside of healing layer, before its completion.

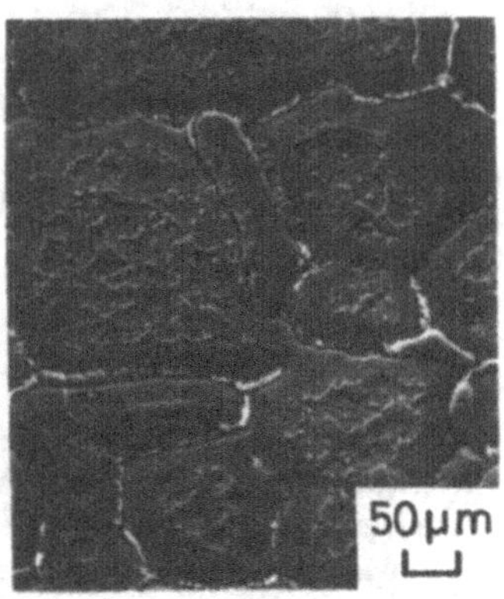

c) Ni–20%Cr, for 30min, underside of healing Cr_2O_3 layer, following completion.

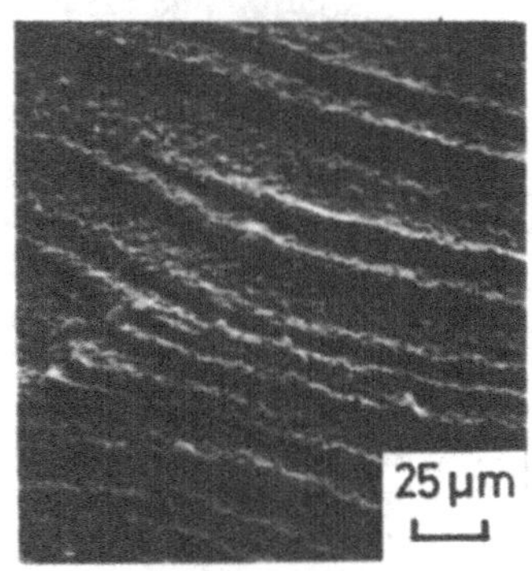

d) Ni–15%Cr, for 1h, detail of healing Cr_2O_3 layer, (under-side).

Fig. 2. Scanning electron micrographs of scale formed on Ni–Cr alloys after oxidation in 1atm oxygen at 1025°C.

following transport of Cr^{3+} ions to the outer surface of the scale and reaction with oxygen (or with the base metal oxides for situations where this oxide is present on the surface). There have been extensive studies of such oxidation processes and assessment of the data has indicated that the process is complex, even for chromium.[8] The growth rate is slow compared to that of the base metal oxides but rate constants at 1000°C varying by as much as four orders of magnitude have been reported. These differences have been attributed to formation of the volatile species, CrO_3, to variations in surface preparation and exposure methods, to the development of convoluted scales, to the purity of the metal and impurity doping of the oxide and to the presence of short-circuit diffusion paths in the scale.

Formation of CrO_3 by reaction of Cr_2O_3 with oxygen is important during oxidation at temperatures above about 900°C in gases of high oxygen pressure since it can result in significant scale thinning by evaporation of the external scale.[9] This can be particularly significant at high gas-flow rates and in the absence of outer scales of the base metal oxide.

It is now widely accepted that, while Cr_2O_3 scales thicken relatively slowly, the growth rates are too rapid to be accounted for by lattice diffusion of Cr^{3+} ions.[10] Although there is no real quantitative evidence from grain-boundary diffusion measurements, qualitative observations are consistent with the importance of short-circuit transport, especially along oxide grain boundaries, in Cr_2O_3 scale growth.[11] This evidence is as follows:

(a) Chromium diffusion coefficients in polycrystalline Cr_2O_3 are much higher than in single crystals[10].
(b) The oxidation rates predicted from measurements in single crystals are over three orders of magnitude less than those observed in practice, for polycrystalline scales, even at 1200°C. Chromium diffusion along dislocations in low-angle grain boundaries cannot account for the observed high rates.[12]
(c) Thin monocrystalline Cr_2O_3 scales grow on some crystals of etched chromium while thick polycrystalline oxides grow on others.[13]
(d) Stress generation, leading to convoluted scales, may result from counter-diffusion along short-circuit paths.[13]
(e) Oxidation of chromium at 800°C is faster than anticipated from extrapolation of rates measured at higher temperatures, while the rate decreases rapidly with time, consistent with a decrease in the number of short-circuit paths as grain growth takes place [14].

Several Cr_2O_3 scale morphologies have been observed as a result of oxidation, ranging from planar scale/alloy interfaces to a detached, convoluted configuration (Fig. 3). Large, oxide-free cavities are often apparent at the scale/alloy interface. Convoluted scales are usually single-layered on alloys,[13,15] but can be multilayered on chromium,[13,14] consistent with repeated cracking and healing at temperature. For many M-Cr alloys, fracture at temperature may result in formation of stratified scales,[16,17] which are much faster growing oxides of the base metal. These scales can establish because the relative alloy interdiffusion coefficients[18] and the diffusion rates of Cr^{3+} ions in Cr_2O_3 are such that selective oxidation leads to considerable depletion of chromium in the substrate at the alloy/scale interface. Failure of the scale exposes this

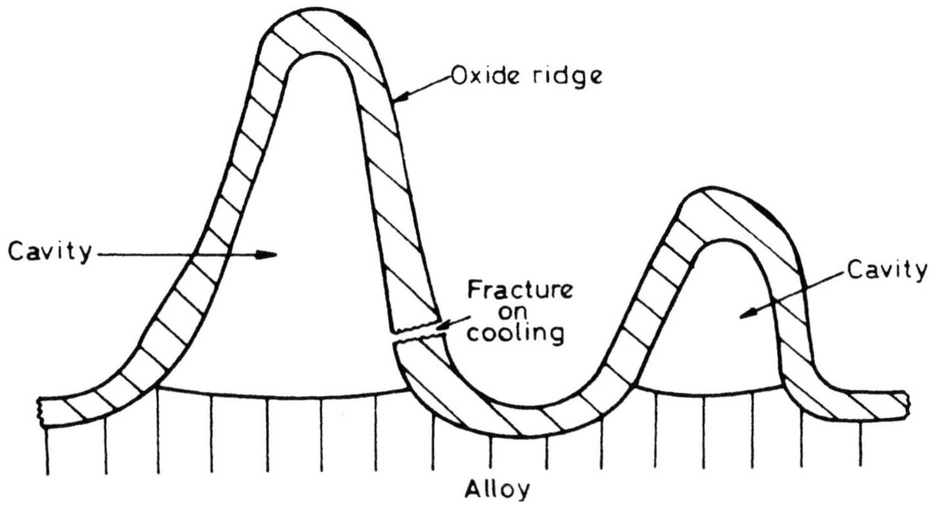

Fig. 3. Schematic representation of a convoluted Cr_2O_3 scale on Fe–28%Cr after oxidation at 1000°C or a convoluted Al_2O_3 scale on Fe–28%Cr–4%Al after oxidation at 1200°C (Transient oxides are omitted for clarity).

chromium–depleted alloy to the environment and the healing Cr_2O_3 layer is unable to be re–established rapidly, allowing the stratified scale to develop. As the scale/alloy interfaces encroaches into the less chromium–depleted regions, the healing layer may eventually be reformed.

The development of convoluted Cr_2O_3 scales has been attributed to the scale–growth process.[13] In particular, fast short–circuit cation diffusion outwards, concurrent with inward diffusion of oxygen along the oxide grain boundaries, results in the formation of new oxide within the grain boundaries of the Cr_2O_3 layer. This causes generation of compressive stresses and wrinkling of the scales by plastic deformation.[13] However, as discussed later for Al_2O_3 scales, lateral growth may also be a factor in determining the convoluted scale morphology.

3. ESTABLISHMENT AND GROWTH OF Al_2O_3 SCALES

The establishment of Al_2O_3 healing layers on binary M–Al alloys at 900°C to 1300°C is similar to that of Cr_2O_3 on the corresponding M–Cr alloys and needs 6 to 12 wt% Al in the alloy.[19–21] As such alloys have poor mechanical properties, it is more usual for Al_2O_3–forming alloys to be developed from the M–Cr–Al systems. Here, the chromium, which forms an oxide of stability intermediate between that of the base metal oxide and Al_2O_3, at a level of 15 to 25 wt%, promotes establishment of an Al_2O_3 healing layer on alloys containing only 3 to 4 wt% Al.[22] The reasons for this effect are not fully understood. The probable explanation is that chromium acts as a getter for oxygen, preventing it from entering the alloy.[1] This allows aluminium to diffuse to the surface to form a healing Al_2O_3 layer at the base of the transient scale. However, the fact that, in many cases, almost no Cr_2O_3 or base metal oxide is detected outside the Al_2O_3 layer and that manganese does not have a similar effect as chromium in this respect suggests that the explanation may not be the whole story.

The development of a healing layer of α–Al_2O_3 at temperatures of 900°C or more on ternary M–Cr–Al alloys is similar to that of Cr_2O_3 on the binary M–Cr alloys,[23,24] although alloy grain boundaries apparently play little role in transporting aluminium to the surface. The various factors influencing how the steady–state scale is established have been reviewed[25] and shown to be similar to those for the corresponding M–Cr alloys. Thus, the ease of development of an α–Al_2O_3 healing layer on M–Cr–Al alloys is in the order Fe–Cr–Al > Ni–Cr–Al > Co–Cr–Al. This is mainly due to the high replenishment rate of aluminium to the surface and the low oxygen solubility and diffusivity in the alloy for the Fe–Cr–Al system. Also, the compositional variation of the Fe–Cr–Al spinel oxide helps to prevent the initially–formed iron oxides from growing rapidly and dominating scale growth.

Following establishment of the healing Al_2O_3 layer, the rate of subsequent oxidation depends on its effectiveness as a protective barrier. The oxide has a very stoichiometric structure, so lattice transport is slow. Although several structures of this oxide are possible, with the α–, θ–, δ– modifications being observed in oxidation studies, the main interest is in behaviour at temperatures from 900° to 1300°C where the α–phase predominates.

There is considerable disagreement regarding the mechanisms of transport during growth of the Al_2O_3 layer at these temperatures. Most evidence suggests that inward grain–boundary diffusion of oxygen is the main diffusion process contributing to scale growth;[26] however, more recent research suggests that aluminium transport may predominate.[27]

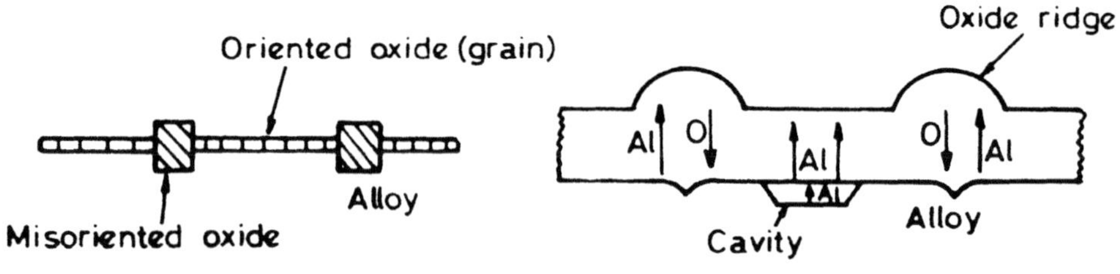

a) Early stages, misoriented oxide

b) Later stages, oxide ridges

Fig. 4. Schematic representation of morphologies of Al_2O_3 scale on NiAl during oxidation in oxygen at 1000°C.[30]

The Al_2O_3 layers formed during oxidation are often non-uniform in thickness (Fig. 4), with protrusions at the scale/gas interface and intrusions at the scale/metal interface.[28–31] Localized thickening may result from rapid diffusion paths for oxygen[28,31], or metal ions[32], or both species[29,30]. The presence of oxide ridges was given as evidence for aluminium transport in oxide grain boundaries[29], arising from a countercurrent of aluminium and oxygen. This leads to outward growth of the oxide ridges at the outer surface and corresponding inward growth at the scale/metal interface. However, as discussed elsewhere[33], such features could also be explained in terms of a stress-assisted diffusional-creep process. In particular, residually-stressed oxide may result from formation of oxide at internal surfaces within the bulk scale.[34] Thickening of Al_2O_3 layers at high temperatures could take place by diffusional creep, involving a compressive flux of metal ions and oxygen from the oxide grain boundaries to the oxide/alloy or oxide/gas interfaces.

Various scale configurations have been observed for Al_2O_3 scales. Interfacial voids[35,36] may result from oxygen vacancy diffusion, with precipitation following vacancy coalescence.[37] Alternatively, the vacancies may be generated by incorporation of metal ions from the substrate into the scale if the latter does not relax back sufficiently to maintain intimate contact with the receding metal.[38] A third suggestion is that a Kirkendall effect in the alloy causes an excess of vacancies to condense out as interfacial voids.[39]

Several investigations have reported much larger interfacial cavities, associated with a convoluted scale configuration[23] and large areas of scale having grown out of contact with the alloy substrate, producing large ridges, similar to those observed for Cr_2O_3 scales on Fe–Cr alloys (Fig. 3). The mechanisms to account for the development of convoluted scales are not well-established. It has been suggested[23,40,41] that inward transport of oxygen down oxide grain boundaries is largely responsible for growth of the scale. Reaction at, or near, the oxide grain boundaries between this species and Al^{3+} ions which diffuse outwards through the scale results in formation of new oxide within the scale. The ensuing compressive stresses cause lateral growth of the oxide layer, leading to sliding of the scale/alloy interface which may cause considerable stress concentrations at wavy interfaces,[34] deformation of the alloy substrate and, eventually, to localized detachment from the metal. Such detached scales are commonly observed for Fe–Cr–Al alloys which are very weak at 1000° to $1200^{o}C$, but are less frequent for the stronger Ni–Cr–Al system.

Detachment of the scale usually initiates at irregularities in the metal surface, such as abrasion markings or emergent grain boundaries, and may be associated with the presence of voids at the interface. These may arise from high local tensions which form in the presence of boundary waviness, especially during interfacial sliding, and may exceed the critical void nucleation stress and permit nucleation of stable voids.[34] Following detachment, continued lateral growth allows progressive coarsening of the convoluted scale morphology. The ridges move laterally and thicken at a similar rate as the attached scale, by vapour phase transport of aluminium across the cavity. The oxide is accommodated into the convoluted configuration either by some intricacies of the mechanism or by deformation.

Another proposal to account for convoluted scales attributes them to growth and impingement of large Al_2O_3 crystals at the underside of the scale following localized detachment.[42] Such crystals result from aluminium evaporation from the substrate and their unrestrained growth causes buckling of the scale. It has also been proposed that convoluted scales are transient features only;[43] detachment of scale results from loss of oxide epitaxy with the metal surface following migration of alloy grain or sub-grain boundaries. Scale detachment should occur only during movement of the boundaries and is thus only likely to be significant in the early stages when metal grain growth may be considerable.

The concept of formation of oxide within a bulk scale is a contraversial subject in oxidation. Analysis shows that it may not be possible when transport involves anion and cation vacancy motion, either in the lattice or in the grain boundaries.[44] However, molecular diffusion of oxygen in oxide grain boundaries may result in chemical potential gradients, consistent with the operation of sites for formation of new oxide in the scale.[34] Theoretical analysis indicates that such oxide can form only if both major and minor defects are interstitials or if one is an uncharged interstitial and the other a charged vacancy.[45] Indeed, enhanced oxide grain-boundary diffusion may involve a neutral species, such as oxygen interstitials.[46]

4. DEVELOPMENT OF STRESSES IN A GROWING OXIDE SCALE

As an oxide scale thickens, the system is subjected to stresses. These include internal stresses which arise from the scale-growth process, thermal stresses which result from changes in the exposure temperature and externally-applied stresses.

There are various causes of internal stresses during isothermal oxidation, as reviewed elsewhere.[33,47] The most significant one is probably formation of new oxide at a constrained site such as at the scale/metal interface, within the alloy matrix or within the bulk scale, particularly at oxide grain boundaries. Specimen geometry can also be important in this respect since, if a scale grows on a curved surface, stresses arise as the scale/metal interface retreats, the magnitude depending on the scale-growth mechanism and the radius of curvature of the surface.[48] However, in general, as the rate of growth of protective scales is slow, time-dependent creep of the oxide or of the metal substrate can usually relieve such internal stresses and isothermal scale failure is rarely observed, although multilayered Cr_2O_3 scales have occasionally been formed during isothermal oxidation of chromium,[13,14] and breakaway oxidation of Fe-Cr alloys has been reported.[16]

However, spallation of Cr_2O_3 and Al_2O_3 scales is a relatively common occurrence during cooling or thermal cycling of high-temperature alloys. Here, thermal stresses are developed because of the differences in thermal contraction/expansion coefficients between the scale and the alloy. Several expressions have been derived to show the magnitude of these stresses, such as equation (1)[49] (for cooling from temperature T_1 to temperature T_2):

$$\sigma = \frac{E_c(T_1 - T_2)(\delta_m - \delta_c)}{1+2\left[\frac{E_c}{E_m}\right]\left[\frac{y_c}{y_m}\right]} \qquad (1)$$

where
E = elastic modulus
δ = thermal expansion coefficient
y = thickness
c,m = subscripts referring to oxide and metal respectively

The stresses are compressive if $\delta_m > \delta_c$ and are often rapidly induced, making relief by deformation difficult and scale failure may result.

5. FACTORS INFLUENCING SCALE FAILURE

Although Cr_2O_3 and Al_2O_3 scales are susceptible to scale failure, especially on thermal cycling, various factors influence the likelihood of such an occurrence taking place, as follows[50]:

(a) the extent of contact between the scale and the alloy at temperature (as discussed earlier),
(b) the scale/alloy bond strength,
(c) the scale/alloy interface configuration,
(d) the possibility of alloy grain-boundary sliding,
(e) the properties of the scale.

The adhesion between a metal and its growing oxide is determined by chemical interactions, as well as Van der Waals and Coulomb-type bonds.[47] However, there is disagreement in the literature on the inherent strength of the scale/metal interface. Recently, and also at the present conference, it has been proposed that segregation of sulphur impurities to the interface results in a weakening of the interfacial bond strength of Al_2O_3 scales on nickel-base

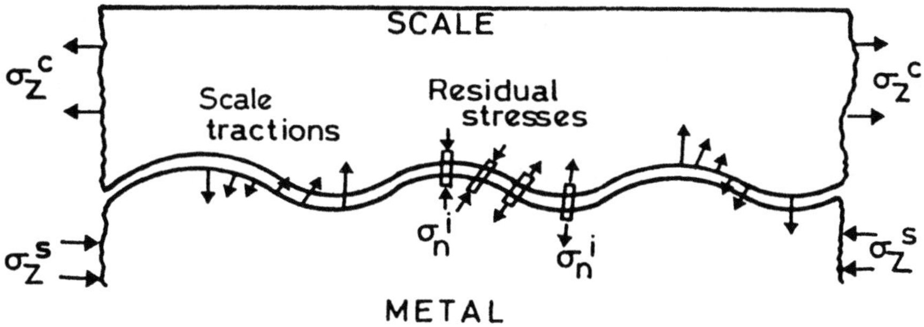

Fig. 5. Schematic representation showing the tractions and residual stresses at a wavy interface for a scale that contracts relative to the substrate on cooling, showing continuity of stress and displacement.[34]

alloys,[51-54] similar to the bond weakening effect of sulphur at metal grain boundaries, as modelled by quantum-mechanical molecular orbital techniques.[55] A similar model may be relevant to the Al_2O_3/metal system, although this has not yet been developed.[56] The converse argument is that the Al_2O_3/alloy interface is inherently weak and sulphur segregation should have little effect on scale-adhesion characteristics. This hypothesis has been supported by experiments in which spalling from a high purity Ni-Cr-Al alloy, with little sulphur segregation, was similar to that from a commercial alloy where considerable sulphur segregation to the interface occurred.[57]

The importance of the scale/alloy interface configuration in influencing scale failure is uncertain. An irregular interface may key the scale to alloy if the wavelength is small and the amplitude is large. In particular, there may be a reduction in growth stress since the interfacial region can accommodate mismatch in specific volumes and in the thermal expansion/contraction coefficients between the alloy and scale. There is also an effective increase in fracture toughness of the interfacial composite region due to increased crack propagation lengths, modified stress fields and an increased capacity for energy dissipation by introduction of plastic alloy protrusions.[50] However, a very irregular interface may cause scale decohesion, not improved adhesion.[58,59] Analysis has shown that an irregular interface can affect the residual stresses induced by the homogeneous mismatch strains arising from differential thermal contraction/expansion effects between the scale and the alloy.[34] It was deduced that alternating regions of residual tension and compression can develop normal to the interface and the peak normal stress can be large, possibly leading to crack propagation and scale failure (Fig. 5).

Alloy grain-boundary sliding which may occur at high temperatures can influence scale adhesion. A scale would accommodate development of slip steps at the scale/alloy interface by fracturing, by separating from the metal or by deforming to maintain contact across the interface. Analysis of the situation for the first two possibilities indicates the importance of the scale thickness in determining whether fracture or loss of contact occurs.[60] Although scale deformation may be able to accommodate such slip steps at high temperature, elimination of grain-boundary sliding is probably beneficial in maintaining adhesion of the scale.

Various properties of the scale may affect its adhesion to the alloy during oxidation. Fracture toughness of the oxide determines the ease of crack propagation in the scale, adjacent to the scale/alloy interface. There has been some study of the toughness of bulk ceramics. For instance, for a given ceramic, a small grain size gives more resistance to crack propagation than a large grain size, while porosity is detrimental to fracture toughness.[61] However, the relevance of these observations to a growing scale is unclear.

6. SPALLATION OF PROTECTIVE SCALES

Following oxidation, Al_2O_3 and, to a lesser extent, Cr_2O_3 scales may spall significantly on cooling. The reasons for the differences between the two oxides may be partly associated with the relative differences in thermal contraction coefficients between the alloy and the scale. The strains caused in the oxide by such effects depend on the alloy substrate as well as on the scale. For instance, the difference in coefficients between chromium or chromium-rich ferritic steels and Cr_2O_3 is reasonably low while it is much larger for an austenitic steel, such as Fe–25%Cr–20%Ni and Cr_2O_3 or for a steel such as Fe–18%Cr–8%Ni–4%Al and Al_2O_3. Hence, the compressive stress developed in the scale during cooling through 500°C is sufficient to exceed the critical compressive fracture strain for Al_2O_3 and Cr_2O_3 scales on the austenitic steels but not sufficient for Cr_2O_3 scales on chromium or the ferritic steel.[62]

The modes of scale spallation, particularly for Cr_2O_3 on 20%Cr–25%Ni–Nb

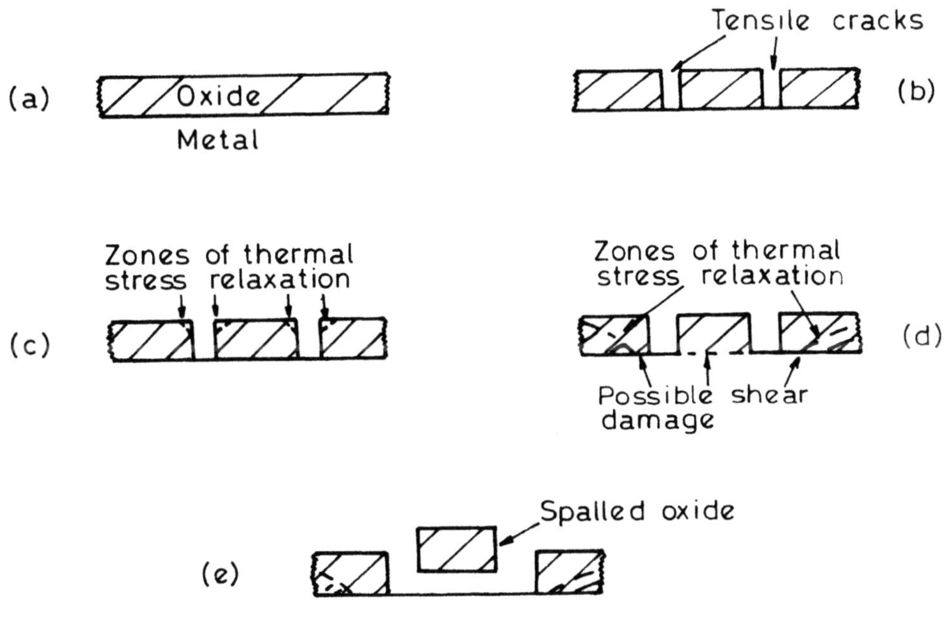

Fig. 6. Schematic representation of cracking, decohesion and blistering and spallation of an oxide scale under tension.[63]

stabilized steel in CO_2, have been studied by Evans and Lobb.[63] It was shown that decohesion of the scale takes place when the strain energy in the layer due to differential thermal contraction effects reaches a critical value which can be equated to the fracture energy of the scale/alloy interface. For tensile stresses (Fig.6), failure would involve cracks, approximately normal to the interface and separation of the scale from the metal should only take place if interfacial sliding occurs, leading to shear fracture at the interface. For compressive stresses (Fig.7), scale failure can occur by one of two processes, depending on the relative fracture strengths of the scale and the interface. If the interface is more resistant to rupture than the scale, compressive shear cracks develop initially, followed by decohesion at the interface (Route 1). If the reverse is true (Route 2), decohesion occurs initially in the absence of through-thickness cracking of the scale, leading to scale buckling during cooling until, eventually, relaxation of the stress and strain energy in the scale takes place by propagation of tensile cracks at the points of maximum curvature of the scale.[63] For the first scale failure mode (Route 1), the initiation of spallation occurs at a temperature change, ΔT:

$$\Delta T = \left[\frac{\gamma_F}{f\ y_c\ E_c\ (\delta_m - \delta_c)^2\ (1-\nu_c)} \right]^{\frac{1}{2}} \quad (2)$$

where γ_F is the fracture energy of the scale/alloy interface
f is the fraction of strain energy used in the fracture process
ν is Poisson's ratio

For the second mode (Route 2), the initiation of unstable buckling and thus spallation occurs at a temperature change, ΔT:

$$\Delta T = \left[\frac{1.052 y_c^4}{D^4} + \frac{1.041 \gamma_F}{E_c\ y_c} \right]^{\frac{1}{2}} \frac{1}{\delta_m - \delta_c} \quad (3)$$

where D is the radius of the area of interface which initially had suffered decohesion.

It has also been proposed that separation of the scale and alloy occurs when either the tensile stress normal to the interface or the interfacial shear stress exceed critical values.[64] It was indicated that through-thickness cracking results in high interfacial shear stresses, but these do not necessarily lead to scale separation since they may be relieved by deformation of the metal or interfacial creep (equivalent to grain-boundary sliding).

Several other authors have also indicated that decohesion followed by buckling is an important process in scale spallation.[65,66] Analysis has shown that, for flat specimens, under the second failure route of Evans and Lobb, a driving force for crack growth at, or parallel to, the interface is generated only if elastic buckling of the scale takes place.[67] Such buckling is initiated when the compressive stress in the scale exceeds a critical value, S, given by equation (4).

$$S = \left[\frac{kE_c}{12(1-\nu_c^2)}\right] \left[\frac{y_c}{b}\right]^2 \qquad (4)$$

where k is a constant
b is radius of separation between the scale and the substrate, prior to buckling.

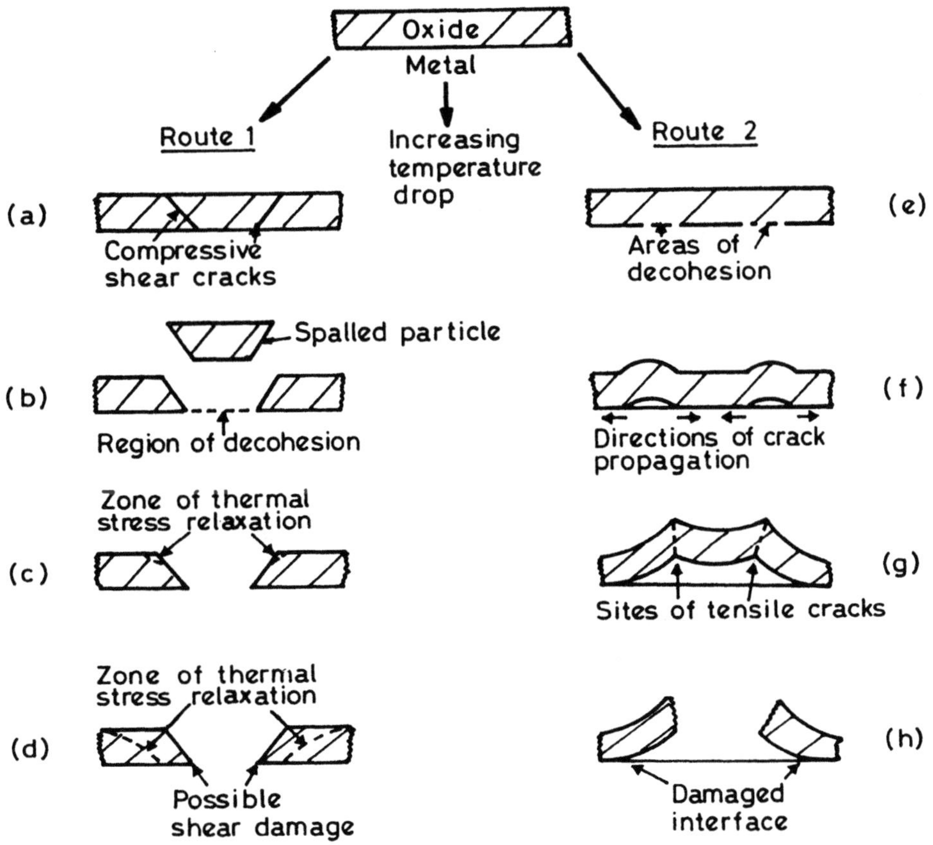

Fig. 7. Schematic representation of cracking, decohesion and blistering, and spallation of an oxide under compression.[63]

Thus, a pre–existing separation is needed for buckling and eventual spallation[34]. If buckling is not present, the interfacial crack does not change the stress state because the prior normal and shear stresses at the interface are zero (Fig. 8(a)). Thus, there is no change in the in–plane compression and no crack driving force. However, if buckling occurs, the in–plane compression is reduced and normal and shear displacements occur at the interface, leading to a stress concentration at the crack tip and, thus, to a crack–growth driving force (Fig.8(b)). The crack grows when this force attains the critical value for the weakest material element, which may be at the interface or in the oxide near the interface. Thus, if the interface has the weakest fracture resistance, K^i_c, a crack will propagate along the interface when the stress exceeds a critical value, σ^c, given by equation (5):

$$\sigma^c = \left[\frac{K^i_c}{y_c^{\frac{1}{2}}}\right] \frac{(1+\nu_c)^{\frac{1}{2}}}{0.8} \qquad (5)$$

Flat surfaces, subject to zero normal stresses, are more resistant to spalling than curved surfaces and, in the former cases, this can only occur when the interface crack reaches a critical size, C_s, given by equation (6) [67]:

$$\frac{C_s}{y_c} = \left[\frac{E_c\, y_c^{\frac{1}{2}}}{K^i_c}\right]^{\frac{1}{2}} \qquad (6)$$

For curved surfaces, there is additional crack driving force and premature spallation may be observed from regions of small radii of curvature.

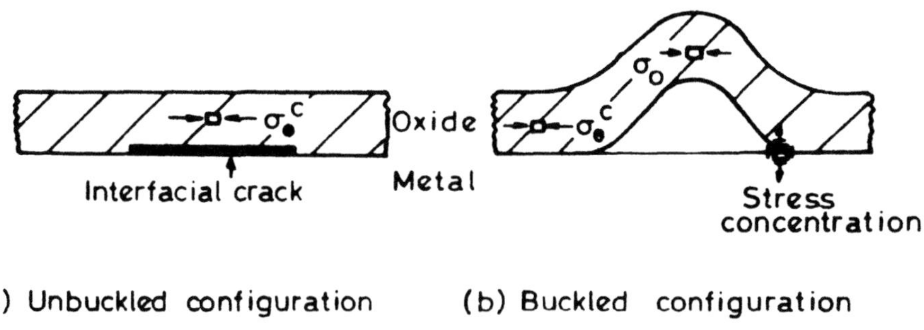

Fig. 8. Schematic representation of the influence of buckling on the crack driving force for an interfacial crack at, or parallel to, the scale/metal interface.[34]

7. CONCLUSIONS

1. Many high-temperature iron-, nickel- and cobalt-base alloys rely on development of external healing Cr_2O_3 or Al_2O_3 layers for protection against the environment. The ease of establishment of such layers is influenced markedly by various alloy and oxide properties as well as by the conditions of oxidation.

2. The subsequent rate of oxidation is determined primarily by the rate of transport of reactants across the healing layer, with short-circuit paths being especially important for these relatively stoichiometric oxides.

3. Mechanical failure of the scale, particularly cracking and spallation on thermal cycling, can be a major life-limiting factor for such alloys. Stresses in an oxidizing system can result from the isothermal oxidation process, the component geometry and differential thermal contraction between the scale and the alloy. Although scale-growth stresses may be relieved by diffusion-assisted creep processes during isothermal exposure, the rapid strain rates developed during thermal cycling can result in scale failure, allowing the environment access to a substrate which may be depleted considerably in the protective element, chromium or aluminium.

ACKNOWLEDGEMENTS

The author is grateful to various colleagues in the Corrosion and Protection Centre, UMIST, particularly G.C. Wood, P.K.N. Bartlett, and F.A. Golightly, for contributions to some of the research discussed in this paper.

REFERENCES

1. G.C. WOOD: Oxidation Metals, 1970, **2**, 11-57.
2. G.R. WALLWORK: Rep. Prog. Phys; 1976, **39**, 401-485.
3. R.A. RAPP: Corros.Sci.,1965, **21**, 382-388.
4. F.H. STOTT and G.C. WOOD: Mat. Sci. and Tech., in press.
5. F.H. STOTT, P.K.N. BARTLETT and G.C. WOOD: J. Mater. Sci. Eng., 1987, **88**, 163-169.
6. F.H. STOTT, P.K.N. BARTLETT and G.C. WOOD: Oxidation Metals, 1987, **88**, 37-55.
7. C. WAGNER: Z. Elektrochem., 1959, **63**, 772-780.
8. H. HINDAM and D.P. WHITTLE: Oxidation Metals, 1982, **18**, 245-284.
9. C.S. TEDMON: J. Electrochem.Soc., 1966, **113**, 766-770.
10. K. HOSHINO and N.L. PETERSON: J. Am. Ceram. Soc., 1983, **66**, 202-203.
11. H.V. ATKINSON: Oxidation Metals, 1985, **24**, 177-197.
12. A. ATKINSON and R.I. TAYLOR: Transport in Non-Stoichiometric Compounds, eds. G. Simkovich and V.S. Stubican; 285-295, 1985, New York, Plenum Press.
13. D. CAPLAN and G.I. SPROULE: Oxidation Metals, 1975, **9**, 459-472.
14. K.P. LILLERUD and P. KOFSTAD: J. Electrochem. Soc., 1980, **127**, 2397-2419.
15. G.C. WOOD and D.P. WHITTLE: J. Electrochem. Soc., 1968, **115**, 126-131.

16. G.C. WOOD and D.P. WHITTLE: Corros. Sci., 1967, **7**, 763–771.
17. O.T. GONCEL, J. STRINGER and D.P. WHITTLE: Corros. Sci., 1978, **18**, 701–720.
18. D.M. POOLE and P.M. THOMAS: J. Inst. Met., 1962, **90**, 228–237.
19. W.E. BOGGS: J. Electrochem. Soc., 1971, **18**, 906–913.
20. G.C. WOOD and F.H. STOTT: Br. Corr. J., 1971, **6**, 247–254.
21. G.N. IRVING, J. STRINGER and D.P. WHITTLE: Oxidation Metals, 1975, **19**, 427–440.
22. F.H. STOTT, G.C. WOOD and M.G. HOBBY: Oxidation Metals, 1971, **3**, 103–113.
23. F.A. GOLIGHTLY, G.C. WOOD and F.H. STOTT: Oxidation Metals, 1980, **14**, 217- 234.
24. C.S. GIGGINS and F.S. PETTIT: J. Electrochem. Soc., 1971, **118**, 1782–1790.
25. G.C. WOOD and F.H. STOTT: High Temperature Corrosion, ed. R.A. Rapp, 227–250, NACE–6, 1983; Houston, Texas.
26. K.P.R. REDDY: Ph.D. Thesis, 1979, Case Western Reserve University.
27. E.W.A. YOUNG and J.H.W. de WIT: Oxidation Metals, 1986, **26**, 351–362.
28. J.L. SMIALEK; Metall. Trans., 1978, **9A**, 309–320.
29. H. HINDAM and W.W. SMELTZER: J. Electrochem.Soc., 1980, **127**, 1630–1635.
30. H. HINDAM and W.W. SMELTZER: Oxidation Metals, 1980, **14**, 337–349.
31. E.J. FELTEN and F.S. PETTIT: Oxidation Metals, 1976, **10**, 189–223.
32. D.J. YOUNG and M. COHEN: J. Electrochem. Soc., 1977, **124**, 769–774.
33. F.H. STOTT: Rep. Prog. Phys., 1987, **50**, 861–913.
34. A.G. EVANS, G.B. CRUMLEY and R.E. DEMARAY: Oxidation Metals, 1983, **20**, 193–216.
35. I.M. ALLAM, D.P. WHITTLE and J.STRINGER: Oxidation Metals, 1978, **12**, 35–66.
36. C.S. GIGGINS, B.H. KEAR, F.S. PETTIT and J.K. TIEN: Metall. Trans., 1974, **5**, 1685–1694.
37. J.L. SMIALEK: NASA Tech. Memor. 81676, 1981, Cleveland, Ohio.
38. G.B. GIBBS: Oxidation Metals, 1981, **16**, 147–158.
39. J.D. KUENZLY and D.L. DOUGLASS: Oxidation Metals, 1974, **8**, 139–154.
40. F.A. GOLIGHTLY, F.H. STOTT and G.C. WOOD: Oxidation Metals, 1976, **10**, 163–187.
41. F.A. GOLIGHTLY, F.H. STOTT and G.C.WOOD: J. Electrochem. Soc., 1979, **126**, 1035–1042.
42. C.S. GIGGINS and F.S. PETTIT: Report ARL 75–0234; 1975, Pratt and Whitney Aircraft, Connecticut.
43. R.A. RAPP: Metall. Trans., 1983, **15A**, 765–775.
44. M.V. SPEIGHT and J.E. HARRIS: Acta Met., 1978, **26**, 1043–1049.
45. A. ATKINSON: Corros. Sci., 1982, **22**, 347–370.
46. F.A. KRUGER: High Temperature Corrosion, ed. R.A. Rapp, 89–94, NACE–6, 1983, Houston, Texas.
47. J. STRINGER: Werkst. u. Korros., 1972, **23**, 747–755.
48. P. HANCOCK and R.C. HURST: in 'Advances in Corrosion Science and Technology', (ed. R.W. Staehle and M.G. Fontana) 1974, New York, Plenum Press.
49. G.D. OXX: Prod. Eng. (Cleveland), 1958, **29**, 61–64.

50. F.H. STOTT: Mat. Sci. and Tech., 1988, **4**, 431–438.
51. C.L. BRIANT and R.A. MULFORD: Metall. Trans., 1982, **13A**, 745–759.
52. V. IKEDA, K. NII and K. YOSHIHARA: in Proc. Symp. JIMIS-3 on 'Oxidation and Hot Corrosion of Metals and Alloys', Tokyo, 1982, Japan Inst. Metals, 207–215.
53. A.W. FUNKENBUSCH, J.G. SMEGGIL and N.S. BORNSTEIN: Metall. Trans., 1985, **16A**, 1164–1175.
54. J.L. SMIALEK: Metall. Trans., 1987, **18A**, 164–175.
55. R.P. MESSMER and C.L. BRIANT: Acta Metall., 1982, **30**, 457–462.
56. J.L. SMIALEK .and R. BROWNING: in Proc. 168th Meeting of Electrochem. Soc., Las Vegas, NV, 1985.
57. K.L. LUTHRA and C.L. BRIANT: Oxidation Metals, 1986, **26**, 397–416.
58. D. DELAUNAY and A.M. HUNTZ: J. Mater. Sci., 1982, **17**, 2027–2036.
59. J.G. SMEGGIL and A.W. FUNKENBUSCH: Report R85-9167564-1, NASA, Cleveland, Ohio, 1985.
60. J.C. GROSSKREUTZ and M.B. McNEIL: J. Appl. Phys., 1969, **40**, 355–360.
61. J. SPRAGGS and N. VASILOS: J. Am. Ceram. Soc., 1963, **46**, 224–235.
62. D.R. HOLMES and R.T. PASCOE: Werkst. u. Korros., 1972, **23**, 859–870.
63. H.E. EVANS and L.C. LOBB: Corros. Sci., 1984, **24**, 209–222.
64. M. SCHÜTZE: Mat. Sci. and Tech., 1988, **4**, 407–414.
65. R.M. CANNON: Am. Ceram. Soc. Bull., 1985, **64**, 449–455.
66. R.M. CANNON, V. JAYARAM, B.J. DAGLEISH and R.M. FISHER: Proc. Ceramic Microstructures'86. 'The Role of Interfaces', Berkeley, California, 1986.
67. A.G. EVANS and J.W. HUTCHINSON: Int. J. Solids Struct., 1984, **20**, 455–466.

OXIDE ADHESION ON PERIODIC TABLE GROUP 4 AND 5 METALS

M. LALLEMANT*, G. BERTRAND*, D. CIOSMAK*,
J.P. LARPIN* and A. SANFELD **

* L.R.R.S., Fac. Sci. Mirande, BP 138. F 21004 Dijon Cedex
** U.L.B. Service de Chimie-Physique, Boulevard du Triomphe, Bruxelles.

ABSTRACT

The oxide scale adhesion on Periodic Table group IV and V metals is evoked through two new concepts in high temperature corrosion : bistability and solute elastic effect.

1 . INTRODUCTION

The topic of this paper concerns the adhesion of oxide scales grown on some pure metals of the group IV and V of the Periodic Table. As these metals are often used as active elements in current steels to improve the adhesion properties of protective oxide scales, it seems interesting to consider their oxidation behavior in some detail.

The oxidation behavior of group IV and V metals has been extensively studied by many authors and rewieved recently by Kofstad.[1] However, in our view, a more careful examination of the morphologies of oxide scales grown on platelet samples has to be performed, because they are quite uncommon and raise many important questions. This paper will describe three types of characteristic oxide scale morphologies.

2 . EXPERIMENTAL RESULTS

The first type of morphology being considered consists of oxide scales exhibiting periodical cracks through the entire thickness, which yield cross-section micrographs of a "millefeuille" pastry appearance. The most representative example is observed in oxidation of titanium, Fig. 1. Under well chosen kinetic conditions, each sublayer is formed in lines of well shaped grains placed side by side, Fig. 2. It is separated from the following by a parallel crack. Sublayer thickness and crack width remain constant through the entire rutile scale thickness. Two others metals, tantalum and niobium, also produce such multilayered oxide scales, Fig. 3.

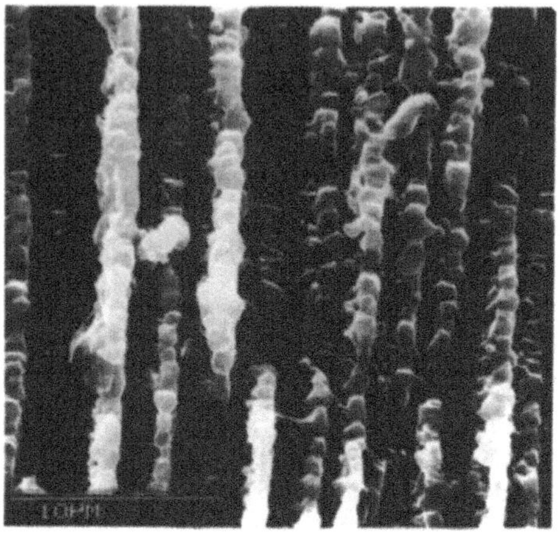

Fig. 1 : Scale morphology of rutile grown on a pure titanium platelet (T = 1173 K, pO_2 = 105 mbar).

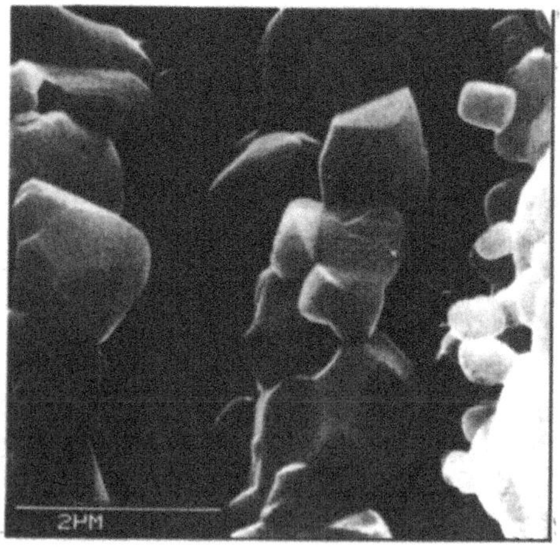

Fig. 2 : Detail of scale shown in Fig.1.

Another exotic oxide scale morphology was found on niobium oxidized at temperatures above 1123 K; it looks like a double egg-cup as shown in Fig. 4. The oxide scale is no longer laminated but has grown only along the edges of the initial platelet. A similar scale morphology was observed during the oxidation of zirconium.

The last example which will be described concerns the oxide scales grown on some titanium-based alloys. Generally, a repetitive variation in chemical composition, which corresponds to alternating sublayer morphologies, is observed through the oxide scale.

a) b)

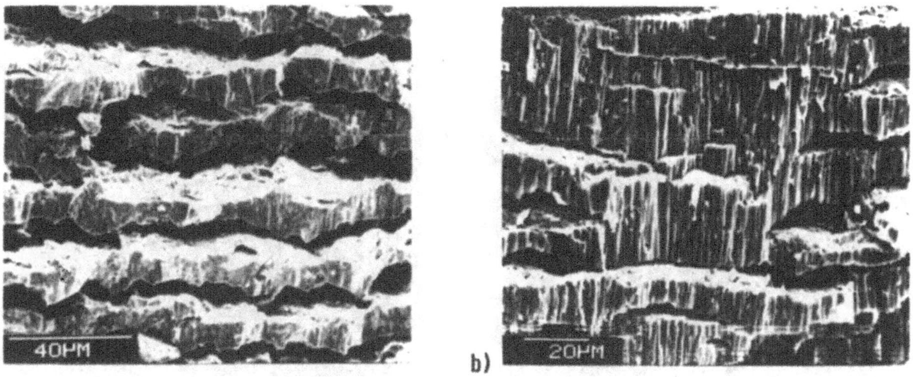

Fig. 3 : Morphology of oxide grown on pure metal platelet :
a) Tantalum (T = 1173 K, pO_2 = 100 mbar, t = 48 h)
b) Niobium (T = 973 K, $pO_2 \cong$ 400 mbar, t = 6 h).

Fig. 4 : Double egg-cup morphology of oxide scale grown on a niobium square section platelet (T = 1173 K, pO_2 = 100 mbar, t = 2 h).

Figure 5 shows the cross-section of an oxide scale grown on TA6V alloy (Ti-6wt%Al-V) which shows this morphological and chemical layering. Similar observations were made for the oxidation of Ti-43wt%Zr, Fig. 9. An alternation of black and white sublayers is clearly seen around the suspension hole. Here, it corresponds to a change in the stoichiometry of the product, $TiZrO_4$ spinel.

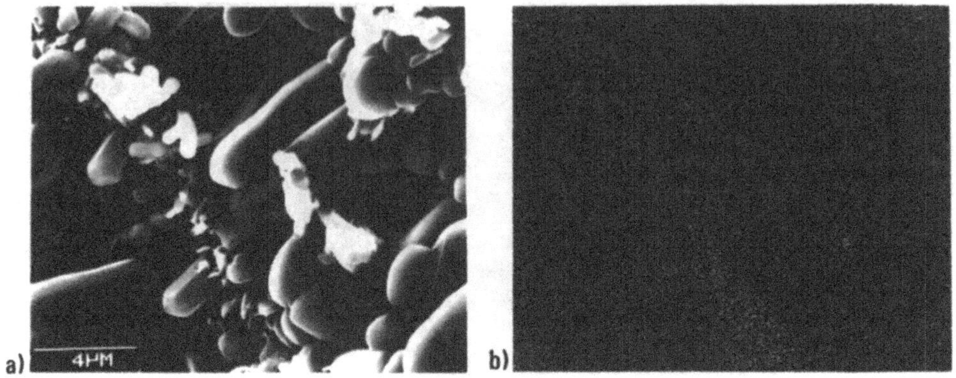

Fig. 5 : Cross section of an oxide scale grown on TA6V alloy (T = 1273 K, pO_2 = 400 mbar, t = 526 h).
a) Morphology; b) Al X-Ray image.

3 . DISCUSSION

The observed unusual morphologies question the classical models of oxidation. How can this self-organization of the corrosion scales occur? What type of processes may be involved? What are the specific properties of the metals considered ?

Two concepts have to be introduced. The first one is <u>bistability</u>, which can be described as the possibility for a dynamic system to evolve following two different regimes for the same parameter range. This is illustrated in Fig. 6.

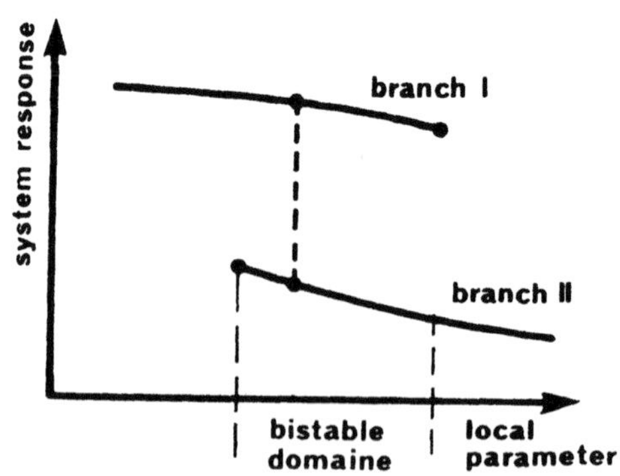

Fig. 6 : Schematic diagram of a bistable system.

If the bistability only exists, without any local perturbations of the considered parameter, no structuring occurs, and the classical high temperature oxidation models are perfectly acceptable. But, supposing that a local parameter, such as interfacial composition or surface temperature, etc... fluctuates in time, due to the dynamics of the chemical reaction, then jumping between the two stable branches of the bistable system is possible.

This concept may be illustrated by two examples of bistable kinetics, schematically represented in Fig. 7, where we presume that the exothermicity of the chemical reaction induces a local increase of the interfacial temperature. The furnace temperature is constant and imposed at a value T_1. Locally, the temperature can increase, eventually up to the limit of the first low-branch, and then a jump on the other high branch occurs. If the shape of the bistable curve corresponds to that plotted in Fig. 7a, no feedback is possible. On the contrary, if the shape of the curve is as plotted in Fig. 7b, where the reaction rate is lower on the second branch, the local temperature will decrease and the reaction regime will join the first branch again, and so on.

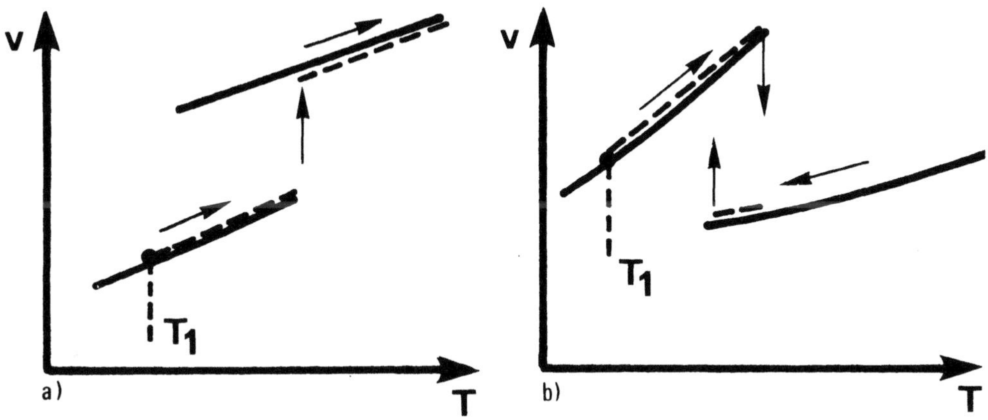

Fig. 7 : Diagrams of bistable systems :
a) No thermal feedback effect possible.
b) A thermal feedback effect exists. Chemical and morphological alternations may be expected.

This kinetic description is exactly what has been observed experimentally during the oxidation of Ti-43wt%Zr alloy in air, as can be seen in Fig. 8. This plots the parabolic rate constant, when kinetics obey the parabolic rate law, versus the furnace temperature for two different air pressures. The shape of this curve corresponds to the second hypothetical bistable diagram in Fig. 7. Oscillations of the surface temperature were measured by

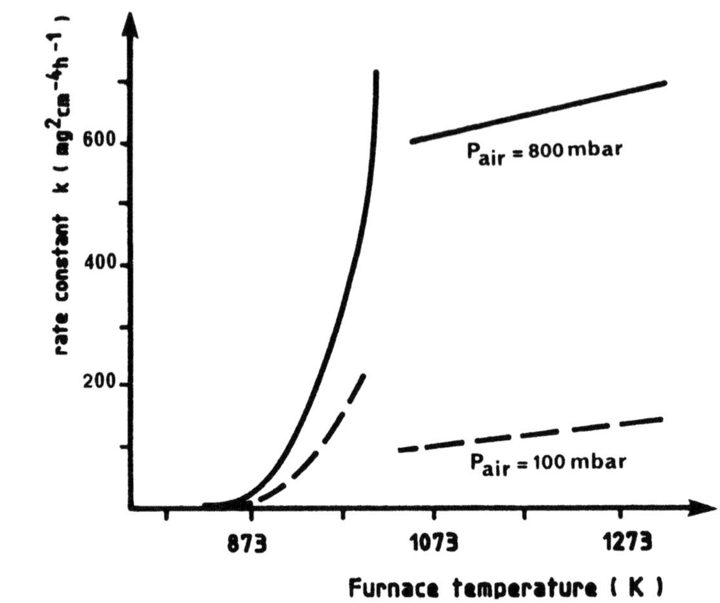

Fig. 8 : The kinetics of oxidation of Ti 43wt%Zr alloy.

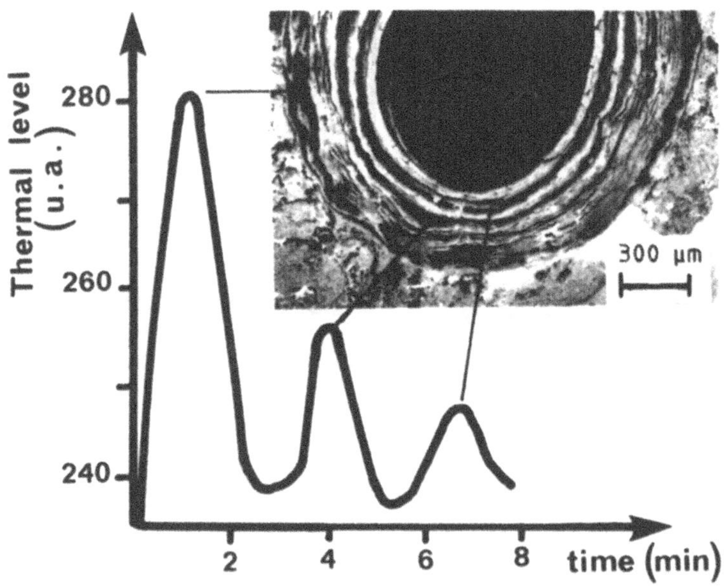

Fig. 9 : Morphology of the oxide scale grown on Ti43wt%Zr alloy and the corresponding experimental thermal excursions measured by I.R. micro-thermography.

infrared micro-thermography. Morphological examination of the oxide scale (Fig. 9) reveals a number of black and white sublayers, composed of non-stoichiometric and stoichiometric $TiZrO_4$ spinel respectively, which corresponds to the number of temperature oscillations.

These results are of considerable importance for the adhesion of oxide scales on metals, and consequently, for metal protection. As can be seen in the previous figures, alternating chemical compositions are often associated with repetitive morphologies of porous and compact sublayers. The jumping between the branches of a bistable system often leads to loss of the protective scale adhesion.

The other interesting concept is related to the solute elastic effect. This process can occur when the local chemical composition of the alloy matrix is modified during the reaction, giving a strong interaction with the stress field resulting from the growth of the oxide scale. This effect is particularly significant in the case of metals which dissolve a large amount of oxygen. It has been shown[2] that a maximum of the stress field may occur in the metal/oxygen solid solution, near the oxide/metal interface. This is due to the opposite fluxes of metal and oxygen ions with different transport properties, together with an interfacial chemical sink. Inserting this description in the mechanical equation of the system, a generalized elasticity law for metallic solution is :

$$\underline{\underline{\sigma}} = -K\left\{\sum_i \alpha_i \Delta c_i\right\}\underline{\underline{I}} + \lambda \operatorname{Tr}\underline{\underline{\varepsilon}} . \underline{\underline{I}} + 2\mu\underline{\underline{\varepsilon}} \qquad (1)$$

where $\underline{\underline{\sigma}}$ is the stress tensor, $\underline{\underline{I}}$ the unit tensor, $\underline{\underline{\varepsilon}}$ the strain tensor, λ and μ are Lamé coefficients, and α_i is the solute expansion coefficient of the ith species. Δc_i is the deviation of the local solute concentration of element i from a reference, the concentration at the interface between oxide scale and metal/oxygen solid solution. Tr is the matrix trace of the strain tensor and K is a coefficient similar to λ and μ.

From the calculated stress in a plane parallel to the interface results that an accumulation of stress occurs at a given depth z in the metal if the following conditions are verified:

$$0 < \frac{D_{ox}^{*} - D_M}{D_M^2} < \frac{k}{V^2} \qquad (2)$$

where:

$$D_{ox}^{*} = \frac{2\,D_{ox}}{1 + \sqrt{1 + 4\,\frac{k\,D_{ox}}{V^2}}}$$

with D_{ox} and D_M the chemical diffusion coefficients of oxygen and metal in the metal/oxygen solid solution, V the growth velocity of the reactive interface due to oxidation and k the reaction rate constant.

These unequalities give two criteria for the observation of periodic cracks. The apparent oxygen diffusion coefficient D^*_{ox} must be greater than the metal diffusion coefficient D_M; while the second criterion relates the diffusion parameters to the kinetic parameters. Such a model is in agreement with the assumption that maximum stress is a necessary condition for cracking and consequently for the loss of adhesion of an oxide scale.

Hence, this approach shows that loss of adhesion may result from exceeding a critical value, as determined by eq. (2). Control of the parameters in such a way that a maximum stress field does not exist, may guarantee adhesion of the protective oxide scale.

In conclusion, in this paper two new concepts were described briefly to provide improved understanding of the adhesion properties of protective oxide scales, and particularly to determine, and consequently to avoid, the range of experimental parameters which may cause a loss of adhesion of the oxide scales.

REFERENCES

(1) P. KOFSTAD. High Temperature Corrosion. Ed. Elsevier Ltd., 1988.

(2) A. SANFELD, M. HENNENBERG, G. BERTRAND and J. C. CANNOT. Corr. Sci., 1988, **28**, 1163-1176.

ELECTRODE POTENTIAL / TEMPERATURE DIAGRAMS AND THEIR APPLICATION TO HIGH TEMPERATURE CORROSION

M. Pourbaix

Cebelcor

Brussels, Belgium

ABSTRACT

Diagrams set up as a function of an electrode potential and temperature allow to represent the influence of temperature, pressure and composition of a gaseous phase on the conditions of stability and metastability of all substances which may exist in a given chemical and electrochemical system. They may be helpful for predicting the conditions of high temperature corrosion of metals and alloys. They may also be a valuable guide for experimental research and technical control.

Examples are given concerning the high temperature behaviour of iron, aluminum, chromium, silicon and carbon, as well as of inorganic coatings, in atmospheres containing oxygen and/or sulphur derivates.

ACTIVE ELEMENTS IN OXIDATION AND THEIR PROPERTIES

G. Beranger, F. Armanet and M. Lambertin

Division Matériaux, U.A. CNRS 849, Université de Technologie

de Compiègne, BP 649, F-60206 Compiègne Cedex, France

ABSTRACT

The beneficial effect of active elements on the oxidation behaviour of high temperature NiCr and NiCrFe alloys has been observed for 50 years and low level additions of active elements are now used with various MCrAl alloys (M = Fe and/or Ni), which include Cr_2O_3- as well as Al_2O_3-forming alloys. Numerous studies on their beneficial influence have been carried out, and different hypotheses, which have been suggested to explain the improvement of the scale adherence and thus of the protective behaviour, will be briefly discussed. Some hypotheses on the benificial effect may be derived from the physico-chemical properties of the active elements (e.g. crystallographic structure, free energy of formation of compounds, electronic structure, electronegativity, radius of ions, etc...). The efficiency of these elements, when added in the bulk or when incorporated near the surface by an appropriate treatment (e.g. ion implantation, coating, ...) will be compared. Further, the resources, consumption and costs of active elements with relation to the design of high temperature alloys will be also considered.

Thus, this paper provides general information concerning active elements to introduce the other more specialized papers to be presented at this Colloquium.

1. INTRODUCTION

In 1937, Griffiths and Pfeil[1] demonstrated that rare earths added as a melt deoxidant to Ni-20Cr alloys had a beneficial effect on their lifetime when used as heating elements under thermal cycling conditions. These empirical observations were applied some years later by Pfeil[2] to develop surface treatments with rare earth salts or oxides in an attempt to increase the corrosion resistance of refractory alloys without affecting their creep resistance.

According to Pfeil[2] the beneficial effect was observed for different levels of doping with chemical elements belonging to the groups II, III, IV, and V of the Periodic Table, whereby the efficiency is decreasing from group II to V but increasing with the atomic weight in a given group. From the classification suggested by Pfeil's patents, only some elements emerged to have pratical importance (Table 1). These studies were the true origin of the later research work, which confirmed and defined the important rôle played by these so-called "Active Elements", (A.E.). Extensive reviews on this topic were made by Whittle and Stringer[3] and by Lees[4].

The purpose of this paper is to collect and compare appropriate physical, chemical, and thermodynamical data of the active elements and their compounds and to point out some common features in these properties, with the objective to provide an introduction to the following more specialized papers.

Table 1

PERIODIC TABLE INDICATING THOSE ELEMENTS WHICH HAVE BEEN FOUND TO ACT AS "ACTIVE ELEMENTS"

IA	IIA	IIIA	IVA	VA	VIA	VIIA	VIII			IB	IIB	IIIB	IVB	VB	VIB	VIIB	O
H																	He
Li	Be											B	C	N	O	F	Ne
Na	Mg											Al	Si	P	S	Cl	Ar
K	Ca	Sc	Ti	V	Cr	Mn	Fe	Co	Ni	Cu	Zn	Ga	Ge	As	Se	Br	Kr
Rb	Sr	Y	Zr	Nb	Mo	Tc	Ru	Rh	Pd	Ag	Cd	In	Sn	Sb	Te	I	Xe
Cs	Ba	La*	Hf	Ta	W	Re	Os	Ir	Pt	Au	Hg	Tl	Pb	Bi	Po	At	Rn
Fr	La	Ac°															

*	Ce	Pr	Nd	Pm	Sm	Eu	Gd	Tb	Dy	Ho	Er	Tm	Yb	Lu
°	Th	Pa	U	Np	Pu	Am	Cm	Bk	Cf	Es	Fm	Md	No	Lw

We shall term as "Active Elements" those which when added in small amount, improve the oxidation resistance, i.e. ensure the integrity of protective scales growing on high-temperature alloys under isothermal or thermal cycling exposure conditions. In this definition are all the elements previosuly mentioned (Table 1), together with silicon and aluminium, for a matter of fact, the adhesion of an alumina scale is improved when alumina particles are finely dispersed in the matrix.

Active elements can be incorporated in or on the metallic substrate in different ways (Fig. 1):

i) by addition of pure A.E. metal or alloys (mischmetal as an example) during alloy melting,
ii) as an oxide dispersion, or
iii) by ion implantation.

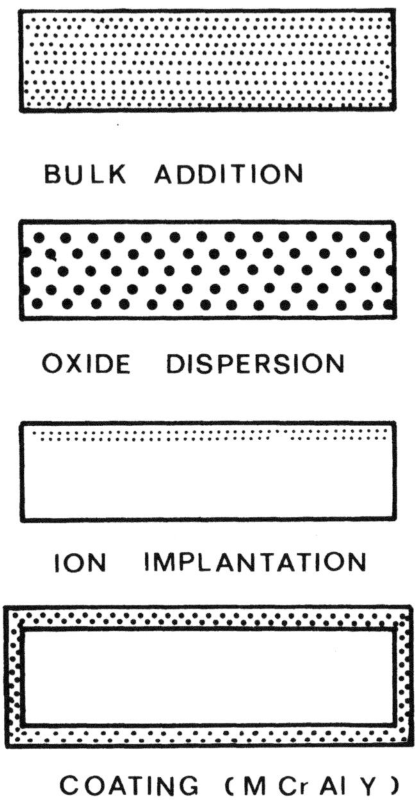

Fig. 1: The different ways to incorporate active elements.

Methods (i) and (ii) influence the mechanical properties of the substrate. Method (iii) is not yet industrially developed and is essentially restricted to laboratory tests, moreover a limited surface zone of the substrate only is potentially protected. An alternative technique consists in applying a MCrAlY-type protective overlay coating to a metallic substrate.

2. EFFECTS AND MECHANISMS

In spite of the numerous studies, which have been devoted to the active elements for 50 years, their benificial effects are still not well understood and even controversial.[3-7] However, the results available indicate that the major effect is an improvement in the scale adherence (Fig. 2). This is more important when the exposure temperature is high or when the oxidation conditions are severe (temperature cycling).

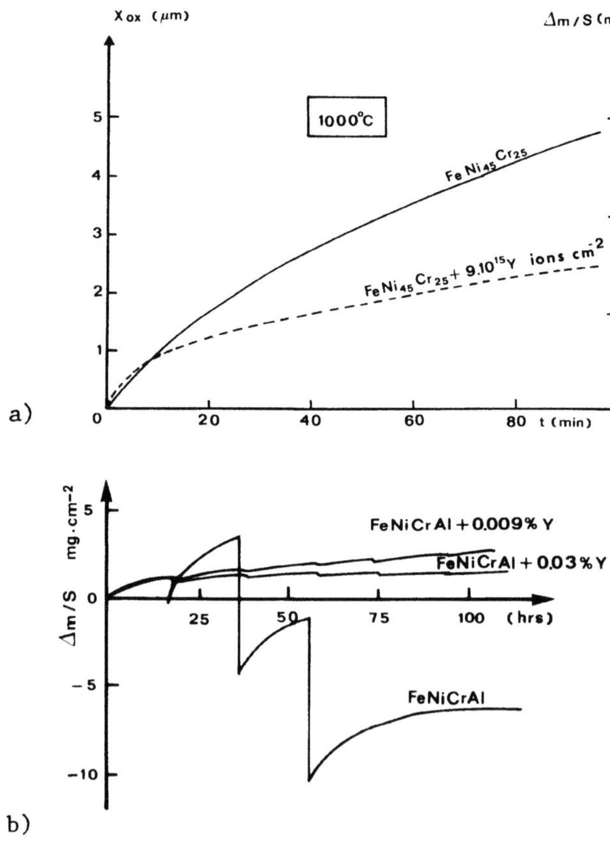

Fig. 2: Oxidation of alloys with and without active elements: a) Isothermal oxidation - b) Oxidation with thermal cycles.[8]

In order to explain the efficiency of these elements, different theories have been proposed.[3, 9-13] It is not the aim of this paper to discuss these models in detail but only to list them and to give short comments:

i) Transient oxidation curtailment: Active elements oxides act as nucleation sites and promote the selective oxidation of the protective scale-forming element, thereby curtailing the transient stage of oxidation (Fig. 3),

ii) Modification of the growth mechanism: The transport mechanism and the growth rate are altered essentially when a Cr_2O_3 scale is formed (there is no real evidence to suggest a similar effect for Al_2O_3 scales). In both

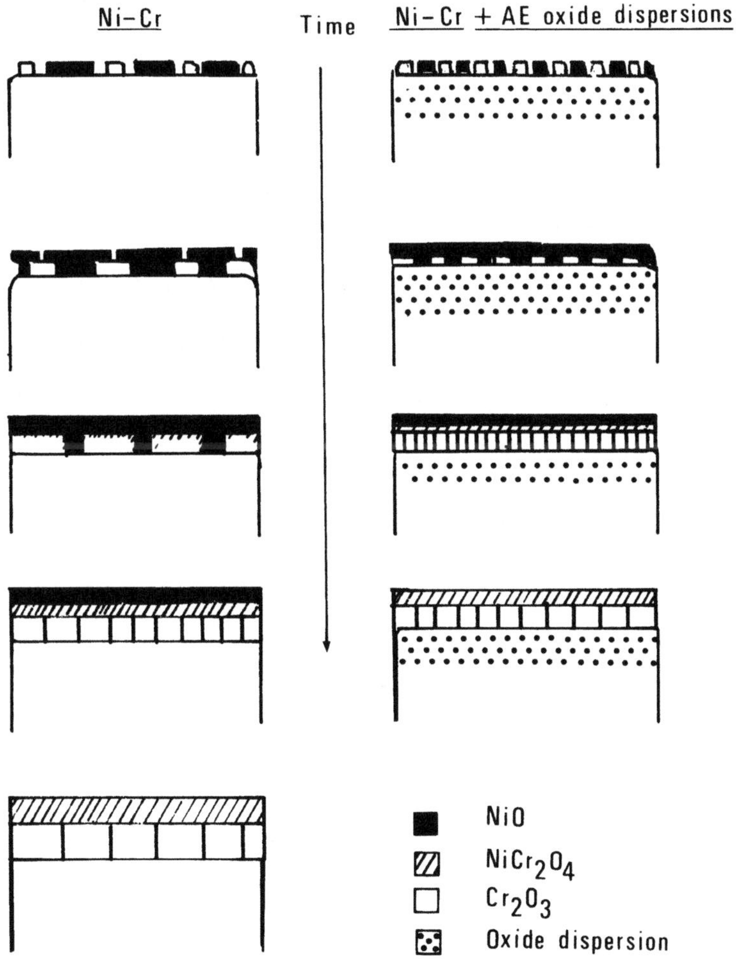

Fig. 3: Influence of active elements on scale nucleation.[3]

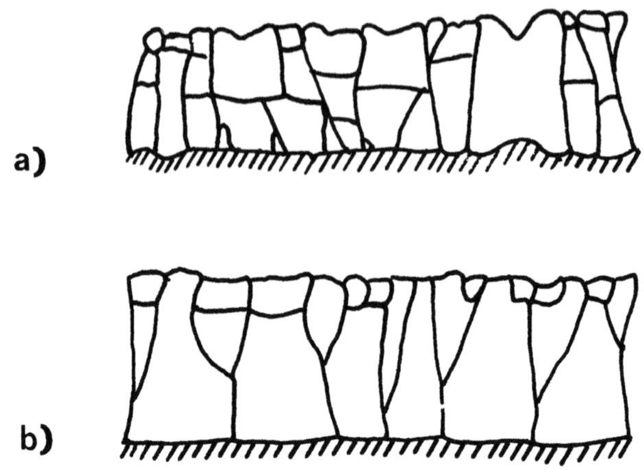

Fig. 4: Influence of active elements on scale structure and morphology. a) Alloy without Y - b) Alloy with Y.[13]

cases, the cation transport is probably reduced or even suppressed.[3]

iii) Modification of scale morphology and structure (Fig. 4): As a consequence of point i), the grain size of the oxide scale is reduced. The texture is modified which can influence scale plasticity and growth stress relief (Fig. 5 & 6). In certain cases, some particular crystalline structures (τ and η Al_2O_3) may become stabilized.[14]

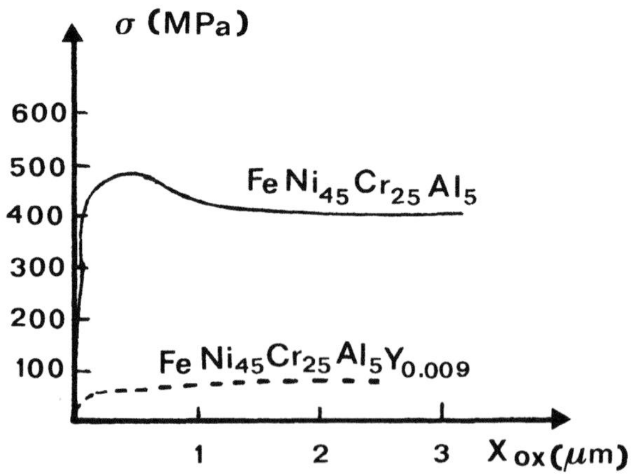

Fig. 5: Influence of active elements on residual growth stresses.[9]

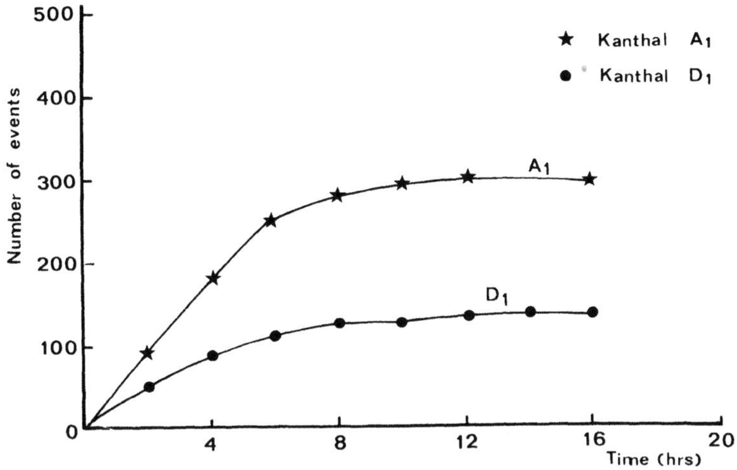

Fig. 6: Acoustic emission during cooling after oxidation of alloys without (A1) and with (D1) active element addition, indicating fewer scale failures for the latter alloy.[15]

iv) Improvement of the scale plasticity can be obtained by a smaller grain structure of the oxide scale. But, if phases such as Y_2O_3, $YAlO_3$ or $Y_3Al_5O_{12}$ (YAG-garnet) precipitate at the grain boundaries, the oxide scale plasticity is reduced by inhibiting grain sliding. The incorporation of doping elements may also modify the concentration and distribution of point defects depending on the valency of the active element cation. Moreover, the ionic radius of the active element cation may play a rôle in terms of size-effects on the lattice distorsion and consequently on diffusivity.

v) Formation of an intermediate scale: An active element oxide layer may develop between the outer scale and the alloy, which has a thermal expansion coefficient intermediate between that of the scale and the substrate. However, very few systems develop such a continous intermediate scale.

vi) Mechanical keying (Fig. 7) is perhaps the most important factor in oxide adhesion during long term exposures. The oxides of active elements, formed by internal oxidation, promote the growth of oxide pegs, leading to enhanced adherence.

vii) Vacancy sink: Active elements, or their oxides, may act

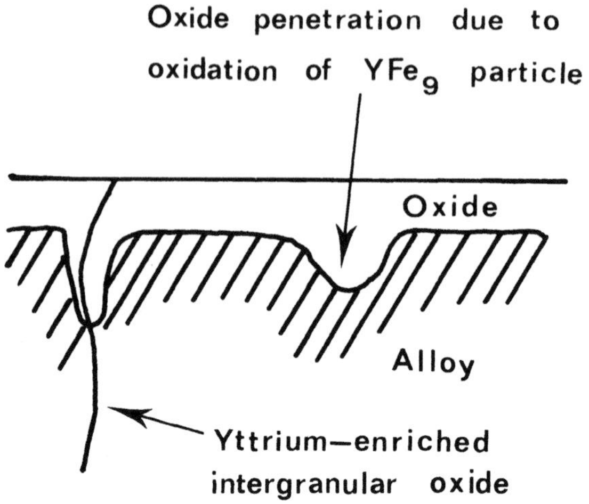

Fig. 7: Mechanical keying of oxide scale.[16]

as vacancy sinks, minimizing the development of voids and porosity at the alloy-scale interface hereby maintaining good scale adherence.

viii) Chemical bonding: Some authors have suggested that active element additions can enhance the chemical bonding at the alloy-scale interface. However, in this case it is difficult to explain the benificial effect of Al_2O_3-dispersions in an alumina-forming alloy.

ix) Prevention of the segregation of impurities: Elements such as sulphur, phosphorus and chlorine can segregate at the scale-alloy interface causing poor oxide-metal adhesion. Addition of active elements or dispersion of their oxides in the substrate are beneficial because they appear to prevent this segregation.

As we can see from this non-exhaustive list of possible effects of active element additions, some seem to be opposite and controversial. However, it does suggest that the main properties, which must be taken in consideration, are:

- free energy or enthalpy of formation of A.E. compounds (oxides, sulphides, oxisulphides),
- crystallographic structure of these compounds,

- Pilling-Bedworth (P.B.) ratio of the compounds,
- solubility of active elements in the main scale constituents and in the metallic substrate,
- diffusivity of active elements in oxide scales and in alloys,
- the main valency of active element ions and their radius,
- the thermal conductivity of oxides which can play a significant rôle in the temperature distribution[17] and consequently on the stress generation in the scale and at the oxide-metal interface,

together with probably several other parameters (like electronic structure, thermal expansion coefficient,...).

From these considerations follows that many properties of active elements have to be considered. Some of the relevant data are readily available and are listed in Table 2 a & b.

3. SOME PROPERTIES OF ACTIVE ELEMENTS

All the elements which are known to improve the scaling behaviour of high temperature alloys have a similar electronic structure for their cations when associated to oxygen in their main oxides. The electronic structures are such that, after ionisation, all the "active elements" possess a similar configuration with eight electrons in their outer shell - ($s^2\ p^6$), like the corresponding rare gas of the preceeding period. For instance:

Y	:	$1s^2\ 2s^2\ 2p^6\ 3s^2\ 3p^6\ 3d^{10}\ 4s^2\ 4p^6\ 4d^1\ 5s^2$
Y^{3+}(Y_2O_3)	:	$1s^2\ 2s^2\ 2p^6\ 3s^2\ 3p^6\ 3d^{10}\ 4s^2\ 4p^6$
Kr	:	$1s^2\ 2s^2\ 2p^6\ 3s^2\ 3p^6\ 3d^{10}\ 4s^2\ 4p^6$

This feature is not so surprising because it is an obvious general rule for elements not belonging to the transition family. Nevertheless, this may explain why some metallic elements such as Mg, Al, Si, Ca, Sc, Ti.. are sometimes considered to be "active".

From a thermodynamical point of view, the main oxides of the active elements have both a high melting temperature (T_f > 2000 K) and a high enthalpy of formation ($-\Delta H_{298} > 1000$ kJ/mole). This implies that these oxides are very stable, and as a matter of fact, more stable than the corresponding sulphides. Some of the sulphides also have high formation enthalpies ($-\Delta H_{298} > 1000$ kJ/mole), mainly when sulphur is associated to oxygen in some oxysulphides, such as Ce_2O_2S, Y_2O_2S or La_2O_2S with $-\Delta H_{298}$ ranging from 1000 to 1700 kJ/mole. This can also

Table 2a

SOME PROPERTIES OF MAIN ALLOYING ELEMENTS AND THEIR COMPOUNDS

	T_f (K)	$-\Delta H$ (kJ)	P.B	α (K^{-1})	E (N/mm^2)	R_{ion} (A)	E_i (eV)	Cryst. Str
Cr	2176			9.4				
Cr_2O_3	2603	526	2.0	9.6		0.64		R a=5.361
CrS		268						α=55.6
Cr_2S_3		222						
Fe	1809			14.6				
FeO	1650	263	1.8	10	25	0.83		CFC a=4.282
Fe_2O_3		543	2.4					
Fe_3O_4	1870	560	2.1	10	25	0.67	24	CSp a=8.394
FeS_2		171						
FeS		200						
Co	1768							
CoO	2070	476	1.7			0.78		CFC a=4.528
Co_3O_4	1220	451	2.0			0.65	25	CSp a=8.110
CoS_2		155						
Co_3S_4		63						
Ni	1725			17.6				
NiO	2223	477	1.68	17.1		0.78	26	CFC a=4.194
NiS		184						
Ni_3S_2		71						
Al	933			25.3				R a=5.127
Al_2O_3	2321	527	1.28	8	39	0.57	53	α=55.16
Al_2S_3	481							

P.B. = Pilling-Bedworth ratio
α = Thermal expansion coefficient
ΔH = Formation enthalpy at 298K
E = Young's modulus
Ei = Ionisation energy

Table 2b

SOME PROPERTIES OF ACTIVE ELEMENTS AND THEIR COMPOUNDS

	T_f (K)	$-\Delta H$ (kJ)	P.B	α (K^{-1})	E (N/mm^2)	R_{ion} (A)	E_i (eV)	Cryst. Str
Si	1685			7.6				
SiO_2	1996	907	2.00	16	10	0.39	102	
SiS		318						
SiS_2		213						
Sc	1810							
Sc_2O_3	2920	1141	1.19			0.83	44	CC a=9.81
ScS		903						
Sc_2S_3		794						
Ti	1933			9.7				
TiO_2	2130	940	1.76	7.5		0.68	91	T a=4.593
TiS		543						c=2.959
Ti_2S_2		418						
Y	1793							
Y_2O_3	2690	1267	1.13			0.89		CC a=10.6
YS		836						
Y_2S_3		836						
Y_2O_2S		1526				1.06	40	
Zr	2126			5				
ZrO_2	2950	1087	1.53	5.5	19	0.80	79	
ZrS_2		577						
Zr_2S_3		669						
Nb	2750							
Nb_2O_5	1783	752	2.37			0.70		M
NbS		447	2.69					O
La	1193							
La_2O_5	2593	744	1.11			1.06	37	R/H a=3.95
La_2S_3		836						c=6.126
LaS		627						
La_2O_2S		1714	1.23			1.22		CC a=11.4
Ce	1077							
CeO_2	2873	1087	1.16			0.90		CFC a=5.41
Ce_2O_3	1960	1254	1.30	13.5		1.20		R/H a=3.9
Ce_2O_2S		1003						c=6.1
Hf	2495			6				
HfO_2	3173	1112	1.60	5.8		0.81	78	C a=5.125
Th	2023			11.2				
ThO_2	3273	1225	1.33	9.5	15	1.10		CFC a=5.596
$ThS_{2.33}$		548						

explain the important rôle played by yttrium (and, may be, by Ce- and La-) in trapping sulphur frequently present in commercial alloys whereby good adherence of the oxide layers is achieved.[18-20]

Some of the active elements, respectively their oxides, seem to be attractive on the basis of the Pilling-Bedworth ratio (Table 2). For example Th, Al, Sc, Y, Ce and La show a rather small value ($1 < P.B. < 1.30$). Another important parameter to be considered is the ratio of the thermal expansion coefficients of oxides and metals α_{ox}/α_{met}.[15] It is desirable to have a ratio close to one, or slightly higher than one, in order to ensure a good mechanical integrity of the oxide layers under thermal cycling conditions. In this respect, the low value for the Al_2O_3/Al ratio ($\alpha_{ox}/\alpha_{met} \ll 1$) is surprising when compared with other systems. From the values compiled in Table 2, it results that there is a lack of data regarding the elastic modulus of many oxides. This is probably an area for future research if mechanical predictions are needed.

It is difficult to draw general conclusions from the available property data collected in Table 2. However the three most relevant active elements Y, La and Ce seem to be characterized by:

- a high stability of their oxides, sulphides and oxysulphides,
- a very low P.B. ratio for their oxides,
- their fairly low melting point, which facilitates plasticity at high temperature.

Despite these conclusions, some elements seem to have an unexpected favourable effect. Thorium and hafnium have a beneficial influence when added as oxide dispersions in spite of the fact that they do not satisfy the previous criteria.[3]

4. RESOURCES - ECONOMIC ASPECTS

Besides the scientific or technical reasons, we have to take into account the availability and cost of extraction and manufacturing of active elements.[21] Moreover their application can be limited by difficulties in metallurgical processing.[22] In this section we shall consider only the main rare earths (R.E.) and related elements (Y). The rare earths are relatively abundant in the earth's crust (Fig. 8).[23] As examples, there is more cerium than nickel and copper and also more yttrium and lanthanum than lead, tin and molybdenum.

The main ores which provide the rare earth metals are

bastnasite, monazite and xenotime (Fig. 9).[24,25] The four most abundant elements are cerium, lanthanum, yttrium and neodymium. The relative abundance depends on the ore and on the mining region (Table 3). The amin countries in which these ores are mined are Australia, China, United States, India, South Africa and Malaysia. The major source of yttrium is xenotime, supplied mainly by Malaysia, Thaïland and China (Table 4, 5, 6).

The development of rare earth demand and the worldwide market in various fields of application are illustrated in Fig. 10. All the metallurgical apllications of rare earths (including the use of R.E. oxide dispersions in ODS alloys) correspond to 20 w% of the total consumption of the three main markets (Fig. 10). The industrial use of rare earths to improve the high temperature oxidation behaviour of refractory alloys or superalloys is relatively low compared to the other metallurgical applications.

As mentioned previously, the price can be a determining factor for the use of active elements, for choosing a particular element and the alloying process. Also, it depends on which form the active element is incorporated into the alloy (pure metal or mischmetal, for example).

For rare earths, the price has remained relatively high during the recent years: 95% Y costs 260 ECU/kg while mischmetal

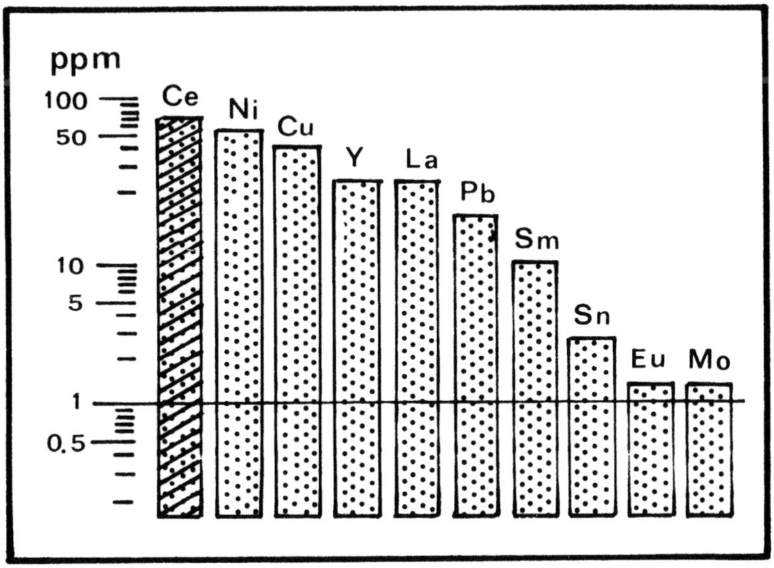

Fig. 8: Relative abundance of rare earths in earth crust.[21]

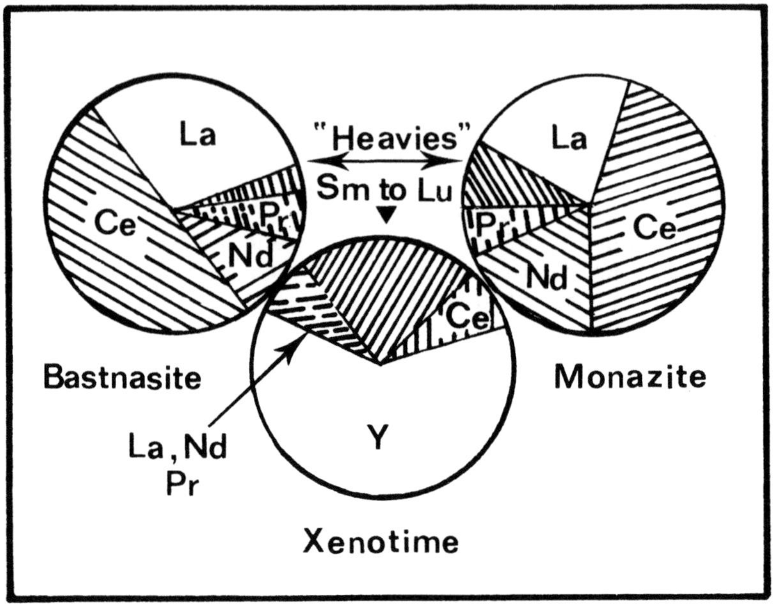

Fig. 9: Composition of the main rare earth ores.[23]

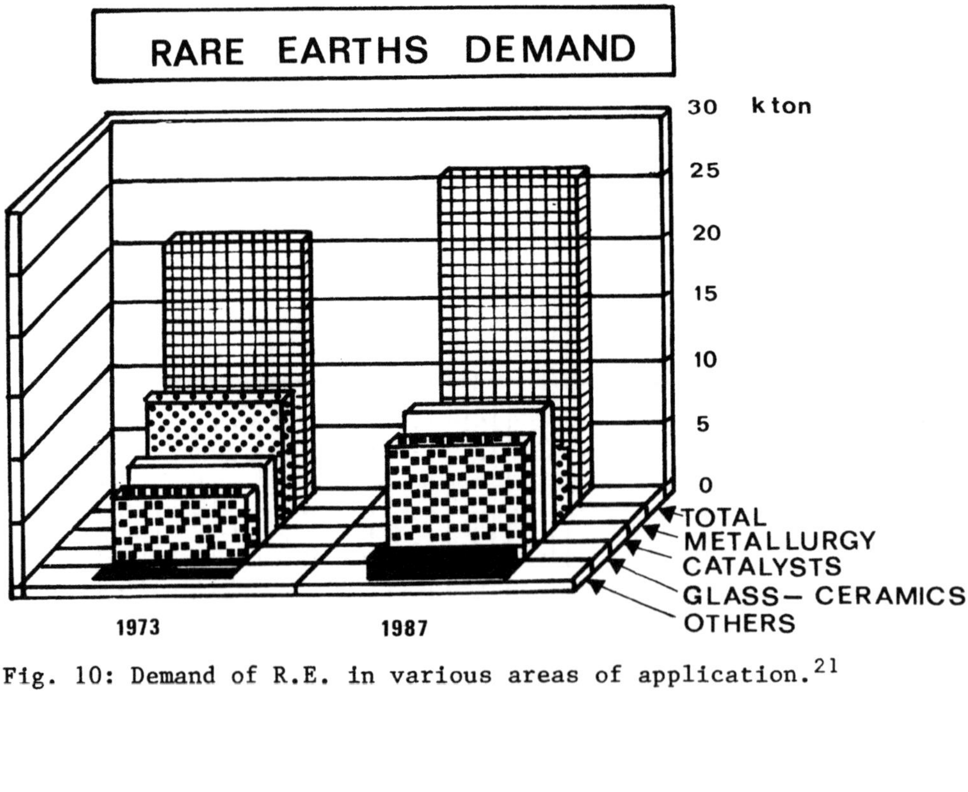

Fig. 10: Demand of R.E. in various areas of application.[21]

Table 3

RARE-EARTH YTTRIUM CONTENTS OF MAJOR SOURCE MINERALS[23]

Rare Earth	Bastnasite		Monazite					Xenotime
	California	China	Eastern Australia	Western Australia	Florida	India	China	Malaysia
● Lanthanum	32.00	27.00	20.20	23.90	17.47	23.00	23.35	0.50
● Cerium	49.00	50.00	45.30	46.03	43.73	46.00	45.68	5.00
Praeseodymium	4.40	5.00	5.40	5.05	4.98	5.50	4.16	0.70
Neodymium	13.50	15.00	18.30	17.38	17.47	20.00	15.74	2.20
Samarium	0.50	1.10	4.60	2.53	4.87	4.00	3.05	1.90
Europium	0.10	0.20	0.10	0.05	0.16		0.10	0.20
Gadolinium	0.30	0.40	2.00	1.49	6.56		2.03	4.00
Terbium	0.01		0.20	0.04	0.26		0.10	1.00
Dysprosium	0.03		1.15	0.69	0.90		1.01	8.70
Holmium	0.01		0.05	0.05	0.11		0.10	2.10
Erbium	0.01	1.00	0.40	0.21	0.04	1.50	0.51	5.40
Thulium	0.02		trace	0.01	0.03		0.51	0.90
Ytrerbium	0.01		0.20	0.12	0.21		0.51	6.20
Lutetium	0.01		trace	0.04	0.03		0.10	0.40
● Yttrium	0.10	0.30	2.10	2.41	3.18	—	3.05	60.80
TOTAL	100.00	100.00	100.00	100.00	100.00	100.00	100.00	100.00

Table 4

WORLD RESERVES OF RARE EARTHS[21]

COUNTRY	Re oxide ores (tons)
AUSTRALIA	753 923
INDIA	1 939 241
UNITED STATES	6 470 911
SOUTH AFRICA	966 820
Other market economy countries	982 984
Pr of CHINA	36 000 000
Other centrally planned economies	500 000
TOTAL	47 613 879

Table 5

THEORETICAL AVAILABILITY OF R.E. ELEMENTS IF THREE MAIN SOURCES WERE MINED AT PRESEND CAPACITY[21]

RARE EARTH OXIDE	PRODUCTION REO (tons/yr)
LANTHANUM	15 000
CERIUM	26 300
PRASEODYMIUM	2 800
NEODYMIUM	8 500
SAMARIUM	670
EUROPIUM	70
GADOLINIUM	410
TERBIUM	27
DYSPROSIUM	140
HOLMIUM	12
ERBIUM	35
THULIUM	7
YTTERBIUM	23
LUTETIUM	10
YTTRIUM	610
TOTAL	54 614

Table 6

MAIN MINING CAPACITIES[21]

ORE / COUNTRY	REO (tons/yr)
MONAZITE/Australia	15 000
BASTNASITE & MONAZITE/China	15 000
BASTNASITE/California	25 000
TOTAL	55 000

is 15 ECU/kg;[22] therefrom could result a driving force to incorporate active elements only near the surface using coatings (MCrAlY) or superficial treatments (ion implantation, laser alloying...). Moreover, taking advantage of the non-equilibrium conditions of these surface treatments, this could also overcome problems of limited solubilities of active elements in high temperature alloys.

5. CONCLUSIONS

The benificial effect of active elements added to high temperature alloys in order to improve their oxidation resistance is now well established. They act essentially on the adherence of the protective oxide scales but detailed understanding of their influence remains to be further explored in spite of the many mechanisms already proposed.

It appears from this short survey that it is difficult to explain the effect of active elements in terms of physical or chemical properties and to make predictions for the design of high temperature alloys. This difficulty is due partially to the lack of appropriate data on active elements and their compounds, e.g. with regard to their solubility in various alloys and oxides, their diffusivity, thermal expansion coefficients, Young's modulus and even some of the thermodynamical values like free energy of formation**.

** Anybody who wishes to contribute, in a cooperative spirit, by supplying such data to built something like a data-bank on active elements is welcome to contact Dr. F. ARMANET (UTC) or Dr. E. LANG (JRC Petten).

ACKNOWLEDGEMENTS

The authors are very grateful to MM. C. Salles and P. Falconnet (Rhône-Poulenc), Mr. J.H. Davidson (Imphy SA) and Mr. Tachikart (UTC) for their help in collecting data on ore resources, metallurgical aspects and thermodynamical values.

REFERENCES

1. W.T. GRIFFITHS and L.B. PFEIL, U.K. Patent n° 459848, 1937.

2. L.B. PFEIL, U.K. Patent n° 574088, 1945.

3. D.P. WHITTLE and J. STRINGER, Phil. Trans. Roy. Soc. London, 1980, A295, 309.

4. D.G. LEES, Oxid. Met., 1987, 27, 75.

5. K.L. LUTHRA and C.L. BRIANT, Oxid. Met., 1988, 30, 257.

6. J.G. SMEGGIL, N.S. BORNSTEIN and M.A. DECRESCENTE, Oxid. Met., 1988, 30, 259.

7. D.G. LEES, Oxid. Met., 1988, 30, 267.

8. J.H. DAVIDSON, P. LACOMBE, A.M. HUNTZ, C. ROQUES-CARMEN, J.C. PIVIN and D. DELAUNAY, Proc. Int. Conference "Behaviour of High Temperature Alloys in Aggressive Environments", Petten, 1979, Commission of the European Communities, p. 209.

9. A.M. HUNTZ, Mater. Sc. and Eng., 1987, 87, 251.

10. C.S. GIGGINS, B.H. KEAR, F.S. PETTIT and J.K. TIEN, Met. Trans., 1974, 5, 1685.

11. J.K. TIEN and F.S. PETTIT, Metall. Trans., 1972, 3, 1587.

12. G.C. WOOD and F.H. Stott, Proc. "6th International Conference on High Temperature Corrosion", San Diego, CA, 1981, NACE, p. 227.

13. F.A. GOLIGHTLY, F.H. STOTT and G.C. WOOD, J. Electrochem. Soc., 1979, 126, 1035.

14. F. ARMANET and G. BERANGER, Ann. Chim. Fr., 1986, 11, 249.

15. F. ARMANET and G. BERANGER, J. Chim. Phys., 1986, 83, 735.

16. I.M. ALLAM, D.P. WHITTLE and J. STRINGER, Oxid. Met., 1978, 12, 35.

17. M. LALLEMANT, G. BERTRAND, D. CIOSMACK, J.P. LARPIN and A. SANFELD, at this conference.

18. J.G. SMEGGIL, Mat. Sc. and Eng., 1987, 87, 261.

19. J.G. SMEGGIL, A.W. FUNKENBUSCH and N.S. BERNSTEIN, Metal. Trans., 1986, A17, 923.

20. J.L. SMIALEK, Metal. Trans., 1987, A18, 164.

21. P. FALCONNET, private communication.

22. J. DAVIDSON, at this conference.

23. B.T. KILBOURN, J. Metals, May, 1988, 22.

24. R.G. BAUTISTA, J. Metals, May, 1988, 21.

25. J.B. HEDRICK, Ceramic Bulletin, 1988, 67, 858.

SESSION 2

EFFECTS OF ACTIVE ELEMENTS IN SYNTHETIC OXIDES

EFFECT OF ACTIVE ELEMENTS ON DIFFUSION PROPERTIES OF SYNTHETIC OXIDES

A. ATKINSON

MATERIALS DEVELOPMENT DIVISION
HARWELL LABORATORY
DIDCOT, OX11 0RA, UK

ABSTRACT

The current understanding of the effect of dopants on diffusion through the lattice and along grain boundaries in oxides is reviewed. The experimental studies for grain boundary diffusion, which is the more relevant to oxidation, show apparently contradictory results and these are discussed critically in terms of the experimental conditions employed. Guided by the results of computer simulations, a possible rationalisation is proposed to explain the results on synthetic oxides and the influence of reactive elements on transport through oxidation scales.

1. INTRODUCTION

The beneficial effects of the so-called reactive elements are currently regarded as being of two kinds:

(i) an improvement in oxide to metal adhesion, and

(ii) a reduction in oxidation rate for some systems.

The first effect occurs quite widely and is particularly important for alloys that form alumina scales. The second effect is not seen for alumina-formers, but is important for scales growing by outward diffusion of cations through the oxide (when the reactive element is absent). This is particularly important for alloys that form chromia scales. In both cases there is much evidence[1,2] to indicate that the reactive element addition modifies the rate at which reactants are transported across the oxide scale; either to slow down the dominant diffusing species (and hence reduce the oxidation rate), or to modify the relative rates of metal and oxygen diffusion through the oxide (which might influence oxide/metal adhesion).

As a result there has been a number of attempts to measure the influence of "reactive elements" on the diffusion properties of oxides typical of those that are formed during high temperature oxidation. The aim of this paper is to review these studies, to compare their results and to draw some general conclusions. There is no universally acknowledged definition of a reactive element, but for the present discussion we will take it to be "an oxide which is more stable than the the host oxide with respect to both low oxygen pressure and high temperature". The experimental systems discussed in this paper are:

(i) Ce, Y and Cr in nickel oxide (because this is a good model oxide for those normally growing by outward diffusion of metal ions),
(ii) Y in chromium oxide (because of its technological importance), and
(iii) Cr in magnesium oxide (because it is similar to Cr in NiO to a first approximation).

We first consider the influence of dopants on bulk lattice diffusion in oxides such as NiO and how this relates to oxidation behaviour. Then the experiments in the systems mentioned above (which concentrate on grain boundary diffusion) are discussed and the results compared with predictions of computer models of the diffusion process. Finally, general conclusions are drawn regarding the relationship between the experiments and the reactive element effect.

2. EFFECT OF DOPANTS ON BULK DIFFUSION

2.1 Homovalent Dopants

The effects of homovalent cation substitutional dopants on diffusion have been studied extensively in the rocksalt-structured oxides NiO, CoO, FeO and MgO. The cations are sufficiently similar in size that these oxides form complete ranges of solid solution. It is found[3] that the influence of the homovalent dopant on self-diffusion of the host cation is small, unless the concentration of the dopant is large. The main effect is on the energy of formation, h_f, of the cation vacancy responsible for diffusion which varies linearly between the two end members of the solid solution i.e.

$$h_f = (1-x)h_{fa} + xh_{fb} \qquad (1)$$

where a and b are the host and dopant and x is the cation fraction of b. This leads to an activation energy which varies linearly with composition. A typical example is Mg doping of CoO where the Mg reduces the diffusion coefficient of Co because it is more difficult to form cation vacancies in MgO than in CoO.[4]

Extensive solid solution is only possible when the ions are similar in size. If the ionic sizes are very different then the solid solution is correspondingly limited and stronger interactions are generated between the dopant and defects on the host lattice. For example, if the dopant is larger than the host it will tend to attract host vacancies to minimise the strain associated with the size mismatch. This attractive interaction can be described by a binding energy between the dopant and the host vacancy. If the dopant is also relatively immobile then the dopant acts as a trap for host vacancies and would thereby be expected to reduce diffusion of the host cation. Typical values for the binding energy between defects due to size mismatch are up to 0.5eV.[5]

2.2 Heterovalent Dopants

In oxides, which are mainly ionically bonded materials, heterovalent dopants carry an effective charge with respect to the host lattice and therefore must be compensated electrically by defects with an opposite effective charge. A typical example is a trivalent dopant, such as Al, in an oxide of a divalent cation, such as NiO. For this particular example one nickel vacancy (with an effective double negative charge) will be created for every two aluminium dopant ions (with effective single positive charge). According to this simple picture the concentration of vacancies is one half the concentration of dopant and the activation energy for Ni diffusion should be the enthalpy of vacancy motion, h_m. Experimental measurements[6] of Ni diffusion in Al-doped NiO have shown that this is not the case (Figure 1). The reason is that the oppositely charged defects have an attractive interaction which binds them together in the same kind of way as described earlier for homovalent oversize dopants. The coulombic interaction between the charged defects is, however, stonger than the strain interaction and hence the binding energy tends to be greater; up to about 1.0 eV. At sufficiently low temperature, or sufficiently high doping, pairs of defects will interact to form larger agglomerates and lead eventually to nucleation and precipitation of a second bulk phase; in this case the spinel, $NiAl_2O_4$. It should be emphasised that these interaction energies are such that most of the defects are bound in pairs, or larger groups, even at 1200°C and a doping level of about 0.5% (as cation fraction).

3. EFFECT OF DOPANTS ON OXIDATION

The growth of an oxide scale by thermal oxidation can be related to the diffusion properties of the oxide using Wagner's theory.[7] The basic theory has been extended[8] to oxidation of alloys where the alloying element is soluble in the host oxide and dopes it. Application of the theory is not straightforward because it is necessary to know the diffusion coefficients of both the host and the dopant as a function of composition in both the alloy and the oxide. Nevertheless the theory has been shown to work well for Co(Fe) and Co(Ni) alloys. It successfully accounts for the dependence of rate constant on alloy composition[9] and the way in which alloy composition varies with depth from the oxide/metal interface. In these successful cases there is extensive solid solution in the oxide and the oxidation rate is controlled by diffusion through the bulk oxide lattice.

When the alloying element becomes a heterovalent dopant in the oxide the theory does not appear to work. This is because in such cases solid solution tends to be very limited and two-layered (duplex) scales are formed whose inner layer grows by inward oxygen transport. An example of this is the oxidation of dilute Ni(Cr) alloys.[10]

A further complication that arises when highly protective oxides are formed is that transport through the oxide is mainly along short circuit pathways, such as grain boundaries, rather than through the oxide crystal lattice.[11] In the following sections we discuss experimental and theoretical studies of how dopants influence diffusion along grain boundaries in oxides and their relevance to the reactive element effect in oxidation.

4. EFFECT OF DOPANTS ON GRAIN BOUNDARY DIFFUSION

There have been several attempts to measure the influence of dopants on diffusion along grain boundaries in oxides and these are summarised in Table 1. From this summary it is apparent that there is no simple pattern emerging and it is therefore necessary to examine the studies in more detail in order to try and understand why they appear to indicate contradictory effects.

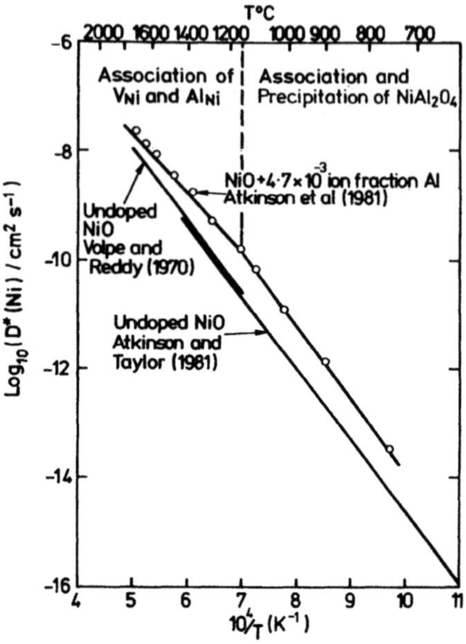

Figure 1: Tracer diffusion of Ni in single crystals of undoped and Al-doped NiO.[6]

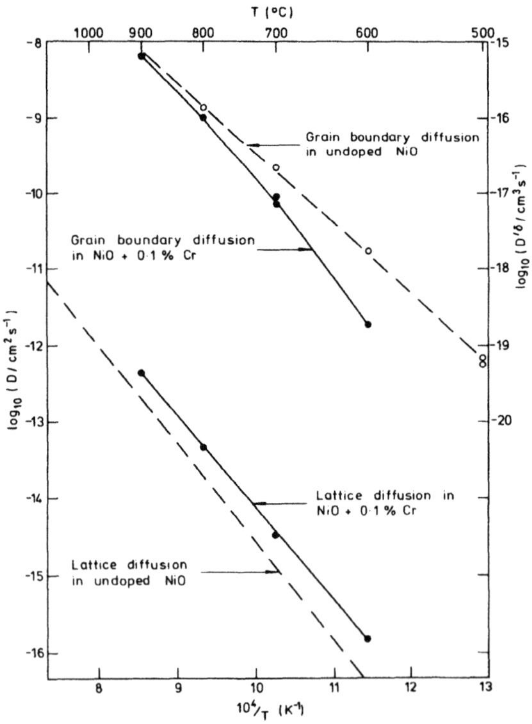

Figure 2: Lattice and grain boundary diffusion of Ni in undoped and Cr-doped (nominally 0.1%) polycrystalline NiO.[10]

Table 1: Attempts to measure the influence of "reactive elements" on grain boundary diffusion in oxides.

Diffusant	Dopant	Host	Effect	Source
Ni	Ce	NiO	none	Chadwick & Taylor,1984[12]
Ni	Y	NiO	none	Moosa et al., 1985[13]
Ni	Cr	NiO	decrease	Atkinson et al., 1987[10]
Cr	Cr	MgO	increase	Osenbach & Stubican,1984[14]
Cr	Y	Cr_2O_3	increase	Park et al., 1987[15]

4.1 Experiments on NiO

The experiments on NiO listed in Table 1 were all carried out using polycrystalline oxide formed by the complete oxidation of Ni, or Ni alloy, foil. The Y and Cr dopants were introduced by alloying additions to the Ni whereas the Ce was introduced as a ceria coating on the Ni surface.

The experiments carried out by Chadwick and Taylor[12] showed that the Ce dopant had no detectable influence on diffusion of Ni along NiO grain boundaries in the temperature range 600-800C. This is in apparent conflict with the observation that Ce causes a large reduction in the rate of Ni oxidation in the same temperature range, by an amount consistent with all grain boundaries being blocked as diffusion pathways for Ni. This apparent difference might be explained by the doping level in the diffusion experiments being insufficient to have an obervable effect. Chadwick and Taylor examined the distribution of Ce using microanalysis techniques. These showed that in the region of the specimens where diffusion was measured the average concentration of Ce was below the limit of detection (<0.02% by weight) whereas the average concentration in the region where the oxidation experiments suggested grain boundary diffusion was blocked was between 0.1 and 1.0%. SIMS imaging of grain boundaries in the diffusion experiments indicated that Ce was present on the boundaries in the diffusion experiments, but subsequent work by Moon[16] revealed that only about 25% of the boundaries had Ce detectable. It would therefore appear that in the diffusion experiments the doping level on most boundaries was too low to have a detectable influence.

Moosa, Rothman and Nowicki[13] also found that Y doping had no effect on the diffusion of Ni along NiO grain boundaries. They concluded that this was consistent with the influence of Y on Ni oxidation, which could be accounted for by a change in oxide grain size; the diffusion along any single boundary unit length being unaffected. The alloy used to fabricate the diffusion specimens and in the oxidation measurements contained nominally 0.1% Y by weight. They examined the distribution of Y in their samples by electron probe microanalysis and found that the distribution was consistent with the Y being present as relatively large precipitates with much of the intervening material having an undetectable level of dopant. It therefore seems likely that in these experiments, as in the previous ones with Ce, most of the grain boundaries were not doped significantly. More recent studies[17] have revealed that Y, like Ce, does have a strong inhibiting effect on the oxidation of Ni at intermediate temperatures.

Atkinson, Smart and Taylor[10] carried out similar experiments with Cr as the dopant. In this study the method of sample preparation was slightly different in that some Cr was additionally supplied through the vapour phase during growth of the oxide. This should encourage more uniform doping of the oxide which would be aided further by the higher diffusivity of Cr relative to Ce[18] (and probably Y). Nevertheless, microanalysis (SIMS) and the observed diffusion behaviour both indicated that the doping was non-uniform, both with depth and laterally from boundary to boundary. Tracer penetration profiles and autoradiography showed that most of the grain boundaries were no longer functioning as fast diffusion pathways as a result of an average dopant concentration in the diffusion zone of about 0.05% (Cr/Ni atomic ratio). From the known[19] segregation coefficient of Cr to NiO grain boundaries at the fabrication temperature the Cr/Ni atom ratio at the grain boundaries was estimated to be about 2%. The residual grain boundary diffusion that could be detected (Figures 2 and 3) was also slower for most of the doped boundaries relative to undoped boundaries; especially at the lower temperatures. The oxidation rate of the alloy from which the diffusion specimens were fabricated (0.1% Cr) was, however, slightly faster than that of unalloyed Ni; indicating that the boundaries had not been blocked by the dopant in the oxidation conditions.

4.2 Cr-doped MgO

Osenbach and Stubican[14] studied the influence of Cr doping on the diffusion of Cr along single boundaries in MgO bicrystals grown from the melt. The bicrystal boundaries were symmetrically tilted about the [100] crystal axis with a misorientation of 15 degrees. They found that the grain boundary diffusion parameter ($sD_{gb}\delta$, where s is the segregation coefficient and δ the boundary width) increased approximately linearly with the bulk Cr concentration (Figure 4). This they interpreted as a dopant increasing the concentration of cation vacancies at the boundary, as for bulk diffusion. At the temperatures of their experiments, the segregation coefficient[20] for Cr to MgO boundaries is about 3 and therefore the maximum concentration of Cr at the boundaries in their experiments would have been about 2% (as Cr/Mg atom ratio).

4.3 Y-doped Cr_2O_3

Recently Park, King and Rothman[15] reported studies of Cr diffusion in undoped and Y-doped chromia polycrystalline specimens fabricated using conventional ceramic sintering routes from powders. The doping level was 0.09% by weight as yttria. The diffusion measurements revealed that grain boundary diffusion of Cr was increased by about a factor of 3 by the Y dopant, which is contrary to the retarding effect Y has on the oxidation of Cr. Furthermore, the diffusivity measured in the undoped samples was too low, by a factor of about a thousand, to account for the growth of chromia during oxidation. The authors characterised their samples using STEM, but were unable to detect any Y at grain boundaries even in specimens doped to 1.35%; although second phase precipitates were located. The concentration of Y at the grain boundaries was thus probably less than about 0.5%.

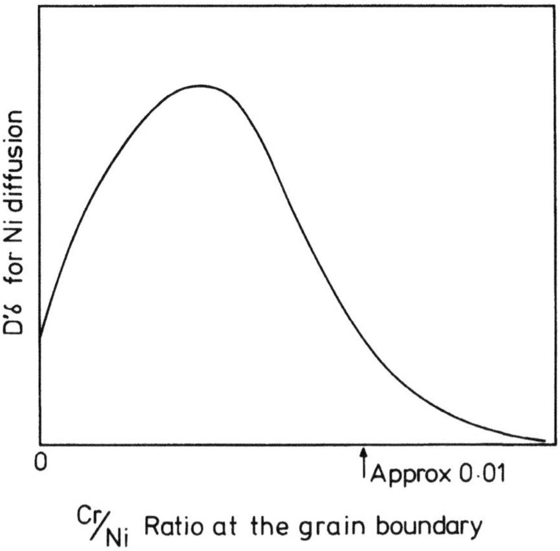

Figure 3: Probable relationship between the grain boundary diffusion coefficient of Ni in NiO and the Cr content in the boundary (schematic).

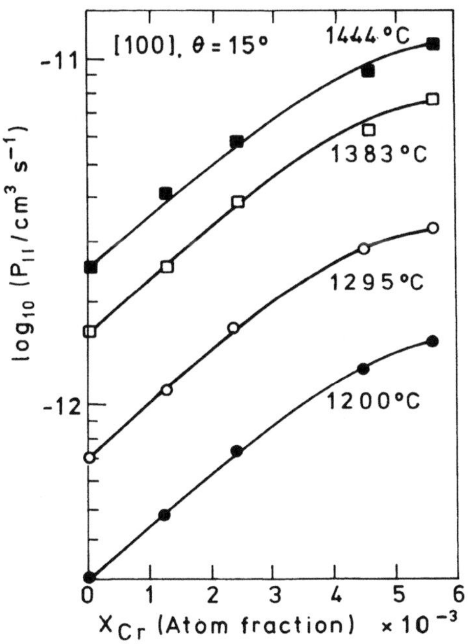

Figure 4: Diffusion parameter ($P = s\,D_{gb}\delta$) for Cr diffusion along [100] tilt boundaries in MgO bicrystals as a function of bulk Cr concentration.[14]

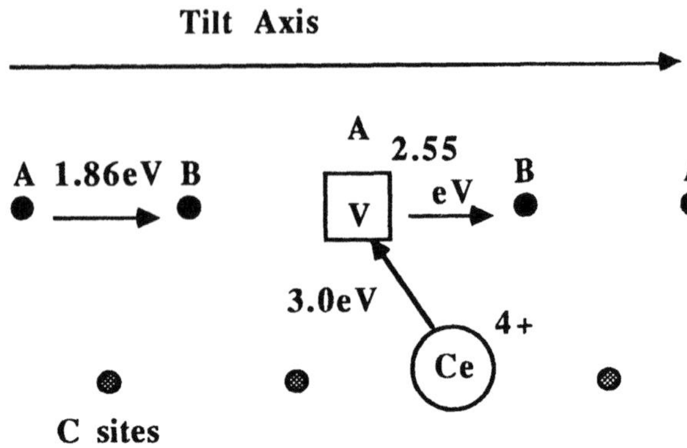

Figure 5: Computed energies for cation-vacancy jumps in the (310)/[001] tilt boundary of NiO

4.4 Computer Modelling of Ce-doped NiO

Duffy and Tasker[21] have carried out extensive computer modelling studies of structure, point defects and diffusion in NiO grain boundaries. In particular, they considered[22] the behaviour of Ce in a symmetrical [100] tilt boundary with a (310) boundary plane. In the undoped boundary the simulations showed that diffusion of Ni along the [100] direction in the boundary took place by vacancies hopping in a linear chain of two crystallographically different sites (labelled A and B in Figure 5) with a migration energy of 1.86 eV. When Ce was introduced into the boundary it was found to interfere with this motion in two ways. In the first, the Ce preferentially occupies the A and B sites and the charge compensating vacancy is strongly bound on a neighbouring site. Since the Ce is much less mobile than Ni,[18] the Ce-vacancy pair blocks the fast diffusivity chain. In the second, the Ce occupies a site which is not part of the high diffusivity chain, but the compensating vacancy is effectively trapped on a neighbouring A or B site because the activation energy for Ni to jump along the chain is increased to 2.55 eV. In both cases vacancies must detour into the crystal lattice to bypass the Ce. For linear diffusion along the chain it was estimated that about 1% of the fast diffusion sites would have to be blocked in order to reduce the effectiveness of the whole chain to being only a factor of two faster than lattice diffusion.

5. OVERVIEW

5.1 Rationalisation of Diffusion Studies

At the present time the different diffusion studies cannot be rationalised if simply taken at face value. Guided by the theoretical studies and the behaviour established for lattice diffusion it is reasonable to assume that the general principles of charge compensating defects and strong interactions between defects also apply to grain boundaries. However the consequences resulting from a given average concentration of dopant are much greater for grain boundary diffusion than for lattice diffusion because:

(i) segregation leads to greatly enhanced concentrations at the boundary, and
(ii) boundary diffusion is almost one-dimensional (pipe diffusion).

Thus for doping by a higher valency cation (e.g. Cr in NiO or MgO) the relationship between boundary diffusivity and dopant concentration is expected to be as shown in Figure 3. At low concentrations and/or high temperatures the increased defect concentration will increase the host diffusion rate (e.g. Cr in MgO), but as the concentration of dopant increases or the temperature is lowered the fast diffusivity pathways become blocked; provided that the dopant has lower mobility than the host (e.g. Cr in NiO). The blocking effect could be regarded as the local formation at the boundary of atomic arrangements that are similar to those in the next phase in the bulk phase diagram (in these examples a spinel) and whose transport properties are qualitatively similar to the bulk second phase in comparison with those of the host. These are embryonic precursors to second phase formation proper. Thus Cr is expected to reduce the diffusion of Ni along NiO grain boundaries because Ni diffusion is slower in $NiCr_2O_4$ than in NiO.[23] The different effect observed in MgO could possibly be accounted for by one, or a combination of, the following:

(i) the temperature was higher in the MgO experiments (which would break up the binding between defects);

(ii) Cr diffusion in MgO will be controlled by a Cr-vacancy interchange (which is different in kind from the Ni-vacancy interchange for Ni diffusion in NiO);

(iii) the ratio of diffusivity of Cr in $MgCr_2O_4$ to that of Cr in MgO is likely to be greater than the ratio for Ni in $NiCr_2O_4$ to that of Ni in NiO.

If this rationalisation of the effects of dopants on grain boundary diffusion is correct then Y doping would be expected to reduce diffusion of Cr along chromia boundaries because: Y has a larger ionic radius than Cr and will trap Cr vacancies as a result; yttria is more refractory than chromia and therefore Cr diffusion is likely to be slower in yttria than in chromia; and Y diffusion in chromia is likely to be slower than Cr diffusion. Therefore the experiments for Ce and Y in NiO and for Y in chromia have given unexpected results on this basis. The characterisation of the samples used suggests that this could have been due to doping of the boundaries being too low and/or too few boundaries doped.

5.2 Relevance to the Reactive Element Effect

It is generally accepted that the reactive elements modify diffusion in chromia scales by reducing outward transport of Cr along grain boundaries and, perhaps, increasing inward transport of oxygen. Modification of diffusion through alumina scales is less clear. The usual reactive elements (Y, Ce, La, Zr, etc.) tend to be those which form highly refractory oxides having significant oxygen diffusion, but very low cation diffusion.[24] Based on the general guidelines outlined above, these elements would be expected to be effective in blocking diffusion of Cr along chromia grain boundaries. They would not be expected to have a similar effect in alumina because cation transport is already very low in alumina and alumina scales may grow by inward transport of oxygen.[2] Thus in order to slow alumina scale growth even further one would require a dopant which forms an oxide with oxygen diffusivity even lower than in alumina itself.

On this same basis Cr, Al and Si (in addition to the ones mentioned above) should be considered as "reactive elements" for the less protective oxides such as NiO and magnetite.

6. CONCLUSIONS

No direct link has been established between the effect of "reactive elements" on grain boundary diffusion in oxides and the RE effect in oxidation.

The balance of evidence indicates that dopants will modify grain boundary host diffusion towards that of the dopant oxide.

Concentrations of at least a few % at the boundary are required.

Contradictory results from diffusion experiments are probably due to lack of control over the dopant distribution.

ACKNOWLEDGEMENT

This review was compiled as part of the Underlying Research Programme of the UK Atomic Energy Authority.

REFERENCES

1 D. P. WHITTLE and J. STRINGER; Phil. Trans. Roy. Soc. Lond., 1980, **A295**, 309-329.
2 A. M. HUNTZ; Mater. Sci. Eng.,1987, **87**, 251-260.
3 V. G. SCHWIER, R. DIECKMANN and H. SCHMALZRIED; Ber. Bunsen. Phys. Chem., 1973, **77**, (6), 402-408.
4 R. DIECKMANN; Solid State Ionics, 1984, **12**, 1-22.
5 W. C. MACKRODT; Solid State Ionics, 1984, **12**, 175-188.
6 A. ATKINSON, A. E. HUGHES, and A. HAMMOU; Philos. Mag. A, 1981, **43**, (5), 1071-1091.
7 C. WAGNER; Z. Phys. Chem. B, 1933, **21**, (1/2), 25-41.
8 C. WAGNER; Corros. Sci., 1969, **9**, 91-109.
9 W. W. SMELTZER; in " Oxidation of Metals and Associated Mass Transport", (eds. M. A. Dayananda et al.), 109-129; 1986, Warrendale PA, The Metals Society.
10 A. ATKINSON, D. W. SMART and R. I. TAYLOR; Werkst. u. Korros., 1987, **38**, 704-709.
11 A. ATKINSON; Rev. Mod. Phys., 1985, **57**, (2), 437-470.
12 A. T. CHADWICK and R. I. TAYLOR; Solid State Ionics,1984, **12**, 343-351.
13 A. A. MOOSA, S. J. ROTHMAN and L. J. NOWICKI; Oxidation of Metals, 1985, **24**, (3/4), 115-132.
14 J. W. OSENBACH and V. S. STUBICAN; J. Amer. Ceram. Soc., 1983, **66**, (3), 191-195.
15 J-H. PARK, W. E. KING and S. J. ROTHMAN; J. Amer. Ceram. Soc., 1987, **70**, (12), 880-885.
16 D. P. MOON; 1987, D. Phil. Thesis, University of Oxford, UK.
17 P. J. GEORGE, M. J. BENNETT, H. E. BISHOP and G. DEARNALEY; AERE Report R12873, Harwell Laboratory, UK, 1988.
18 A. ATKINSON and R.I. TAYLOR; J. Phys. Chem. Solids, 1986, **47**, 315-320.

19 B. BENDER, D. B. WILLIAMS and M. R. NOTIS; J. Amer. Ceram. Soc., 1980, **63**, (9/10), 542-546.

20 M. F. YAN, R. M. CANNON and H. K. BOWEN; J. Appl. Phys., 1983, **54**, (2), 764.

21 D. M. DUFFY and P. W. TASKER; Philos. Mag. A, 1986, **54**, 759.

22 D. M. DUFFY and P. W. TASKER; Report TP 1155, Harwell Laboratory, UK,1986.

23 C.E. BIRCHENALL; in "Mass Transport in Oxides", (eds. J.B. Wachtman and A.D. Franklin), NBS Special Publication 296 (Washington, 1968), 119-127.

24 Hj. MATZKE; Advances in Ceramics, 1986, **17**,1-54.

EFFECTS OF IMPURITIES ON THE MECHANICAL PROPERTIES OF OXIDES

J. Castaing

Laboratoire Physique des Matériaux

C.N.R.S. Bellevue, F-92195 Meudon-Cédex

ABSTRACT

The amount of work on the mechanical properties of oxides is limited, except for Al_2O_3. The influence of impurities on these properties has been seldomly examined in detail.

In this survey, we briefly summarise the mechanisms by which a solid oxide responds to an applied stress: motion of extended defects (cracks, twins, dislocations ...) and/or of point defects (Nabarro creep ...). By presenting typical examples, mostly for Al_2O_3, we illustrate the different behaviours. The rôle of impurities will be discussed for each kind of mechanism.

1. INTRODUCTION

When a metal or an alloy is submitted to a heat treatment in an oxidizing atmosphere, it is readily covered by an oxide layer. If this layer maintains its integrity, the oxidation kinetics decrease, and the metallic substrate becomes protected by a barrier to further oxidation. This phenomenon displays large variabilities depending on the dopants incorporated in the substrate, or on the preliminary surface treatments. It is also clear that one of the important parameters is the deformation behaviour of the protective layer under stress in the presence of impurities and dopants. Although there is a substantial amount of investigations on the mechanical properties of oxides, the influence of foreign atoms on the various mechanisms of deformation has not received much attention. When submitted to a stress, a solid responds in various ways: elastic deformation, dislocation glide or climb, transport of matter, twinning, crack propagation... being the basic mechanisms of deformation.

Depending on temperature and stress, one of the above mentioned mechanisms may be dominant. In oxidation processes, they all can be encountered due to the great variability of temperature and stress conditions.

In the following, we will not go into the details of the mechanical behaviour of oxides. There are several reviews, covering various aspects: Cannon et al.[1] have recently examined the creep mechanisms of ceramics; Bretheau et al.[2] have dealt with the deformation of binary oxides emphasizing the case of single crystals; the plasticity of ceramic materials has been reviewed in the proceedings of the Penn State conference[3] and many results on the properties of MgO and Al_2O_3 can be found in Ref.4.

The examples which we consider in the following, are often taken from these references. We will almost exclusively examine results concerning Al_2O_3 and this for two reasons: (i) alumina is encountered in the oxidation of many high-temperature alloys; (ii) the behaviour of alumina under stress has been extensively investigated, from single crystals to fine-grained ceramics, involving a great variety of deformation mechanisms.

2. DEFORMATION MECHANISMS

2.1. Dislocation Motion

Activation of dislocation sources can be achieved in NaCl-type oxides, even at low temperatures where single crystals exhibit plastic deformation down to T = 4.2 K, Fig. 1.[5] On the contrary, alumina exhibits a brittle behaviour below 1273 K for tests performed at normal pressure, Fig. 1. Such a situation is typical for covalent single crystals; it is also likely to be encountered for Cr_2O_3 single crystals, since this compound has the same crystal structure and chemical bonds with an ionicity factor smaller than for Al_2O_3.[8]

The brittleness of those materials is even greater for polycrystals. Indeed, for a solid-state body with randomly oriented grains, deformation with no crack or cavity formation is achieved by dislocation glide provided the grains possess five independent slip systems (Von Mises' condition).[9] This condition is generally not met for oxides when considering the easy glide systems viz. $\langle 1\bar{1}0 \rangle \{110\}$ for NaCl-type crystals, basal and prism plane slip for alumina, which correspond to the deformation data displayed in Fig. 1.

Failure of ceramic oxides will not occur when an additional plastic deformation mechanism is activated, e.g. by an unusual slip system, twinning, diffusional processes... Temperature is one parameter which will affect the balance between all these mechanisms, impurity or doping being another one.

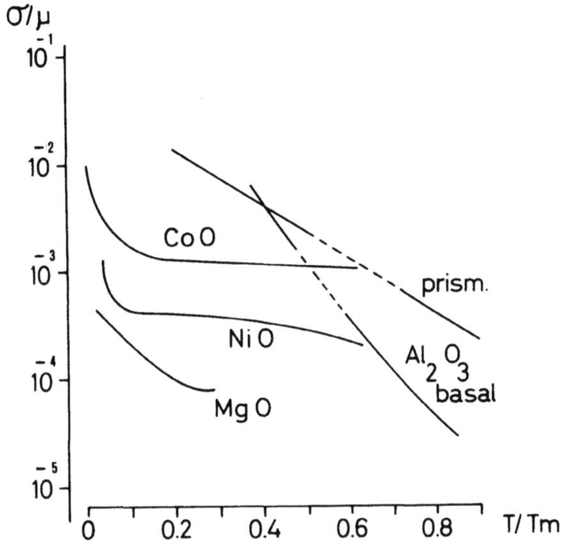

Figure 1: Typical dependence of flow-stress σ on temperature for various oxide single crystals (μ = elastic shear modulus; T_M = melting temperature). MgO, NiO, CoO display plastic behaviour down to very low temperatures. Alumina on the contrary is brittle below 0.6 T_M; to achieve plasticity, hydrostatic pressures have to be superimposed.[6]
The following constants have been used:

	Al_2O_3	CoO	NiO	MgO
μ (10^{11} Pa)	3,6	0.16	1	1,3
T_M (K)	2320	2075	2235	3098
Reference	6	5	7	2,8

In the following, we describe the influence of foreign cations on the glide of dislocations in the various slip systems of Al_2O_3 single crystals. In a first approximation, the crystal structure can be described as hexagonal; detailed studies have been performed for basal and prism plane slip, which have

$^1/_3$ <11$\bar{2}$0> and <10$\bar{1}$0> Burgers vectors, respectively.[10] Rhombohedral slip is more difficult to activate and received less attention.[10]

Mechanical data have been collected, mostly through constant strain rate compression of single crystals. The results are plotted as stress-strain curves. Typical examples are given in Fig. 2 for various iso- and aliovalent doping, the basal slip being activated.[11]

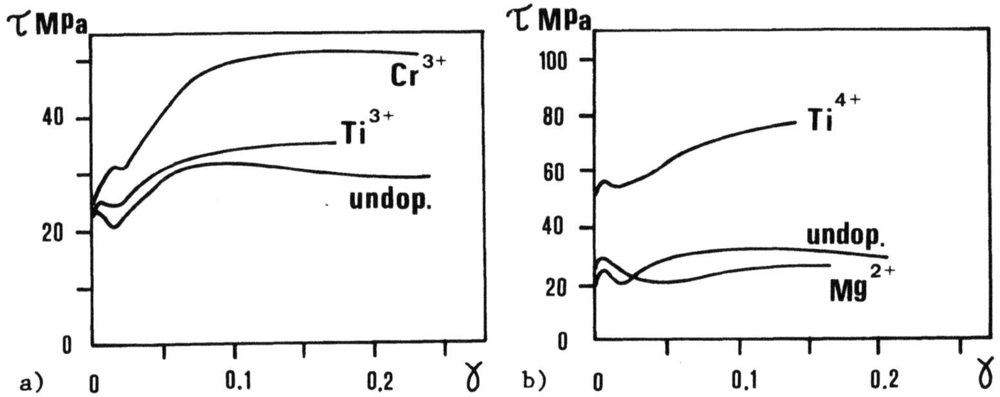

Figure 2: Shear stress τ vs shear strain γ curves for undoped and doped Al_2O_3 single crystals.[12]
a) the undoped and the 0.047% Ti^{3+} doped crystals were deformed at 1793 K and the 0.2% Cr^{3+} doped crystal at 1773 K.
b) Aliovalent doped crystals were deformed at 1793K. 0.002% Mg^{++}; 0.0047% Ti^{4+}.

After a yield drop at the beginning of plastic deformation, a steady state flow is achieved, followed by some strain-hardening. Impurities cause different changes in the flow stress depending on the stage of the stress-strain curve, Fig. 2. The critical resolved shear stress (CRSS), taken as the minimum after the yield drop, is strongly influenced by the concentration c of Cr^{3+} and Ti^{4+}, with hardening rates d(CRSS)/dc of $\mu/30$ and $\mu/2$, respectively (μ = elastic shear modulus).[12] These experimental results can be related to the average strengthening of obstacles to dislocation glide, due to the presence of impurities. They are explained on the basis of elastic interactions even for aliovalent impurities; the tetragonal strain field associated with defect complexes (aliovalent impurities and charge compensating defects) is a stronger obstacle than that resulting from substitutional isovalent impurities.[13]

The influence of foreign ions on the steady state deformation may be quite different from the one on the CRSS. For instance, while Mg^{++} hardens Al_2O_3 at the beginning of the deformation, it has a softening effect at large strains (Fig. 2b). The stress in these conditions is due to an equilibrium between strain-hardening and recovery. For basal slip in Al_2O_3, the increasing amount of dislocation loops impedes further deformation of the crystals. The equilibrium stress is controlled by the loop annihilation, which depends on the diffusivity of the species; the relation between impurities and flow stress is, therefore, a consequence of the effect of impurities on diffusion.[11,14] This link between flow stress and impurities is even more direct for pure diffusional creep (see Section 2.2.).

The influence of impurities has also been investigated for prism plane and rhombohedral slip.[15] A general trend is softening due to the presence of Ti^{3+}, Cr^{3+} and Ti^{4+} impurities, an effect particularly marked for the rhombohedral slip.[15] The origin of this softening is not clearly established.

Since titanium increases the flow stress for basal slip and decreases that for other systems, it could make alumina polycrystals easier to deform by providing five independent slip systems. Full plasticity of polycrystals could also be achieved in the presence of water, which has been shown to dramatically decrease the strength of alumina;[16] further studies of this phenomenon should be undertaken.

2.2. Deformation by Transport of Matter

As already mentioned in the previous section, oxide polycrystals are generally brittle, except at high temperatures ($T > 0.6\ T_m$). In these conditions, dislocations can glide in a number of slip systems larger than five, or they can climb away from their glide plane, a process directly related to diffusion, which controls the deformation rate at large strains ($\varepsilon > 0.05$). A number of models have been developed to describe such steady state creep of materials involving dislocation motion.[2,17] Deformation can also be achieved by transport of matter (pure diffusional creep) without dislocation activity,[2,17] a process which predominates in many ceramics, such as alumina. Nabarro and Coble creep mechanisms correspond to deformation produced by diffusion in the bulk and along grain boundaries; they both correspond to a situation where each grain undergoes the same strain as the whole specimen. Frequently, it is observed that the grains are not deformed, the global strain being provided by grains switching positions (Ashby and Verrall creep).[17] The last three mechanisms are characterized by a linear strain-rate ($\dot{\varepsilon}$) vs stress (σ) relation; large strains ε up to 100% or more can be reached, which encouraged many authors to use the term "superplasticity".[18,19]

To summarise, at high temperatures, large deformations of oxide polycrystals can be achieved at a steady state strain rate $\dot{\varepsilon}$; the relation between $\dot{\varepsilon}$ and the parameters of the mechanical test can be written in a general form:

$$\dot{\varepsilon} = A \frac{\mu b}{kT} \left(\frac{b}{d}\right)^p \left(\frac{\sigma-\sigma_o}{\mu}\right)^n D \tag{1}$$

where
- A, p and n = constants, characteristic for creep mechanism,
- μ = elastic shear modulus,
- b = Burger's vector,
- k = Boltzmann's constant,
- T = temperature,
- d = grain size,
- σ_o = threshold stress,
- D = diffusion coefficient of the rate-controlling mechanism for transport of matter.

The presence of impurities can essentially affect the diffusion coefficient D, the grain size d by altering grain growth at high temperature, and the constants A, p, n if the involved creep mechanism is changed.

Examples of parameters for viscous flow ($\sigma_o = 0$, $n = 1$): Nabarro creep p = 2, A = 5 to 15, D for volume diffusion; Coble creep p = 3, A = 141, D for diffusion in a grain boundary of thickness b. Frequently, experiments give values for p lying between those of Nabarro and those of Coble creep; this is associated with diffusional creep being rate-limited by a combination of grain boundary and lattice diffusion with $2 < p < 3$. Deformation maps are drawn to depict the various possible deformation mechanisms and their limits (Fig. 3). The change from one mechanism to another may be induced by foreign atoms in the lattice or at grain boundaries.

The complexity of the diffusion processes has to be considered to understand the mechanical behaviour of oxide polycrystals. To account for the various diffusion paths, one can write:

$$D = D_v \, (1 + f \, D_{ss}/D_v) \tag{2}$$

where D_v is the volume diffusion coefficient, relevant to deformation, and D_{ss} is the diffusion coefficient for a short circuit path with a fraction f of atom sites. For a dislocation pipe of cross-section area b^2 and a density of dislocations proportional to σ^2, f becomes $10(\sigma/\mu)^2$; for a grain boundary of thickness δ, f is $\pi\delta/d$.[20] For the deformation of polycrystals, eq. (2) is written as:

$$D = D_V \left(1 + \frac{\pi\delta}{d} D_{GB}/D_V\right) \tag{3}$$

where D_{GB} is the diffusion coefficient in the grain boundary.

In a compound AB, the transport of matter requires the motion of atoms A and B, which has to be related to D in eq. (3); the fluxes J of matter are such that the global composition of the material is preserved:

$$J^A = J^B \tag{4}$$

To maintain this requirement locally in the various diffusion paths, the following conditions have to be fulfilled:[21]

$$\begin{aligned} J_V^A &= J_V^B \quad \text{(volume diffusion)} \\ J_{GB}^A &= J_{GB}^B \quad \text{(grain boundary diffusion)} \end{aligned} \tag{5}$$

Alternatively, one can assume local variations of the composition, giving a condition:

$$J_V^A + J_{GB}^A = J_V^B + J_{GB}^B \tag{6}$$

This procedure allows to relate the coefficient D to the four self-diffusion coefficients of A and B.[21] The mathematics are heavy and different results are obtained depending on whether condition (5) or (6) are used, the second one being most commonly applied in the literature.[21]

At this point, we have shown that the steady state strain rate $\dot{\varepsilon}$ is directly related to self-diffusion phenomena, as discussed by A. Atkinson at this conference. Deformation by transport of matter requires also the creation and annihilation of point defects in the grain boundaries, a process which may be rate-controlling when diffusion is fast; two parameters have to be introduced, in addition to the four diffusion coefficients, in order to describe the creep kinetics. They correspond to the interfacial rate constants for diffusion of A and B related to point defects.[22,23] The various ways to combine these elementary processes lead to complex features; an example is given in Fig. 3 for the deformation of alumina at 1773 K and for various grain-sizes.[22] Similar plots can be made for other temperatures. Creep can be controlled by oxygen diffusion in the volume, D_V^O, or in the grain boundaries, D_{GB}^O, and by aluminium diffusion, D_V^{Al} or D_{GB}^{Al}. The following relations are generally accepted for Al_2O_3 to explain creep data (Fig. 3):[22]

$$D_V^O < D_V^{Al} < D_{GB}^{Al} < D_{GB}^O \qquad (7)$$

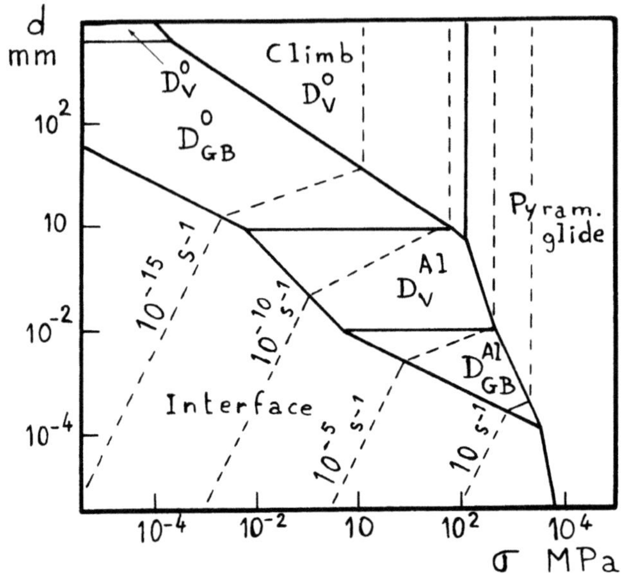

Figure 3: Deformation map of alumina polycrystals at 1773 K plotting grain size α vs stress σ. Most of the fields correspond to pure diffusion creep, the rate-controlling self-diffusion coefficient being indicated. Exceptions are "climb" for the pure dislocation climb field controlled by oxygen diffusion D_V^O, "interface" for creation-annihilation of point defects controlling the rate of deformation and "pyramidal glide". Dashed lines indicated values of strain rates $\dot{\varepsilon}$.[22]

These relations have not been verified by direct tracer measurements, neither for undoped nor for doped alumina, for which only oxygen lattice diffusion has been studied in detail.[10,14]

Extensive data on deformation of Al_2O_3 have been collected; divalent and tetravalent cations accelerate creep when cation volume diffusion is rate-controlling.[23,24] For grain boundary diffusion control, some impurities affect the creep rate,[22] e.g. Fe enhances the creep rate while Mg does not;[22] Y added to Mg hardens alumina.[19] Dramatic increases result from co-doping which can affect both volume and grain boundary diffusion, e.g. Fe + Ti or Mn + Ti.[22]

These early studies have been complemented by a recent study on the relation between creep and electrical properties of

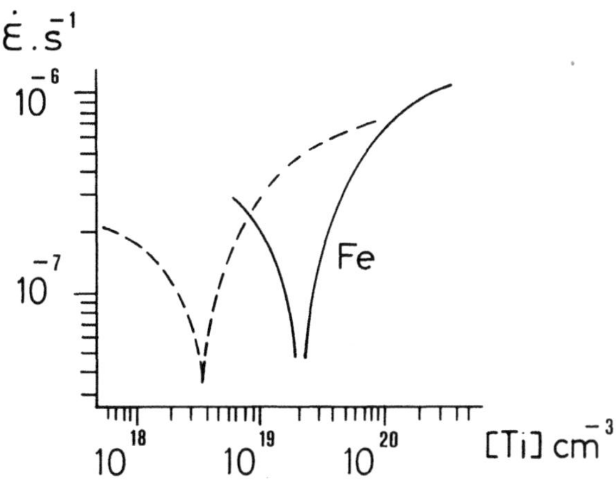

Figure 4: Creep rate $\dot{\varepsilon}$ as a function of Ti concentration for alumina doped with 500 ppm Fe and without Fe. T = 1683 K; d = 20 μm; σ = 30 MPa. After Tsauer and Kröger.[25]

alumina-doped with various amounts of Ti alone or together with 500 ppm Fe.[25] The results are summarized in Fig. 4; the minimum creep rate corresponds to a concentration at which acceptor domination (by residual Fe impurities) changes to Ti donor domination. Fig. 4 clearly illustrates how co-doping can increase or decrease creep rates. All these results have been interpreted in terms of modification of cation and anion diffusion in alumina, and their relation to point defects and impurities or dopants.[23,24,25] For that purpose, creep rates are adjusted by fitting the six parameters already mentioned, which are balanced according to the needs to explain the data. In most studies, the deformation mechanisms are deduced from the parameters p and n of eq. (1), microstructural observations being very seldom. Grain boundary segregation and solubility limits depend on temperature-time cycles, while various grain sizes are obtained by grain growth during high temperature annealings. Conditions may be such that redistribution of impurities and dopants in the specimen can occur, leading to uncertainties in the determination of the parameters of the creep equation. This, together with the lack of tracer diffusion measurements makes many of the results on creep mechanisms in alumina somewhat speculative; nevertheless, we believe that, basically, diffusion is the rate-controlling mechanism.

In this section, we have examined the deformation of ceramics by pure diffusion creep. Such a process assumes perfect accommodation of grains; actually, cavitation may occur, in particular at triple points. This gives additional deformation and explains why tensile strain rates are larger than compressive ones,[19] leading to fracture at small strains in the first case.

2.3. Other Deformation Mechanisms

Besides dislocation motion and transport of matter, oxide specimens may respond to stresses by other mechanisms; crack propagation, twinning etc., which are typical of low temperature behaviours. Indentation tests have been often used because they allow easy observations at various temperatures[26] and for a large number of specimens.[27] Fracture of alumina single crystals has been studied at various temperatures for the different cleavage planes,[28] as well as twinning.[29] We do not attempt to evaluate the influence of impurities on these processes, which are covered by very limited experimental observations.

3. CONCLUSIONS

We have examined some mechanical properties of oxides, essentially of Al_2O_3, which were obtained by conventional tests, such as tension and compression at constant strain rate or stress on compounds in thermodynamic equilibrium. These conditions are far from the actual situation of oxide layers on metals or alloys. The link can essentially be made through the understanding of the mechanisms of deformation and by exploring the rôle of doping on the stress effects in oxidation. However, there is a lack of studies on the basic properties of oxides, even in the case of the most well-known ones, such as alumina, where hardening and softening have been observed for deformation controlled by dislocation motion, as well as by diffusion. Cr_2O_3 received little attention;[30] its properties must have many features in common with Al_2O_3.

REFERENCES

1. W.R. CANNON and T.G. LANGDON; J. Mater. Sci. 1988, 23, 1-20.

2. J. BRETHEAU, J. CASTAING, J. RABIER and P. VEYSSIERE; Adv. Phys. 1979, 28, 829-1014.

3. Deformation of Ceramics II, Edited by R.E. Tressler and R.C. Bradt; Plenum Publishing Corp., 1984.

4. Structure and Properties of MgO and Al_2O_3 Ceramics, Advances in Ceramics, vol. 10; Edited by W.D. Kingery, The American Ceramic Society, 1984.

5. A. DOMINGUEZ-RODRIGUEZ, J. CASTAING, H. KOISUMI and T. SUZUKI, Revue Phys. Appl. 1988, 23, 1361-1368.

6. J. CASTAING, J. CADOZ and S.H. KIRBY, J. Phys. Colloque C3, 1981, 42, 43-47.

7. F. GUIBERTEAU, A. DOMINGUEZ-RODRIGUEZ, M. SPENDEL and J. CASTAING, Revue Phys. Appl. 1986, 21, 87-92.

8. J. CASTAING, Ann. Phys. Fr. 1981, 6, 195-221

9. A. KELLY and G.W. GROVES in: "Crystallography and Crystal Defects" 1970, Longman.

10. A.H. HEUER and J. CASTAING, pp. 238-257 in Ref. 4.

11. B.J. PLETKA, T.E. MITCHELL and A.H. HEUER, Acta Metall. 1982, 30, 147-156.

12. B.J. PLETKA, T.E. MITCHELL and A.H. HEUER, phys. stat. sol. (a) 1977, 39, 301-311.

13. T.E. MITCHELL and A.H. HEUER, Mater. Sci. Engin. 1977, 28, 81-97.

14. K.P.D. LAGERLOF, T.E. MITCHELL and A.H. HEUER, pp. 199-212 in Ref. 3.

15. D.M. KOTCHIK, Ph.D. thesis, 1978, The Pennsylvania State University.

16. A.H. HEUER, R.F. FIRESTONE, J.D. SNOW and J. TULLIS in: "Ceramics in Severe Environments" (Ed. W.W. Kriegel and Hayne Palmour III) 1971, 331-340, Plenum Press, New York.

17. J.P. POIRIER, "Creep of crystals", 1985, Cambridge University Press.

18. W.R. CANNON pp. 741-749 in Ref. 4.

19. C. CARRY in: Proc. Symp. on Superplasticity of the Int. Meet. on Adv. Mat., Tokyo, June 1988, Materials Research Society.

20. M.F. ASHBY and J.H. FROST in "Constitutive Equations in Plasticity" (Ed. A.S. Argon) 1975, 117-147, The MIT Press, Cambridge, USA.

21. M. JIMENEZ-MELANDO, A. DOMINGUEZ-RODRIGUEZ, J. CASTAING and R. MÁRQUEZ, Scripta Met. 1986, 20, 739-742.

22. R.M. CANNON, pp. 818-838 in Ref. 4.

23. R.S. GORDON, pp. 418-437 in Ref. 4.

24. R.S. GORDON and Y. IKUMA, pp. 341-352 in Ref. 3.

25. J.M. TSAUER and F.A. KRÖGER, J. Mat. Sci. 1987, 22, 2332.

26. H.M. CHAN and B.R. LAWN, J. Amer. Ceram. Soc. 1988, 71, 29-35.

27. B. MUELLER, P. GRAU and G. KLUGE, Phys. Stat. Sol.(a) 1984, 83, 499-506.

28. M. IWASA and R.C. BRADT pp. 767-779 in Ref. 4

29. W.D. SCOTT and K.K. ORR, J. Amer. Ceram. Soc. 1983, 66, 27-32.

30. B. BURTON and P.L. REYNOLDS, J. Mat. Sci. 1978, 13, 219-221.

SESSION 3

EFFECT OF ACTIVE ELEMENTS ON OXIDATION OF METALS AND ALLOYS

EFFECT OF ACTIVE ELEMENTS ON THE OXIDATION BEHAVIOUR OF Al_2O_3-FORMERS

A.M. Huntz

Laboratoire de Métallurgie Structurale, ISMA, CNRS UA 1107, Université Paris XI, 91405 ORSAY Cedex, France

ABSTRACT

The main mechanisms suggested in the literature for interpreting the effect of active elements on the growth and adhesion of alumina scales are reviewed and discussed, with particular emphasis on some of the more recent contributions.

First, the various active elements will be considered on whether they can detrimentally affect alumina scales, show varying effects or improve the protective character of such scales.

Then, the discussion will be focused on some of the more important mechanisms developed during the recent years and which account for the improvement of the properties of Al_2O_3 scales, i.e. effect of active elements on:

- the strengthening of the alloy-scale bonding with relation to segregation phenomena;
- the scale morphology and microstructure;
- the diffusion mechanisms;
- the scale plasticity.

The discussion will be supported by results obtained on doped massive alumina.

1. INTRODUCTION

The effect of active elements on the oxidation behaviour of alloys which develop an alumina scale has been extensively studied during the recent twenty years, the effect being empirically discovered already in 1937 by L.B. Pfeil (U.K. Patent n° 459848, 1937). In addition to an extensive literature available on this subject, several review papers[1-6] discussed the phenomena, the possible mechanisms by which active elements can influence the oxidation behaviour and the theories accounting for these effects. Nevertheless, at this date, the mechanisms are not clearly defined and studies on this subject are continuing.

Amongst all the mechanisms suggested in the literature to account for the active element effect, the most frequently mentioned ones are the following:

(1) Active elements can promote nucleation of alumina and consequently reduce the aluminium content necessary to form a continuous protective Al_2O_3 scale, compared with that in addition-free alloys.

(2) They can form an intermediate inner oxide layer which acts as a diffusion barrier. There is no sufficient experimental evidence for such an assumption.

(3) They can reduce the accumulation of voids at the oxide-alloy interface. In a recent review paper, Stott et al.[4] discussed this mechanism.

(4) They can act by a mechanical keying effect, as was impressively shown by some authors.[3]

(5) They can improve the metal-scale bond, as recently discussed by Stott et al.[4] This improvement can be associated with the elimination of impurities at the alloy-scale interface, i.e. with segregation phenomena as suggested by studies of Smeggil.[7,8]

(6) They can modify the morphology and the microstructure of the scale, as discussed in ref. 4.

(7) The previous point is also related to the effect of active elements on transport phenomena in alumina which was extensively studied indicating that active elements can modify the Al_2O_3 growth mechanism, as discussed in ref. 5.

(8) Finally, active elements can modify the mechanical properties of the scale.[1,6]

The present paper will be focused on the more recent contributions from the literature dealing mainly with the four last points.

Before, an attempt will be made to define and classify the "active elements".

2. ACTIVE ELEMENTS FOR ALUMINA SCALES

For obvious reasons (namely to improve industrial applications of Al_2O_3-forming alloys), most of the workers interested in high temperature alloys have orientated their activities towards the beneficial effect of active elements. Nevertheless, some literature data indicate that certain elements can also detrimentally affect the protective character of alumina scales or show variable effects depending on their amount, chemical state, etc...

For instance, titanium was found to have a detrimental effect[9] on the alumina scale adherence owing to the presence of TiO_2 precipitates. This was confirmed by studying the oxidation behaviour of alloy IN100 (NiCoCrAl with 5.2% Ti)[10,11] and of high purity Fe-37Ni-20Cr-5Al alloys either undoped or doped with 0.23% Ti.[12] However, a detrimental effect of titanium is not always observed: when Ti is added together with Y and Ta to NiCoCrAl alloys[10] or with Y to FeNiCrAl alloys,[12] no significant effect of this element on the oxidation behaviour is observed. Besides, it was shown that the effect of Ti depends on its amount: an addition of 0.2% Ti to a FeNiCrAl alloy does not significantly alter the oxidation kinetics, while an addition of 0.54% Ti to the same alloy significantly increases the oxidation weight gain rate at 1200°C (Fig. 1).[12] It must also be noted that Ti

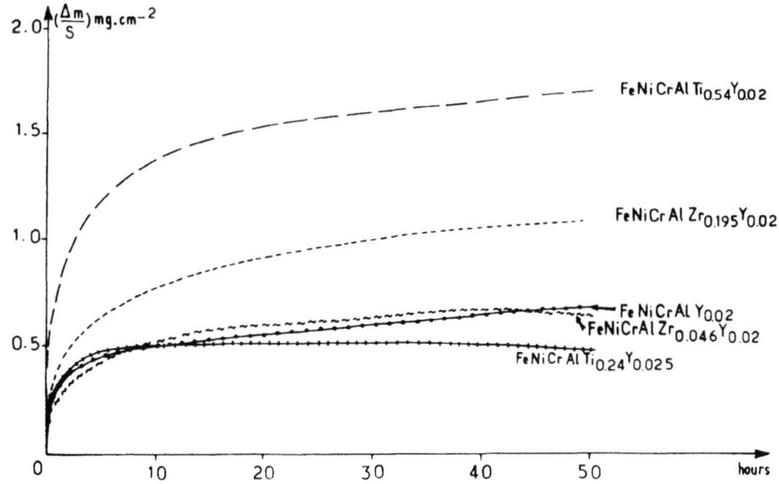

Fig. 1: Weight gain during oxidation at 1200°C of Fe-37Ni-20Cr-5Al alloys doped with Y, Y + Ti or Y + Zr.[12]

appeared to have in certain cases a benificial effect on the corrosion behaviour in sulphidizing atmospheres.[13,14]

In a similar way, variable effects were observed for zirconium additions. In FeNiCrAl alloys,[12] additions of Zr alone have a detrimental effect, while, if this element is added together with Y, no effect is observed, provided the amount of Zr is small (0.05%); with 0.2% Zr the weight gain is greater than for the zirconium-free alloy (Fig. 1). Detrimental effects of Zr-additions were also observed for the isothermal oxidation and sulphidation behaviour of aluminide coatings on stainless steel:[15] the weight gain rate increased with the amount of Zr in the coating. However, scale adherence was improved by a mechanical keying effect, provided the amount of Zr was limited. The authors suggested that, at lower concentration in the coating, this element could have a beneficial effect, in agreement with studies by Kahn et al.[16] who showed that the oxidation rate constant kp of a NiCrAl alloy achieves a minimum value for ~0.038 at% Zr. This important effect of the amount of Zr-addition was also observed more recently in the oxidation of FeCrAl alloys by Pandey et al.[17]: an addition of 0.2% Zr was found to be most effective in improving the isothermal and cyclic oxidation behaviour of those alloys. A beneficial effect of zirconium was further observed for the sulphidation behaviour of FeCrAl alloys.[14,18] In the case of NiCoCrAl alloys, it appears that addition of Zr has no beneficial effect in contrast to that of yttrium (and hafnium), Fig. 2.[19]

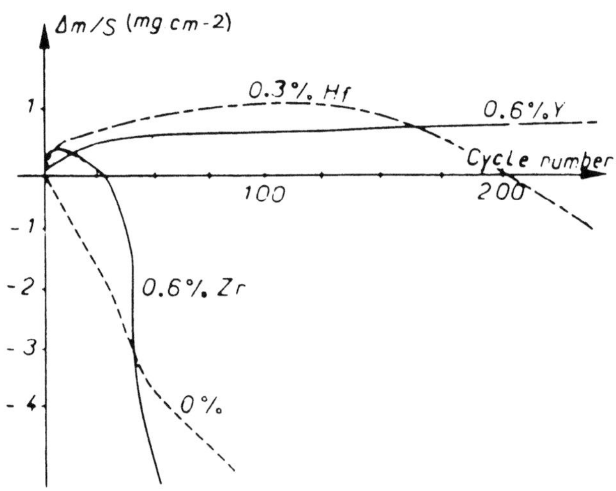

Fig. 2: Weight gain during cyclic oxidation at 1100°C in air of a NiCoCrAl alloy without and with additions of Y, Hf or Zr.[19]

Thus, it appears that for titanium and zirconium additions, variable effects can be observed depending on the amount of the addition and the type of alloy. Similar observations were made for hafnium and yttrium additions to CoCrAl[20] and FeNiCrAl alloys[21], but in the case of yttrium, the amount necessary to induce an improvement of oxidation resistance is now well established.

Regarding now an element such as carbon, which is present in most of the industrial alloys, its effect is not clear at all. The influence of carbon appears to depend on its chemical state in the alloy and on phenomena occuring during the oxidation treatment, i.e. carbide precipitation or dissolution.[22] As an example, Fig. 3 shows that, when the carbides are dissolved in the alloy prior to the oxidation treatment, then the weight gain rate at 900°C is higher than at 1000°C. This is related to the competition between oxidation and carbide precipitation which occurs at 900°C: some of the carbides which progressively precipitate are incorporated in the scale, react with oxygen and induce an enrichment of the alumina scale in Cr and C. Carbon in alumina creates irreversible defects which increase the growth rate. This effect of C was verified on synthetic FeCrAlY alloys without and with an addition of 50 ppm C.[23] But it is not easy to assess the effect of carbon because this element simultaneously affects the mechanical properties of the alloy and probably of

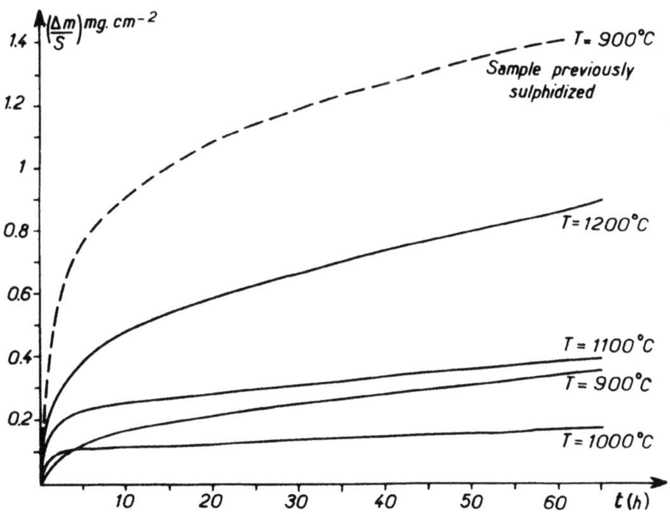

Fig. 3: Weight gain curves of Fe-23Cr-5Al samples, pre-annealed at 1200°C, and then oxidized in 1 atm O_2.[22]

the scale. It is well known that alumina scales on carbon-free alloys are not at all adherent, and under this aspect it can be argued that carbon appears to be a beneficial active element.

In contrast, there is a general agreement on the detrimental effect of sulphur on the protective character of alumina scales. Many examples could be presented which show that sulphur, provided either by the atmosphere or from the alloy, induces rapid corrosion of the alloy. Fe-23Cr-5Al samples which had been previously sulphidized and then oxidized, for instance at 900°C, show an oxidation weight gain which is about three times greater than for unsulphidized samples (see Fig. 3).[22] At my knowledge, no studies on the chemical state of sulphur in the Al_2O_3 scale have been carried out. But it is interesting to note that in Cr_2O_3 scales developed on a NiCrFe alloy, sulphur was detected (by XPS and EXAFS analysis) as $S^=$ ions on oxygen sites of the oxide lattice.[24] In such a state in alumina, sulphur could increase the aluminium vacancy concentration according to:

$$1/2\ S_2 + O_o^x \rightleftarrows S_o^\circ + \frac{1}{3} V_{Al}''' + 1/2\ O_2 \qquad (1)$$

or the oxygen interstitial defects:

$$S_2 + O_o^x \rightleftarrows 2S_o^\circ + O_i'' \qquad (2)$$

Both phenomena could account for the observed increase of the alumina growth rate. Further studies have to be conducted in this domain. Besides, another effect of sulphur appears to be very important for the adhesion of alumina scales. Segregation of sulphur to the metal-oxide interface was experimentally observed[7-8,25-27] and was assumed to be responsible for both the poor adherence of the scale and the fast transport in the scale. The poor adherence was attributed to a weakening of the cohesive bond between the scale and the adjacent metal in analogy with theoretical calculations made on the effect of sulphur segregated at metal grain boundaries.[28] The fast transport was explained by sulphur segregation in the oxide grain boundaries where it would promote cation transport.[26] Anyway, all authors agree upon the detrimental effect of sulphur.

On the other hand, it is generally agreed that rare earth elements, including yttrium, or oxide dispersoïds of such elements have a beneficial effect on the high temperature oxidation resistance of alumina-forming alloys, provided the amount and distribution of these elements is adequately chosen. The elements most frequently mentioned are Y, Hf, Ce, La ... or oxide dispersoïds of these elements. Publications on this topic are so numerous that they will not be considered here in detail; we refer to some comprehensive reviews in ref. 1-6 and 29. It should be underlined however that, as discussed before for Ti and

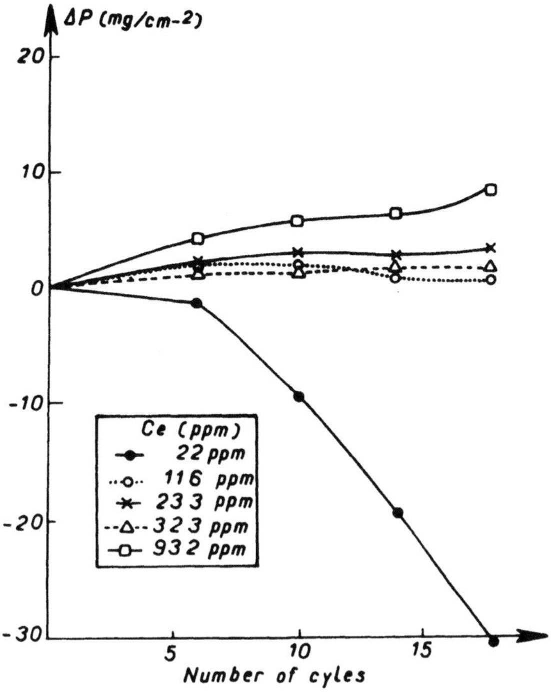

Fig. 4: Effect of Ce on the cyclic oxidation behaviour of Fe-45Ni-25Cr-5Al alloys, in air. (cycle: 15h at 1300°C, then air-cooled to room temperature).[30]

Zr, also the effect of Y and other rare earth metals depends on their amount and on the presence of other "active" elements. As an example, Fig. 4 shows the effect of Ce on the oxidation behaviour of a FeNiCrAl alloy.[30]

This review of active elements is not exhaustive but only an attempt to give an idea of the complexity of the problems. An "active" element can have a detrimental or a beneficial influence; but most important is that synergic effects can occur. Hence, a given element can have two opposite rôles depending on its amount, its distribution, the type of alloy to which it is added, the content of other impurities, etc... It is for these elements as for persons: their behaviour depends on their neighbourhood (see "trading places").

3. INFLUENCE OF ACTIVE ELEMENTS ON THE GROWTH AND ADHESION OF Al_2O_3 SCALES: RECENTLY SUGGESTED MECHANISMS FOR THEIR BENEFICIAL EFFECT

Some of the mechanisms proposed in the literature to explain the beneficial effect of active elements and referred to in the Introduction, were experimentally verified for specific alloys doped under controlled conditions.

For instance, several papers presented evidence for a pegging-mechanism to be responsible for good scale-substrate bonding,[3,15,19,29,31-33] (Fig. 5), although, in other cases, for very low yttrium levels in the alloy, the oxide-metal interface was observed to be reasonably flat and with no evidence of pegs but yet providing good adhesion of the scale.[34] On the other hand, even when macropegs had formed, reduced scale adhesion was observed.[35] Besides, for all studies where the active element was added by implantation, a mechanical keying effect can be excluded.

Other authors showed that the accumulation of voids at the oxide-alloy interface was reduced by the presence of active

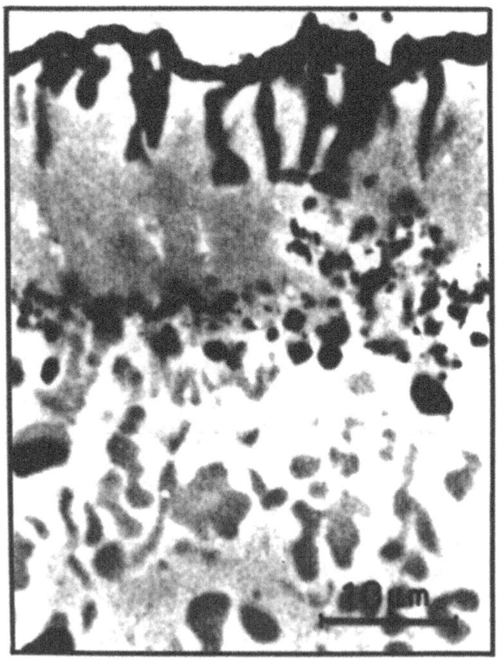

Fig. 5: Oxide pegs developed on a ZrAl pack-cemented coating on an austenitic stainless steel during oxidation for 2h at 900°C in 1 atm O_2.[15]

elements acting as vacancy sinks in the alloy, thus reducing the oxide detachment.[4,20,35-41] But, in a study on the rôle of Zr or Y in a NiCrAl alloy,[42] the amount of voids was found to be relatively independent on the reactive element addition.

There is little experimental evidence for the rôle of active elements in enhancing nucleation of alumina. Nevertheless, it was suggested by Pivin et al.[21,43] that this effect would be much more efficient for Cr_2O_3-forming alloys than for Al_2O_3-formers.

Also, the formation of an intermediate inner oxide layer (such as Y_2O_3, ZrO_2 ...), which would allow to accommodate the stress between the alumina scale and the metallic substrate, was not really observed.

In the following the discussion will be focused onto the more important mechanisms developed during the recent years.

3.1. Strengthening of the Alloy-Scale Bonding and Segregation Phenomena

The assumption that active elements prevent sulphur and other harmful elements to segregate at the metal-oxide interface appears to be of great interest. On one side, a theoretical study developed by Anderson et al.[44] predicts that a substantial enhancement of bond strength can be provided by a few monolayers of yttrium between the Al_2O_3 scale and the metallic substrate. On the other hand, interesting and important experimental investigations were carried out on segregation phenomena.[25,28,45-48] First, it was shown that elements such as sulphur (perhaps also chlorine and phosphorus) have a detrimental effect on the scale-metal bonding (in Ni-base superalloys).[46] Using quantum chemical molecular orbital techniques, Messmer and Briant[28] claimed that sulphur segregation induces a bond weakening. In analogy to this model one might also presume a weakening effect of S to the Al_2O_3 scale-metal interface. Then, Smeggil et al.[25,45] showed the crucial rôle of yttrium additions to NiCrAl alloys whereby the segregation of sulphur to the interface was eliminated. This is illustrated by Fig. 6 where it can be seen that the disastrous effect of sulphur (30-40 ppm) on the oxidation behaviour of normal purity NiCrAl disappears when Y is added: the Y-doped sample behaves as the high purity sulphur-free sample.[8] However, such a mechanism was excluded by Briant et al.[27] who did not find any correlation between the resistance to scale spalling and sulphur segregation, and further observed that the presence of yttrium in the alloy did not eliminate sulphur segregation at the scale-substrate interface.

On the other hand, many experiments showed segregation of yttrium or zirconium at the Al_2O_3-metal interface.[26,48] This

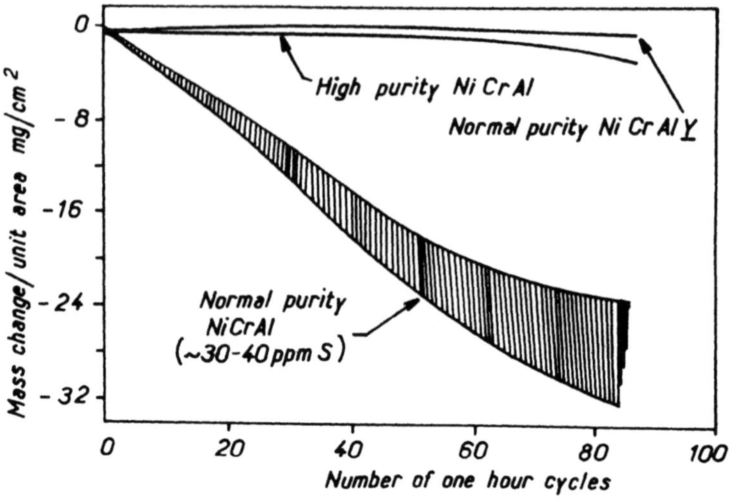

Fig. 6: Effect of sulphur content on the 1180°C cyclic oxidation behaviour of NiCrAl, normal purity NiCrAl: 30-40 ppm S, normal purity NiCrAlY: S trapped by Y, high purity NiCrAl: S < 1-2 ppm.[8]

phenomenon could also induce a bond strengthening between the scale and the metallic substrate, in agreement with the theoretical analysis of Anderson.[44]

3.2. Scale Morphology and Microstructure

Many authors also investigated, during the more recent years, the effect of active elements on the microstructure and the morphology of alumina scales. In most cases, scales developed on alloys without reactive element additions are non-uniform in thickness, convoluted, with protrusions at the scale-gas interface and/or intrusions at the scale-metal interface.[4,49-52] It has been proposed that localized thickening results from rapid-diffusion paths and counterflow diffusion of aluminium and oxygen.[49] Sometimes whiskers[19,33,39,50,51] and voids in Al_2O_3 grains[19,47] are observed. Generally, the oxide grains are small (~ 0.1 to 1 µm) and both zones with equiaxed grains and columnar grains are observed. The columnar grains are localized towards the interface and the little equiaxed grains towards the outer surface[19,33,38,53] as schematically shown in Fig. 7. This is not always the case: Ramanarayanan et al.[54] observed only equiaxed grains in the alumina scale formed on a FeCrAl alloy. Also in the case of oxidized β NiAl, the scale does not consist of columnar grains.[49,50,55-57]

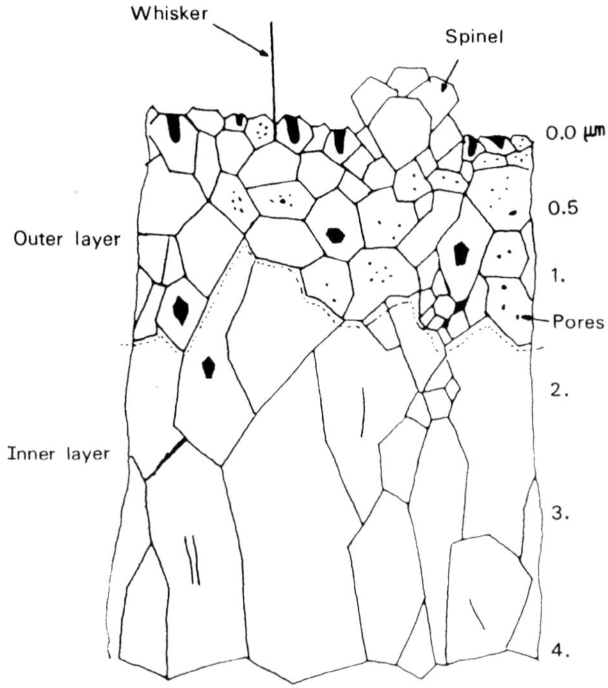

Fig. 7: Schematic representation of the microstructure of the alumina scale formed on a NiCoCrAl alloy oxidized for 100h at 100°C.[54]

There is indeed some scattering in the results obtained on alloys doped with active elements. On a NiCoCrAl alloy doped with Y, Hf or Zr, Choquet[19] always observed an outer equiaxed grain zone and an inner zone composed of bigger columnar grains without pores. Golightly et al.[34] made the same observation on FeCrAl (+Y) alloys and noted that the grains in the scale are more columnar on samples containing yttrium than on yttrium-free samples. They proposed that the outward transport of aluminium is suppressed by yttrium. More recently, Smialek et al.[42] did not observe on NiCrAl (+Y) alloys an unequivocal correlation between the active element additions and the scale microstructure, and similar observations were made by Pivin et al.[21] on FeNiCrAl alloys. On β NiAl, the influence of implanted active elements is rather spectacular:[51,55,57] on yttrium-free samples, the scale surface is convoluted, with large ridges and whiskers, (Fig. 8), whereas on doped samples, the scale surface is smooth without any whiskers (Fig. 9). Spectacular modifications were also observed on FeCrAl alloys by Ramanarayanan et al.[54] In contrast to the Y-free samples, the scale on the Y-doped samples has a distinct columnar grain morphology. The average width of these columnar grains is about 0.5 µm whereas the grain size of

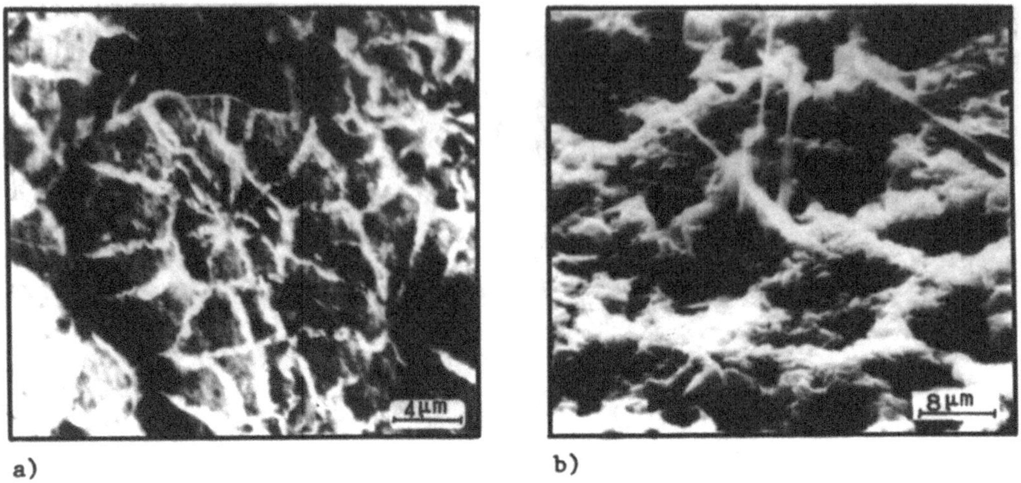

Fig. 8: Surface morphology of the alumina scale on β NiAl after 10h (a) and 4 days (b) oxidation at 1100°C in O_2.[55]

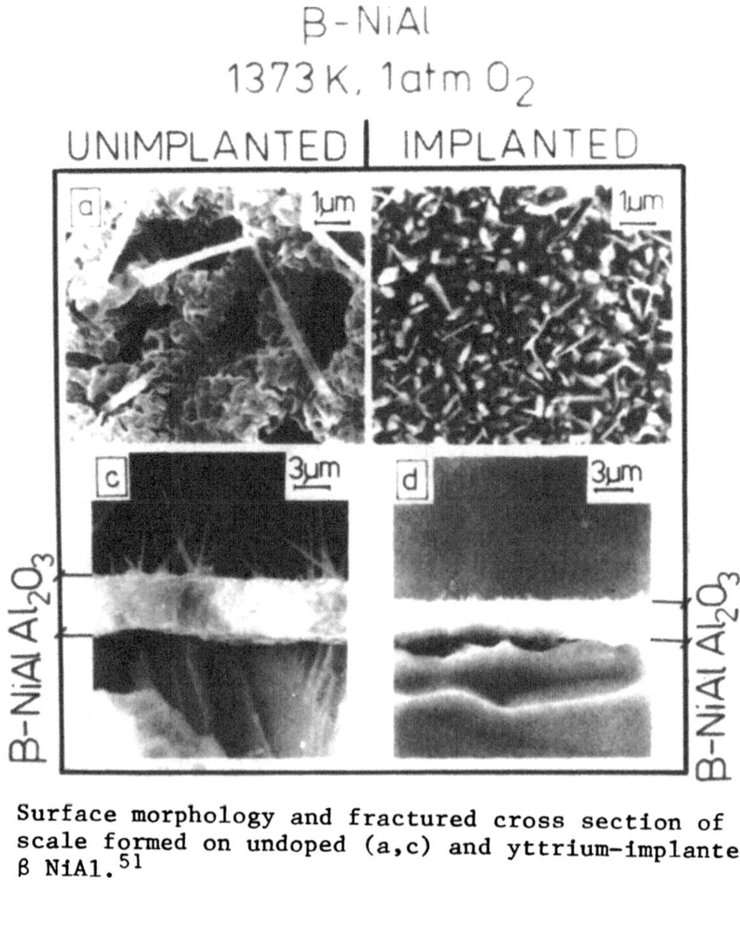

Fig. 9: Surface morphology and fractured cross section of the scale formed on undoped (a,c) and yttrium-implanted (b,d) β NiAl.[51]

a)

b)

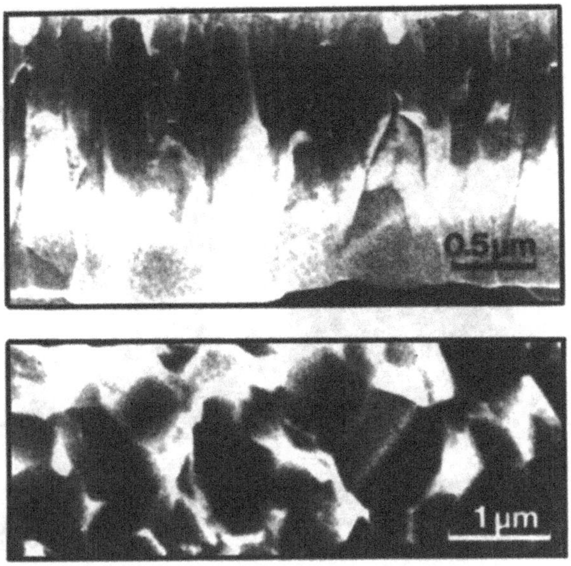

Fig. 10: Facture cross sections of alumina scales formed on a) FeCrAlY alloy, b) FeCrAl alloy.[54]

the equiaxed grains on the undoped material is about 1 µm (Fig. 10). These authors found that the yttrium-addition restricted the grain growth in the alumina scale. They also observed grain boundary microvoids and precipitates of $Y_3Al_5O_{12}$ in the grain boundary regions.

At my knowledge, until now, only one study was focused onto the orientational relationship between alumina and a NiCrAlY alloy.[58] No preferential growth texture was detected for the columnar grains of Al_2O_3: the alumina c-axis direction is randomly orientated. However, heterophased interfaces (between α-Al_2O_3 and γ NiCrAlY) have a "special" character, the {111} planes of the γ-phase being, within an accuracy of $\pm$ 4°, parallel to the $\{11\bar{2}3\}$ planes of the α-alumina, (Fig. 11). The frequency of such boundaries corresponds to 15% of the boundaries examined. However no comparison was made with undoped NiCrAl, so the effect of yttrium cannot be deduced from this interesting work.

The disagreement regarding the influence of active elements on the morphology of alumina scales is probably related to the transport processes in alumina which will be considered in the following. Nevertheless, it can be concluded that, when an effect of active element is observed, it consists in a decrease

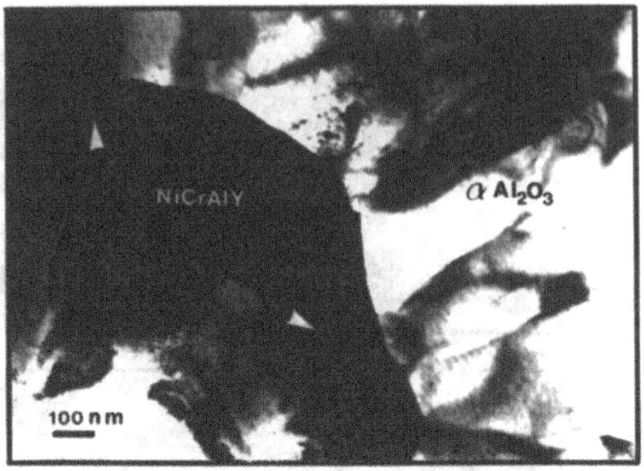

a)

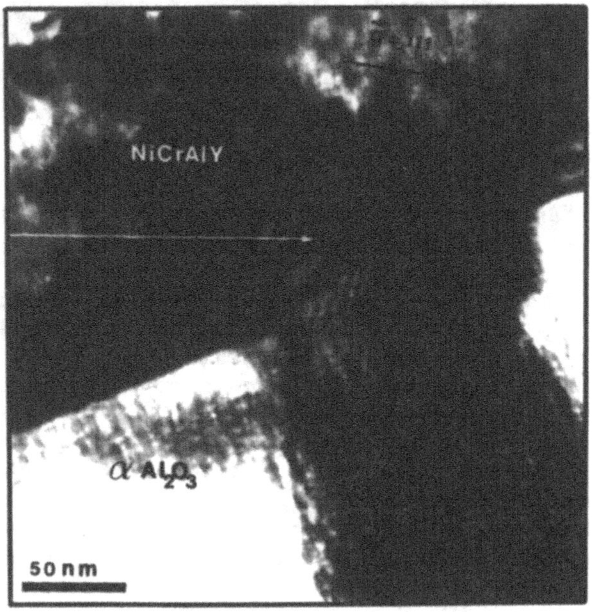

b)

Fig. 11: Interface between NiCrAlY (γ-phase) and columnar α-alumina.[58]

a) a $Y_3Al_5O_{12}$ garnet is observed at G, and many dislocations appear in the metal near the interface.

b) coïncidence between {111} planes of γ-phase and $\{11\bar{2}3\}$ planes of α-alumina.

of scale convolutions, in inducing orientated rather than randomly distributed oxide grains, and in a decrease in grain size. The active element is incorporated in the scale either as precipitate or segregated at the grain boundaries.[54] These effects of
active elements are in agreement with similar observations on massive alumina:

* The solubility of yttrium in alumina is very small[59,60] and Y-Al-rich precipitates are observed according to the equilibrium diagram.[61]

* Yttrium, added at low concentration, segregates to the grain boundaries of alumina[62] and induces a decrease in grain size.[60]

* Recently, it was observed that on alumina monocrystals, doped with 300 ppm Y_2O_3, ytrrium segregation occured at the surface (Fig. 12).[63] This would account for the previously discussed strengthening mechanism of the metal-scale interface bond.

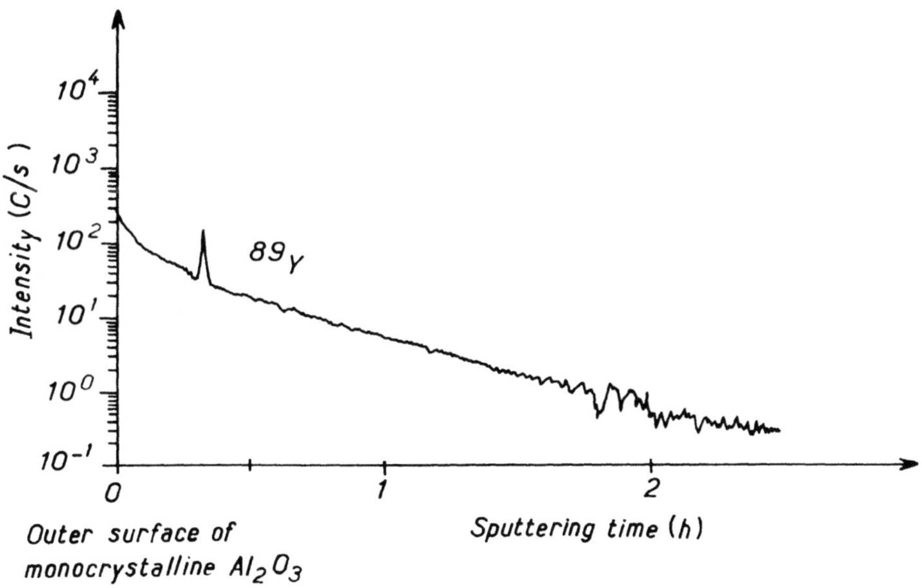

Fig. 12: Y-profile (SIMS) on an alumina (+ 300 ppm Y_2O_3) monocrystal, treated for 72h at 1720°C; 1h sputtering ~0.14 µm.[63]

3.3. Diffusion Mechanisms

In the literature, there is general agreement that active elements influence the transport processes in Al_2O_3 scales, but the situation is not as clear as for Cr_2O_3 scales, and there is considerable disagreement regarding the detailed mechanism. This is mainly due to the fact that the transport mechanism in undoped alumina has always been subject of controversy: many authors suggest that the preponderant diffusing species is oxygen[1,19,36,39,49,64,65] while others estimate that Al_2O_3 scales grow by outward diffusion of aluminium[52,55,66] or by diffusion of both oxygen and aluminium.[34,50,57] The beneficial effect of active elements is assumed to be due to a decrease of either oxygen or aluminium diffusion. Thus, Reddy et al.[64] indicated that, at low levels, active element additions decrease grain boundary diffusion of aluminium while, at high levels, they increase grain boundary diffusion of oxygen. Golightly et al.[36] postulated that Y-additions suppressed aluminium diffusion in the bulk lattice. Ramanarayanan et al.[65] also suggested that Al transport in the oxide scale is suppressed by yttrium and that the scale grows exclusively by inward oxygen transport. From studies on β NiAl, Young et al.[66] estimated that Y reduces the oxygen diffusion and increases the aluminium diffusion in agreement with results by Jedlinski et al.,[50,51,57] whereas, our own work suggested that yttrium in FeCrAl promotes diffusion of oxygen.[52]

All these results, (and others not mentioned), clearly indicate that, in contrast to the situation for chromia scales,[67] active elements do not inverse the growth mechanism of alumina scales, although they influence the transport processes. This complex situation is probably due to several parameters:

* depending on the type and amount of active element addition and on the oxidation conditions (T, atmosphere...), the element incorporated in the alumina scale can be either in solid solution, segregated or precipitated.

  The effect of the atoms in solid solution will depend on their donor or acceptor nature: if present as donor, such as Zr^{4+}, Ti^{4+} ... or Y^{3+} as suggested for by El Aiat et al.[68] the active element would promote aluminium vacancy or interstitial oxygen defects or decrease the amount of interstitial aluminium or oxygen vacancy defects. The inverse situation would be obtained for acceptor elements. In alumina, it seems that Schottky defects are more favourable than Frenkel defects (Table 1).[69] Thus, one could imagine that a donor element would promote the lattice aluminium diffusion and decrease the lattice oxygen diffusion in alumina, and vice versa for an acceptor.

Table 1: Calculated values of formation and migration enthalpy (in kJ/mol) of defects in alumina.[69]

defects	V_{Al}	V_o	Al_i	O_i
formation enthalpy	879	338	1042	1012
migration enthalpy	366	280	460	231

With regard to grain boundary segregation, the limited work in this domain, (sintering, conductivity... see ref. 5), suggests that an element such as yttrium would increase the transport in alumina. A recent study of S. Lartigue[70] indicates that yttrium in polycrystalline alumina modifies the relative proportion of general and special grain boundaries, increases the intergranular dislocation network and induces a "hard" behaviour of grain boundaries, i.e. impedes the absorption of lattice dislocations in grain boundaries.

These results suggest that segregated yttrium in alumina must have an effect on the intergranular transport processes but further experimental studies have to be carried out in order to point out the specific influence of segregated yttrium on intergranular diffusion in alumina.

If the active elements are precipitated in the grain boundaries, it seems very difficult to assess the general consequences on the transport mechanisms. Nevertheless, by considering oxygen diffusion data in various phases, see Table 2, it appears that oxygen lattice diffusion in $Y_3Al_5O_{12}$, Y_2O_3 and ZrO_2 is faster than in alumina, at least if an extrapolation to 1100°C is valid.[19] Moreover, such precipitates in alumina create new interfaces which may act as short-circuit paths.

The few results obtained on diffusion in doped alumina agree with these considerations: cation diffusion is enhanced by the presence of yttrium.[71,72]

* In undoped polycrystalline alumina, the diffusion coefficients of oxygen and aluminium are of the same order of magnitude. Extrapolation of the results of Kingery et al.[73,74] gives:

$$D^{O}_{Al_2O_3} = 6.6\ 10^{-18}\ cm^2\ s^{-1}$$

$$D^{Al}_{Al_2O_3} = 2.34\ 10^{-17}\ cm^2\ s^{-1}$$

Table 2: Diffusion coefficients of oxygen in various oxides.[19]

	OXIDES	T (°C)	D_o ($m^2.s^{-1}$)	Q ($kJ.mol^{-1}$)	D* 1100°C (m^2s^{-1})	reference
Al_2O_3	(m)	< 1600	$6.3\ 10^{-12}$	241	$4.1\ 10^{-21}$	74
Al_2O_3	(m)	> 1450	$1.5\ 10^{-3}$	572	$2.4\ 10^{-25}$	85
ZrO_2 (0.15 at% Ca)	(m)	700 - 900	$1.0\ 10^{-5}$	117.1	$3.4\ 10^{-11}$	86
ZrO_2 (16 mol% MgO)	(m)	1050 - 1300	$1.0\ 10^{-8}$	62.8	$4.1\ 10^{-11}$	87
$Y_3Al_5O_{12}$	(m)	1060 - 1550	$5.2\ 10^{-7}$	325	$2.1\ 10^{-19}$	88
Y_2O_3	(m)	1050 - 1240	$6.1\ 10^{-10}$	81.8	$4.6\ 10^{-13}$	89
Y_2O_3	(p)	1000 - 1500	$7.2\ 10^{-4}$	245	$3.3\ 10^{-13}$	90
Y_2O_3	(m)	1100 - 1500	$7.3\ 10^{-10}$	191	$3.8\ 10^{-17}$	91

m = monocrystal
p = polycrystal
* in some cases, D* was obtained by extrapolation

Moreover, very recent studies on monocrystalline alumina indicate that oxygen and aluminium have fairly similar diffusion rates in Al_2O_3.[63] Further, due to the fact that alumina is a stoichiometric compound (for instance calculations from Table 1 indicate that the Schottky defect concentration at 1727°C is about 0.01 ppm), impurities incorporated in alumina scales will dominate the transport behaviour. Hence, it is not surprising that from one alloy to another the growth mechanism can alter from rate-controlling aluminium to preponderant oxygen diffusion.

* Nevertheless, a growth mechanism mainly by intergranular diffusion is generally accepted for Al_2O_3 scales. Indeed, Choquet[19] recently calculated the diffusion coefficients in alumina on the basis of the parabolic rate constant Kp determined for a NiCoCrAlY alloy oxidized for 100h at 1100°C:

assuming $D_L >> D_{gb}$, he found at 1100°C

$$D_L = 2.10^{-14}\ cm^2\ s^{-1}$$

then, assuming $D_{gb} >> D_L$, he obtained

$$\delta\ D_{gb} = 9.5\ 10^{-19}\ cm^3\ s^{-1}$$

(δ being the grain boundary width).

The calculated value for D_L is much higher than values found at 1100°C in polycrystalline alumina, while the $\delta\ D_{gb}$ value agrees with values collected by Kröger[75] and Atkinson.[76] Considering all Kp-values available from the literature, it can be concluded that in all cases the growth of alumina scales is controlled by (oxygen or aluminium) diffusion along grain boundaries.

* The question is now why an element such as yttrium has a beneficial influence? Perhaps it is not related to its effect on the transport processes, although such effects were observed. Indeed, it can be noted that:

  i) the effect of yttrium on the growth kinetics of alumina scales is very small compared to its effect on the growth kinetics of chromia scales.[67]

  ii) In many cases, yttrium decreases the oxidation rate of alumina-forming alloys, but in some cases also an increase is observed, for instance for FeCrAl alloys at $T > 1373$ K.[77] Consequently, the predominant mechanism by which yttrium, or other active elements, improve the oxidation resistance of Al_2O_3-forming alloys is probably not directly related to their influence on transport properties but rather associated with some effect on the scale plasticity.

3.4. Scale Plasticity

There is no direct evidence of an improvement of the plasticity of alumina scales by active elements, though there exists an extensive number of results which show that active elements considerably improve the alumina scale adhesion.

Nevertheless, as discussed by Castaing[78] at this conference, it is well known that active elements have an influence on the deformation behaviour of massive alumina. For instance, a recent study of Carry[79] conducted on Mg-containing alumina samples, indicated that doping with yttrium decreases cavitation during compressive creep tests and that a longer time is needed to reach a given strain level. This study, combined with TEM observations,[70] suggested that, in Mg and Y doped alumina, the decrease in creep rate and cavitation is related to the intergranular segregation of yttrium, which hinders the accommodation processes.

Recently some techniques, based on deflection measurements, high angle X-ray diffraction, acoustic emission or microstructural observations on thin foils, were developed to obtain informations on the stress or strain in alumina scales.[80]

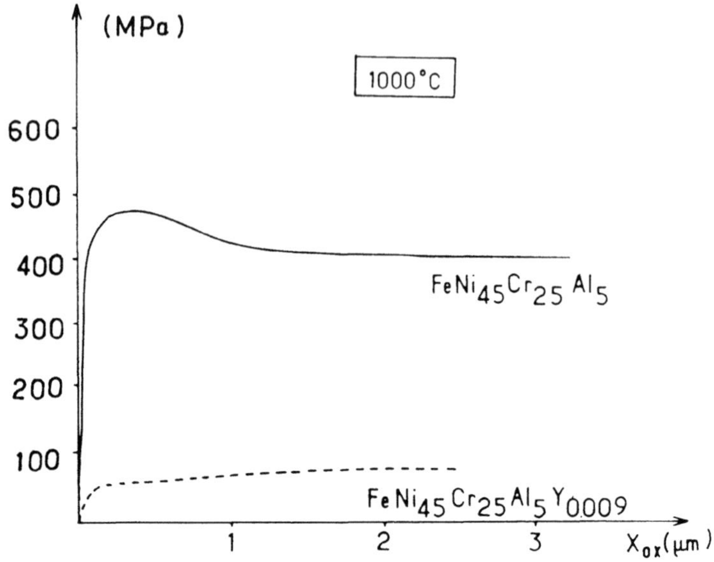

Fig. 13: Stress in alumina scales on FeNiCrAl samples without and with yttrium addition, as determined from deflection tests.[81]

From deflection tests the oxide stress σ_d, accommodated by the sample bending, can be determined according to:

$$\sigma_d = \frac{E_m X_m^2 D}{3 L^2 X_{ox}} \qquad (3)$$

where

E_m = Young's modulus of the substrate material,
X_m, X_{ox} = substrate and oxide thickness, respectively,
L = sample length,
D = deflection.

From oxidation tests on FeNiCrAl alloys[81] it was found that Y-additions significantly reduced the stress level accommodated by sample bending, Fig. 13, but from these measurements it is not possible to decide whether yttrium decreases the total stress (accommodated + residual stress) in the scale or whether Y present in the scale reduces the stress accommodation.

Attempts were also made to calculate the residual stress (at room temperature) in alumina scales from strain measurements by X-ray diffraction.[10,19,82] For instance, Choquet[19] determined on a NiCoCrAlY sample oxidized at 1100°C, after cooling to room temperature, a compressive stress value of 5700 ± 450 MPa.

According to:

$$\sigma_{\Delta T} = \int_{Ti}^{Tf} \frac{E_{ox}}{1 - \nu_{ox}} \frac{(\alpha_m - \alpha_{ox})\, dT}{1 + \frac{X_{ox}}{X_m} \frac{E_{ox}}{E_m} \frac{1 - \nu_m}{1 - \nu_{ox}}} \qquad (4)$$

where

$\sigma_{\Delta T}$ = thermal stress,
ν = Poissoin coefficient,
α = thermal expansion coefficient,

the thermally induced compressive stress can be estimated to be 5180 MPa. These results indicate that the isothermal growth stress is small compared to the thermal stress developed during cooling and that yttrium is probably responsible for the low stress level generated by the oxide growth. Unfortunately, no experiments were performed on undoped samples. Luthra et al.[82] did not find compressive stresses in alumina scales developed on similar alloys, doped with Y, whereas they determined high compressive stresses in alumina scales grown on FeCrAl alloys which were doped either with yttrium or hafnium. Thus, results obtained in this domain so far do not yet provide clear ideas about the effect of active elements on the stress in alumina scales.

However, some information and suggestions can be obtained from microstructural observations. In most cases yttrium causes a columnar fine-grained oxide scale microstructure[54,65] and segregates at the grain boundaries. This will affect the plastic deformation behaviour. It has been shown that in polycrystalline alumina with a small grain size ($\leqslant$ 15 µm), diffusional creep is the most important mechanism for plastic deformation.[83] Besides, the creep strain rate is extremely sensitive to grain size and increases with decreasing grain size. Plastic deformation will occur more easily in doped scales than in undoped ones. The growth stresses will be relieved by plastic flow during oxidation. That is perhaps the reason why the stress measured by the deflection tests during oxidation and related to the stress accommodated by the sample bending, is small in presence of Y (case of FeNiCrAlY for instance), according to the fact that the total stress in the scale σ_{ox} is given by

$$\sigma_{ox} = \sigma_d + \sigma_{pf} + \sigma_r \qquad (5)$$

where σ_r = residual stress at the oxidation temperature,
σ_d = stress accommodated by deflection,
σ_{pf} = stress relieved by plastic flow.

In samples doped with active elements, σ_{pf} would adopt a high value, compared to undoped materials, and for a given σ_{ox} the value of σ_d would be much smaller than σ_{pf}. The same consideration can explain why the residual stress σ_{res}, measured at room temperature, is of the same order of magnitude as the thermally induced stress $\sigma_{\Delta T}$. Indeed,

$$\sigma_{res} = \sigma_{ox} - \sigma_{pf} + \sigma_{\Delta T} \tag{6}$$

and it can be assumed that $\sigma_{pf} \sim \sigma_{ox}$.

Besides, using the elastic strain-energy criterion for oxide scale failure, it has been shown[53,84] that

$$\varepsilon_c = (B\ \gamma\ /\ X_{ox}\ E_{ox})^{1/2} \tag{7}$$

where ε_c is the critical elastic strain level for oxide scale spalling, γ is the fracture surface energy of the delaminating interface, B is a constant factor (usually about 4) and E_{ox} is the Young modulus of the scale. In the case of alumina scales, and in contrast to chromia scales (see ref. 67), X_{ox} is of the same order of magnitude for doped and undoped scales. But one could presume that due to the improvement of the substrate-oxide bond strength by yttrium segregation, or by elimination of sulphur segregation at the oxide-metal interface, γ has a greater value for doped samples than for undoped ones. This would mean that the critical strain-energy for scale spalling would be greater in the case of the yttrium-containing alloys, implying a better adhesion of the alumina scale.

Anyway, it appears difficult to give general rules for the effect of active elements on Al_2O_3-scale plasticity. This is probably due to the fact that, in many cases, the chemical state of the active element (in solid solution, segregated, precipitated) is unknown and that to date no clear ideas have been developed to explain the influence of an element on the alumina plasticity in relation with its localization and chemical state (see ref. 78).

4. CONCLUDING REMARKS

It appears that, despite the extensive literature on the subject, there is still a remarkable controversy regarding the effect of active elements on the growth mechanism and the adhesion of Al_2O_3 scales.

Firstly, in many cases, there exists a controversy about the beneficial or detrimental effect of certain elements. This appears to depend on their amount, their chemical state in the scale, the alloy nature, the neighbourhood... For some specific elements, clear ideas are available: sulphur is recognized as a detrimental element, whereas yttrium (in limited amount) has a beneficial effect on the scale adhesion. But for most other elements, such as Ti, Zr, Hf, C..., varying effects are being reported in the various investigations.

Secondly, further studies have to be carried out in order to explain the exact mechanism by which active elements affect the growth and adhesion of alumina scales. The more recent investigations emphasize the following principal ideas:

- Active elements induce a strengthening of the substrate-oxide bonding, (although this idea is contradicted by some authors), by segregating at the metal-oxide interface and/or by preventing detrimental elements (S, Cl...) to segregate to this interface.

- Active elements modify the microstructure of Al_2O_3 scales, particularly by decreasing the grain size, but also by inducing intergranular segregation or second phase precipitation. This should obviously have an influence on both the transport processes and the plasticity of the alumina scale:

  * The effect of active elements on the transport processes in alumina scales is being estimated to be of minor importance - although this is a subject of controversy.

  * The major recognized effect of active elements for Al_2O_3-formers is the improvement of the scale adhesion. It is generally suggested that active elements affect the scale plasticity. But there is no direct evidence for this point, although some observations (decrease in grain size, stress measurements...) appear to support this idea. It is probably in this domain where more data on both doped massive alumina and alumina scales are needed.

- It cannot be excluded that other mechanisms, such as mechanical keying, elimination of interface voids, etc... are also relevant, and it is probable that several mechanisms may simultaneously interact.

REFERENCES

1. D.P. WHITTLE and J. STRINGER, Phil. Trans. R. Soc. London, Ser. A, 1980, 295, 309.

2. G.C. WOOD and F.H. STOTT, in High Temperature Corrosion, NACE-6, 1983, 227.

3. H. HINDAM and D.P. WHITTLE, Oxid. Met., 1982, 18, 245.

4. F.H. STOTT and G.C. WOOD, Mat. Sci. and Eng., 1987, 87, 267.

5. A.M. HUNTZ, Mat. Sci. and Eng., 1987, 87, 251.

6. D.P. MOON, UKAEA Report n° AERE R 12930, 1988.

7. J.G. SMEGGIL, A.W. FUNKENBUSCH and N.S. BORNSTEIN, Met. Trans., 1986, 17A, 923.

8. J.G. SMEGGIL, Mat. Sci. and Eng., 1987, 87, 261.

9. A.R. NICOLL and G. WAHL, Superalloys 1984, Met. Soc. of AIME Pub., 1984, 805.

10. A. BOUMAZA, G. MOULIN and A.M. HUNTZ, Oxid. Met., 1988, 30, 3-4, 141, and
A. BOUMAZA, Thesis, University Paris XI, Orsay, France, 1987.

11. R. MEVREL, ONERA, private communication.

12. M.H. LAGRANGE, A.M. HUNTZ and J.H. DAVIDSON, Corr. Sci., 1984, 24 (7), 613.

13. D. FELLMAN, Y. BIENVENU and M. FOUCAULT, Mem. Sci. Rev. Met., 1983, 80, 37.

14. S. TANIGUCHI and T. SHIBATA, 10th ICMC, Madras 7-11 Nov. 1987, Ed. by Oxford and IBM Publishing Co, PVT, LTD, Vol IV, 3385.

15. P. PAUWELS, Thesis, ENSMP-ARMINES, France, 1988.

16. A.S. KAHN, C.E. LOWELL and C.A. BARRET, J. of the Electrochemical Soc., 1980, 127 (3), 670.

17. J.L. PANDEY, S. PRAKASH, and M.L. MEHTA, 10th ICMC, Madras, 7-11 Nov. 1987, Ed. by Oxford and IBM Publishing Co, PVT, LTD, vol IV, 3507.

18. D. SAXENA, S. PRAKASH, M.L. MEHTA and I.P. SARASWAT, 10th ICMC, Madras, 7-11 Nov. 1987, Ed. by Oxford and IBM Publishing Co, PVT, LTD, vol IV, 3495.

19. P. CHOQUET, Doctor Thesis, University Paris XI, Orsay, France, 1987 and
P. CHOQUET and R. MEVREL, Proc. Int. Conf. on Metallurgical Coatings, San Diego, 17-21 April 1989., D. DIOT, P. CHOQUET and R. MEVREL, Proc. ICRS-2, 23-25 Nov. 1988, Nancy.

20. I.M. ALLAM, D.P. WHITTLE and J. STRINGER, Oxid. Met., 1978, 12 (1), 35.

21. J.C. PIVIN, D. DELAUNAY, C. ROQUES-CARMES, A.M. HUNTZ and P. LACOMBE, Corr. Sci., 1980, 20, 351.

22. G. BEN ABDERRAZIK, G. MOULIN, A.M. HUNTZ and R. BERNERON, J. of Mat. Sci., 1984, 19, 3173.

23. A.M. HUNTZ, G. MOULIN and G. BEN ABDERRAZIK, Ann. Chim. Fr., 1986, 11, 291.

24. J. ROUSSELET, Doctor thesis, University Paris XI, Orsay, France, 1988, and
Proceedings of Euroanalysis VI, 7-11 Sept. 1987, Paris, Ed. GAMS, Paris.

25. A.W. FUNKENBUSCH, J.G. SMEGGIL and N.S. BORNSTEIN, Met. Trans., 1985, 16A, 1164.

26. D.G. LEES, Oxid. Met., 1987, 27 (1-2), 75.

27. C.L. BRIANT and K.L. LUTHRA, Met. Trans., 1988, 19A, 2091 and 2099.

28. R.P. MESSMER and C.L. BRIANT, Acta Met., 1982, 30, 457.

29. H. HINDAM and D.P. WHITTLE, J. Electrochem. Soc., 1982, 129, 1147.

30. J.H. DAVIDSON, P. LACOMBE, A.M. HUNTZ, C. ROQUES-CARMES, J.C. PIVIN and D. DELAUNAY, in: Behaviour of high temperature alloys in aggressive environments. The Metals Society, London, 1980, 209.

31. D.P. WHITTLE and D.H. BOONE, Proc. 8th ICMC, Mainz, 1981, Dechema, Frankfurt-Main, 718.

32. C.S. GIGGINS and F.S. PETTIT, Report ARL 75-0234, 1975.

33. I.A. ALLAM, D.P. WHITTLE and J. STRINGER, Oxid. Met., 1979, 13, 381.

34. F.A. GOLIGHTLY, F.H. STOTT and G.C. WOOD, J. Electrochem. Soc., 1976, 126, 1035.

35. D. DELAUNAY and A.M. HUNTZ, J. Mat. Sci., 1982, 17, 2027.

36. F.A. GOLIGHTLY, F.H. STOTT and G.C. WOOD, Oxid. Met., 1976, 10, 163.

37. C.S. GIGGINS, B.H. KEAR, F.S. PETTIT and J.K. TIEN, Met. Trans., 1974, 5, 1685.

38. J.D. KUENZLY and D.L. DOUGLASS, Oxid. Met., 1974, 8, 139.

39. J.K. TIEN and F.S. PETTIT, Met. Trans., 1972, 3, 1587.

40. A. KUMAR, M. HASRALLAH and D.L. DOUGLASS, Oxid. Met., 1974, 8, 227.

41. T. AMANO, S. YAJIMA and Y. SAITO, Trans. JIM, 1979, 20, 431.

42. J.L. SMIALEK and R. GIBALA, Met. Trans., 1983, 14A, 2143.

43. J.C. PIVIN, C. ROQUES-CARMES, J. CHAUMONT and H. BERNAS, Corr. Sci., 1980, 20, 947.

44. A.B. ANDERSON, S.P. MEHANDRA and J.L. SMIALEK, J. Electrochem. Soc. 1985, 132, 1695.

45. J.G. SMEGGIL, A.W. FUNKENBUSCH and N.S. BORNSTEIN, Proc. Spring Electrochemical Society Meeting, Cincinnati, Ohio, 1984.

46. C.L. BRIANT and R.A. MULFORD, Met. Trans., 1982, 13A, 745.

47. J.L. SMIALEK, NASA, TM 81676, 1981, and TM 100 209, 1987. J.L. SMIALEK and R. BROWNING, NASA TM 87168, 1985.

48. J.L. SMIALEK, Met. Trans. 1987, A18, 164.

49. H. HINDAM and W.W. SMELTZER, J. Electrochem. Soc., 1980, 127, 1630 and Oxid. Met., 1980, 14 (4), 337.

50. J. JEDLINSKI, 10th ICMC, Madras, 7-11 Nov. 1987, Ed. by Oxford and IBM Publishing Co, PVT, LTD, vol VI, 3709.

51. S.S. MROWEC, A. GIL and A. JEDLINSKI, Werkstoffe und Korrosion, 1987, 38, 563.

52. A.M. HUNTZ, G. BEN ABDERRAZIK, G. MOULIN, E.W. JOUNG and J.H.W. DE WIT, Applied Surface Science, 1987, 28, 345 and Solid State Ionics, 1987, 22, 285.

53. E.J. FELTEN and F.S. PETTIT, Oxid. Met., 1976, 10, 189.

54. T.A. RAMANARAYANAN, R. AYER, R. PETKOVIC-LUTON and D.P. LETA, Oxid. Met., 1988, 9 (5-6), 445.

55. D. NICOLAS-CHAUBET, Thesis, University Paris XI, Orsay, 1989.

56. J.K. DOYCHAK, T.E. MITCHELL and J.L. SMIALEK, Mat. Res. Soc. Symp. Proc., 1985, 39, 475.

57. J. JEDLINSKI and S. MROWEC, Mat. Sci. and Eng., 1987, 87, 281.

58. G. TREMOUILLES, Doctor Thesis, University Paris VI, France, 1988, and
G. TERMOUILLES, J.L. DEREP and R. PORTIER, Advances in Thermal Spraying, ITCS'86, Pergamon 1986, 465.

59. J.D. CAWLEY, AND J.W. HALLORAN, J. Amer. Cer. Soc., 1986, 69 (8), C 195-196.

60. G. PETOT-ERVAS, D. DEWEIRDER, M. LOUDJANI, B. LESAGE and A.M. HUNTZ, Advances in Ceramics, 1987, 23, 125.

61. I.A. BONDAR, Izv. Akad. Nauk. SSSR, Neorg. Mat., 1984, 20 (2), 257.

62. D. BOUCHET, University Paris XI, Orsay, private communication.

63. D. PROT, B. LESAGE, C. MONTY and A.M. HUNTZ, unpublished work.

64. K.P.R. REDDY, J.L. SMIALEK and A.R. COOPER, Oxid. Met., 1982, 17 (5-6), 429.

65. T.A. RAMANARAYANAN, M. RACHAVAN and R. PETKOVIC-LUTON, Oxid. Met., 1984, 22 (3-4), 83, and J. Electrochem. Soc., 1984, 131 (4), 923.

66. E.W.A. YOUNG and J.H.W. DE WIT, Solid State Ionics, 1985, 16, 39 and Oxid. Met., 1986, 26, 351.

67. M.J. BENNET and D.P. MOON, this conference.

68. M.M. EL AIAT and F.A. KRÖGER, J. Am. Ceram. Soc., 1982, 65 (6), 280.

69. G.J. DIENES, D.O. WELCH, C.R. FISHER, R.D. HATCHER, D. LAZARETH and M. SAMBERG, Phys. Rev., 1975, B11 (8), 3060.

70. S. LARTIGUE, Doctor Thesis, University Paris XI, Orsay, France, 1988 and Phys. Rev., 1975, B11 (8), 3060.

71. M.H. LAGRANGE, A.M. HUNTZ and J.Y. LAVAL, Ann. Chim. Fr., 1987, 12, 9.

72. A.M. HUNTZ, G. MOULIN and B. LESAGE, Proc. 9th ICMC, Toronto, June 3-7, 1984, voll II, National Research Council of Canada, Toronto, 1984, 400.

73. A.E. PALADINO and W.D. KINGERY, J. Chem. Phys., 1962, 37 (5), 957.

74. Y. OISHI and W.D. KINGERY, J. Chem. Phys., 1960, 33 (2), 480.

75. F.A. KRÖGER, Solid State Ionics, 1984, 12, 189.

76. A. ATKINSON, this conference.

77. J. NOWOK, Oxid. Met., 1982, 18 (1-2), 1.

78. J. CASTAING, this conference.

79. C. CARRY and A. MOCELLIN, Ceramics International, 1987, 13, 89.

80. A.M. HUNTZ et al., Mat. Sci. and Techn., 1988, 4 (5), 470 and 1988, 4 (12), 1079.

81. D. DELAUNAY, A.M. HUNTZ and P. LACOMBE, Corr. Sci., 1980, 20, 1109.

82. K.L. LUTHRA and C.L. BRIANT, Oxid. Met., 1986, 26 (5-6), 397.

83. R.M. CANNON, W.H. RHODES and A.H. HEUER, J. Am. Ceram. Soc., 1980, 63, 46.

84. M.I. MANNING, in: Corrosion and Mechanical Stress at High Temperature, eds. V. Guttman and M. Merz, Applied Science Publishers Ltd., London, 1981, 323.

85. J.D. CAWLEY, J.W. HALLORAN and A.R. COOPER, NASA T.M. 83622, 1984.

86. W.D. KINGERY, D.C. HILL and R.P. NELSON, J. Amer. Cer. Soc., 1960, 43, 473.

87. Y. OISHI, H. KOIZUMI, Y. SAKKA and K. ANDO, J. Met. Sc. Letter, 1985, 4 (2), 176.

88. H. HANEDA, Y. MIYASAWA and S. SHIRASAKI, J. of Crystal Growth, 1984, 68 (2), 581.

89. M.F. BERARD, J. Amer. Cer. Soc., 1968, 51, 648.

90. M.F. BERARD, J. Amer. Cer. Soc., 1967, 50, 113.

91. K. ANDO, J. Amer. Cer. Soc., 1983, 66, C-222.

EFFECT OF ACTIVE ELEMENTS ON THE OXIDATION BEHAVIOUR OF Cr_2O_3-FORMERS

M.J. Bennett and D.P. Moon

Materials Development Division, Harwell Laboratory, Didcot, Oxfordshire. OX11 ORA

ABSTRACT

Active elements with a high affinity for oxygen, such as yttrium, cerium, lanthanum and other rare earths, have been known for half a century to improve the oxidation behaviour of high temperature alloys whose protection derives from the formation of Cr_2O_3 scales. It is now clear that the active elements can promote the formation of the protective scale, can reduce the rate of continuing scale growth and can inhibit scale failure processes. Various mechanisms, involving modifications to transport properties and mechanical behaviour of scales, have been proposed to explain the observed beneficial effects. These mechanisms will be reviewed in the light of available experimental evidence, including the dependence of protection upon the individual active element, its concentration and initial distribution and source (i.e. whether an alloying addition, oxide dispersion, coating or surface implant).

1. INTRODUCTION

The protection of technological alloys in oxidising environments at elevated temperatures relies upon the formation and maintenance of a coherent, crack-free and adherent oxide scale of low diffusivity, such as chromia or alumina. For over half a century it has been known[1] that the behaviour of both types of scale can be improved by the addition of the so-called active elements, such as yttrium, cerium and other rare earths, which have a higher affinity for oxygen than the base alloy constituents. The active elements may be incorporated in the bulk alloy as metallic or oxide dispersoid constituents or be added as surface layers by coating or ion-implantation. The crucial rôle of the active elements has been extensively researched with the resultant generation of numerous mechanistic theories.

This review will concentrate on the rôle of active elements in improving the oxidation behaviour of Cr_2O_3-formers. It will not be exhaustive in summarising the evidence for and against the various theories advanced, as this has been covered recently[2], but rather the review will provide a considered view of current understanding. It has to be recognised that a technological Cr_2O_3 scale is not a definable, unique entity and that the Cr_2O_3 scales formed on chromium metal, on binary iron/nickel/cobalt - chromium alloys and on chromium-bearing technological alloys will almost certainly show considerable compositional and microstructural differences. Nevertheless all such scales have one important common property, that of providing oxidation resistance to the underlying substrate. Therefore, all observations concerning Cr_2O_3scales are considered relevant to this review. Essentially, oxidation protection involves the initial establishment of a Cr_2O_3 scale and then its steady-state growth, while a major cause of loss of protection derives from scale cracking and spallation. As a consequence it seems logical to discuss the rôle of active elements with respect to these three aspects separately. Finally, factors controlling active element effectiveness will be summarised.

2. THE EFFECT OF ACTIVE ELEMENTS UPON Cr_2O_3 SCALE ESTABLISHMENT

Active element additions can promote the selective oxidation of chromium in an alloy[3] and thereby the earlier development of a protective Cr_2O_3 scale to curtail transient oxidation of the base metal constituents. As an example, this has been observed to be a consequence of the addition of 0.1 wt.% Ce or CeO_2 to Fe-20Cr upon its oxidation behaviour at 1000°C.[4] Additionally, protective scales may be formed on alloys with a lower chromium concentration than normally would be required to form a continuous Cr_2O_3 layer.[5,6] For the oxidation at 1000°C of Co-25Cr, which is a borderline Cr_2O_3-former, a 1% yttrium alloying addition or the implantation of 4×10^{17} yttrium ions cm^{-2} induced a continuous Cr_2O_3 scale. An internal oxidation pretreatment additionally shortened the transient stage of oxidation (with less cobalt oxide being formed) and reduced the minimum implant dosage for selective chromium oxidation.[6]

Of the various mechanisms proposed to explain the rôle of active elements, modification by the active element of alloy diffusion properties and the formation of an active element oxide diffusion barrier lack general applicability. The most probable mechanism[7] is that dispersed active element oxide particles at the

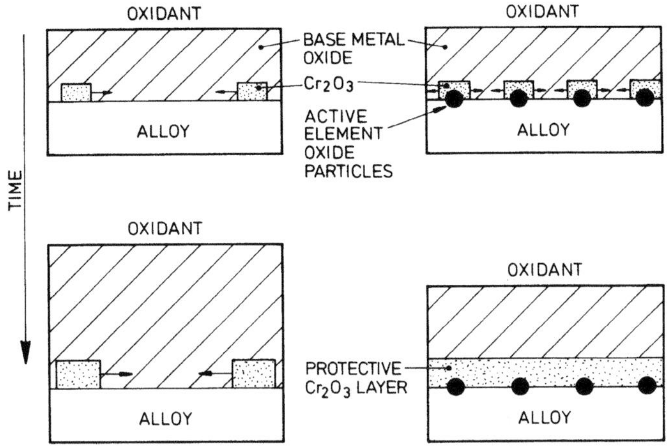

Fig. 1 Schematic representation of the rôle of active element oxide particles in an alloy surface acting as heterogeneous sites for Cr_2O_3 nucleation and facilitating subsequent lateral growth processes which link the nuclei to form a protective layer.

alloy surface act as heterogeneous nucleation sites for the first-formed oxides, thereby decreasing internuclei spacings. As a consequence, less time will be required for subsequent lateral growth processes to link the nuclei to form a protective Cr_2O_3 scale. This is shown schematically in Figure 1.

This mechanism is consistent with further experimental observations. The promotion of Cr_2O_3 formation by active element oxide dispersions[7], coatings[8] and surface ion implants[9] is independent of the chemical nature of the active element. Also, smaller active element oxide particle spacings favour the selective oxidation of chromium. This was demonstrated for an Fe-20Cr-0.02Ce alloy internally oxidised to produce a distribution of CeO_2 particles whose number decreased and size increased with distance into the alloy.[10] Samples were ground to different depths and subsequently oxidised at 1000°C. Increased spacings of CeO_2 particles at the alloy surface gave progressively less Cr_2O_3, with increasingly more time being required to form a continuous protective oxide layer. Eventually Fe_2O_3 was the predominant oxide formed. The particle spacing effect appears to saturate at high particle concentrations. Variation in the volume fraction of ThO_2, Y_2O_3 and CeO_2 dispersoids in the range 1-3 vol.% in Ni-20Cr had little effect upon initial oxidation.[7] For Fe-20Cr-25Ni stainless steel implanted with cerium or yttrium (which are considered to oxidise internally), the time required for protective scale formation was independent of ion dose above a threshold value.[11] Comparison with the corresponding influence of Ni implantation confirmed that this behaviour was a chemical effect of cerium or yttrium and not merely a consequence of sputter or irradiation damage arising from the implantation. The saturation particle spacing for a given alloy might be governed by an inter-dependence with the nature and concentration of alternative surface discontinuities resulting from surface cold work, heat treatment, surface finish and purity.

3. THE EFFECT OF ACTIVE ELEMENTS UPON Cr_2O_3 SCALE GROWTH

The rate-controlling process in the growth of Cr_2O_3 scales is believed to be cation diffusion along high-angle grain boundaries[12] or other short-circuit paths[13], since lattice and low-angle boundary diffusivities are too slow to account for the scale growth rates. Chromium vacancies are probably the predominant mobile species in scales formed at high oxygen partial pressures.[14] Active element additions significantly reduce the thickening rate of Cr_2O_3 scales. As an example the Arrhenius plots of the parabolic rate constants for the growth of a duplex spinel/Cr_2O_3scale on cerium or yttrium implanted and on unimplanted Fe-20Cr-25Ni stainless steel[15] are shown in Figure 2.

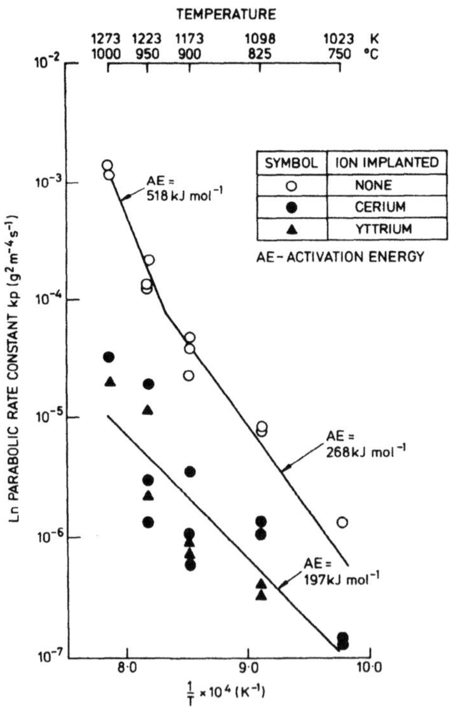

Fig. 2 Arrhenius plot of the parabolic rate constants for the oxidation of cerium, yttrium and unimplanted Fe-20Cr-25Ni stainless steel at 750-1000°C.

The main effect of the active element implantation was to reduce, by an order of magnitude, the pre-exponential (geometric) factor. The apparent reduction in the activation energy for the oxidation of the implanted steel may be less than indicated owing to scatter in the data, but in any case was only of secondary importance.

The active elements were effective for this and most other Cr_2O_3 scales in the temperature range 0.4-0.6 T_m, where T_m is the absolute melting temperature of the scale. Since short-circuit diffusion is considered to control scale growth under the same conditions, it would seem logical that the active element additions inhibit scale growth by reducing transport along high-diffusivity pathways. Further understanding should follow from establishing the exact form and location of the active element within the scale. More detailed microstructural characterisation of Cr_2O_3 scales has become possible during the last few years as a result of developments in analytical instruments, such as the field emission gun scanning transmission electron microscope (FEG-STEM), and in specimen preparation

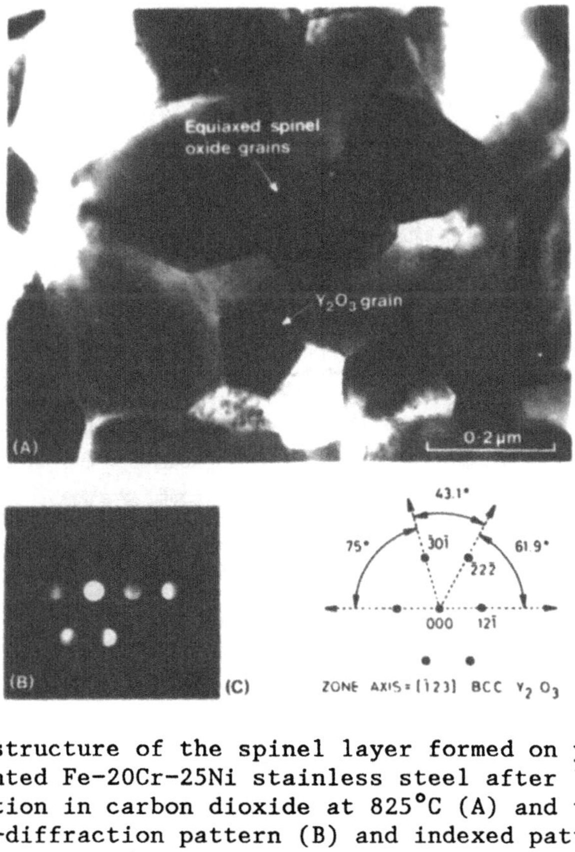

Fig. 3 Microstructure of the spinel layer formed on yttrium implanted Fe-20Cr-25Ni stainless steel after 775 h oxidation in carbon dioxide at 825°C (A) and the micro-diffraction pattern (B) and indexed pattern (C) of an incorporated Y_2O_3 grain.

procedures. In all of the Cr_2O_3 scales so far examined, the active element was located within a continuous band, which almost certainly provided a marker of the original metal surface. These bands occurred either at or more precisely just beneath the gas interface, as in the case of yttrium implanted into Cr[16], Ni-33Cr[17] and Co-45Cr[18], or near the alloy interface as on Ni-20Cr[19]. For the duplex scale formed on yttrium or cerium implanted Fe-20Cr-25Ni stainless steel, there was a continuous active element bearing band within the outer spinel layer at the interface with the inner Cr_2O_3 layer.[15,20] Analytical electron microscopy (AEM) revealed Y_2O_3[20] (Figure 3) and $YCrO_3$[16,18] particles at grain boundary junctions within the bands in the spinel and Cr_2O_3 layers. $YCrO_3$ particles, formed by Y_2O_3-Cr_2O_3 solid-solid reaction, were also found within a Cr_2O_3 scale formed on the oxide dispersion-strengthened alloy MA 754 (Ni-20Cr-0.6Y_2O_3).[21] The composition and structure of these particles were established by energy dispersive X-ray analysis (EDX) and micro-diffraction, as illustrated in Figure 3. The Y_2O_3 and $YCrO_3$ particles acted as the sources of yttrium ions which segregated along the oxide grain boundaries in the scales formed on yttrium implanted Fe-20Cr-25Ni stainless steel (Figure 4a) and yttrium

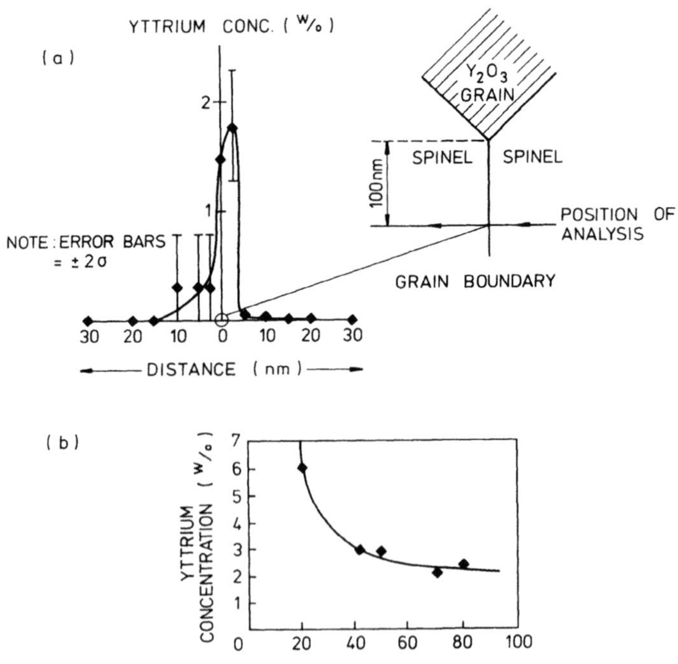

Fig. 4 Energy dispersive analysis using a field emission STEM of the yttrium concentration profile across a spinel-spinel grain boundary at 100 nm from a Y_2O_3 grain (a). Variation of the yttrium concentration along a spinel-spinel grain boundary with distance from a Y_2O_3 grain (b).

implanted Cr (Figure 5). Segregant concentration decreased with increasing distance from each particle (Figure 4b). It has been demonstrated that other effective active elements, such as Ce and La, are present within Cr_2O_3-based scales as oxide particles,[9,22] and by inference probably also occur as grain boundary segregants, but this has yet to be established.

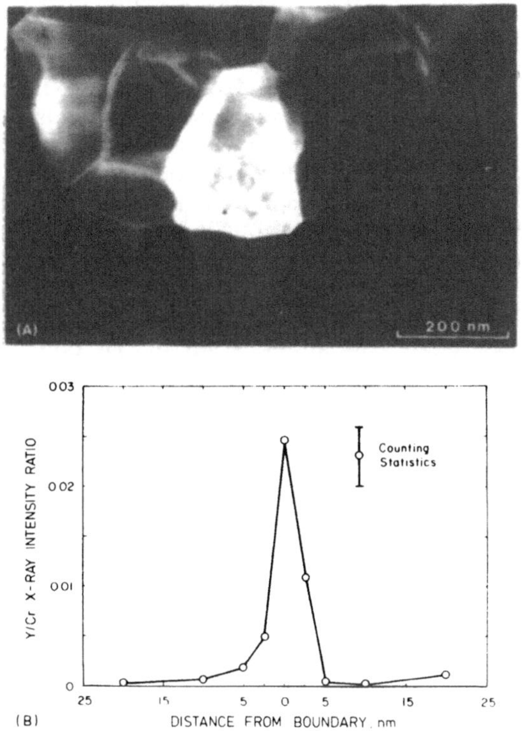

Fig. 5 Dark field STEM image of a parallel section through the scale formed in 20 h at 900°C in O_2 at 10^5 Pa on Cr implanted with 2 x 10^{16} Y ions cm^{-2} (A) and a plot of the Y:Cr X-ray intensity ratio determined by EDX in a STEM across one of the grain boundaries in the parallel section (B).

AEM also indicated that the crystallite sizes of the main scale constituents (Cr_2O_3 or spinel) within the band where the yttrium was located were considerably smaller than those outside the band or within the corresponding scales formed on the metal or alloy without active element additions (e.g. Ref. 20). This implies that the initially formed oxide grains, whose nucleation could have been promoted by the active element addition, were also prevented from growing during subsequent oxidation by the incorporated active element oxide particles and grain boundary segregants.

Diffusion through these active element bearing bands controlled continuing scale growth, as was demonstrated for the formation by cation diffusion of a NiO scale on CeO_2 coated Ni.[23] The crucial microstructural feature responsible for the reduction in scale growth rate is undoubtedly the grain boundary segregation of the active element. In contrast, the decrease in grain size should have increased grain boundary diffusion, while the fraction of boundary area occupied by active element oxide particles was insufficient to block diffusion physically. Of the various mechanisms suggested[2] for the inhibition of grain boundary diffusion the most likely candidate would appear to be complex defect formation by vacancy-active element association, increasing the activation energy for diffusion along segregated short-circuit paths.[24] This mechanism is supported by recent atomistic modelling,[25] which indicated that the activation energy for nickel vacancy diffusion near an isolated cerium ion on a nickel site in a (310)/[001] tilt boundary in NiO exceeded that for lattice diffusion. For this case at least, the lattice was a more favourable pathway than the segregated grain boundary. A high concentration of segregated active element is required to inhibit grain boundary diffusion,[26,27] which would be favoured by the provision of small, closely spaced and uniformly distributed active element oxide particles to act as sources.

On yttrium implanted Cr[16] and Co-45Cr[18], the yttrium segregants completely inhibited cation diffusion so that continuing growth of the Cr_2O_3 scales occurred entirely by anion movement. Sequential oxidation in oxygen containing normal and enriched concentrations of ^{18}O has also established that the initial growth of the Cr_2O_3 scale on Fe-25Cr coated with 40 Å CeO_2 resulted primarily from inward transport (Figure 6), as the ^{18}O peak is at the scale-alloy interface.[28] In addition, isotopic exchange was apparent with the oxygen in the CeO_2 particles. It is not clear whether this inward oxygen transport was comparable with that which might have occurred concurrently with cation movement in the development of the unmodified scale, or whether it was promoted by the incorporation of the active element within the Cr_2O_3 scale. Initial growth by oxidant transport has been observed also on Co-25Cr[6] and on the Fe-20Cr-25Ni stainless steel[9], but in both systems subsequent scale growth occurred by cation transport. This is shown for yttrium implanted Fe-20Cr-25Ni stainless steel in Figure 7.

Rutherford backscattering (RBS) analyses of the yttrium distribution within scales of increasing thickness indicated that the maximum yttrium concentration, which may be taken as a marker of the original surface, initially was near the gas interface but with increasing exposure time became incorporated at a greater depth within the scale. During extended exposures, the active element effect on scale growth would be controlled essentially by a geometric factor, i.e. the proportion of boundaries effectively blocked by active element segregants. This would be consistent with the order of magnitude reduction in the pre-exponential factor of the parabolic rate constant for Fe-20Cr-25Ni stainless steel oxidation by cerium or yttrium ion implantation (Figure 2).[15]

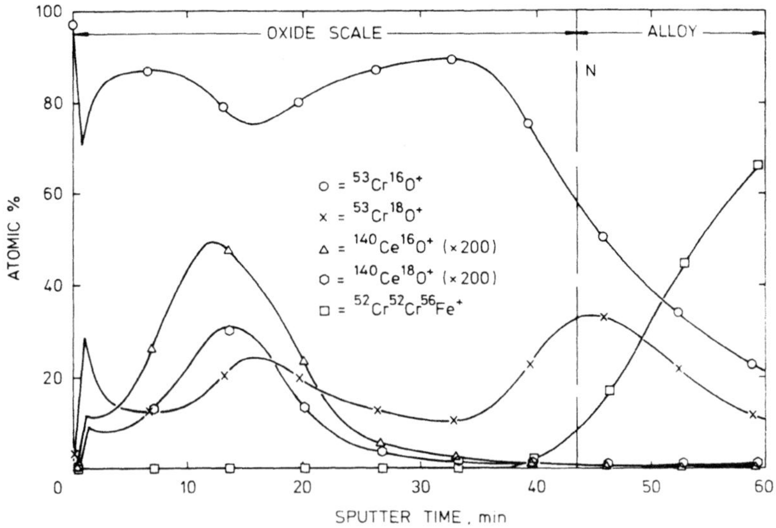

Fig. 6 SIMS depth profile through the Cr_2O_3 scale formed on Fe-25Cr, coated with 40 Å CeO_2, after oxidation at 1100°C for 1 h in 10 Pa $^{16}O_2$ and then for 1 h in 0.66 Pa $^{18}O_2$.

At temperatures lower than ∿ 0.4 T_m, active elements appear to exert little influence on the oxidation rate. This could be due to a decrease in active element solubility at scale grain boundaries to a level at which diffusion blocking is insignificant.

A consistent picture of any possible effect of active elements on Cr_2O_3 scale growth rates at temperatures above ∿ 0.6 T_m has not yet emerged due to a lack of experimental observations. It could be argued that lattice diffusion would increasingly dominate scale growth. The increased solubility of active elements in the oxide lattice at these temperatures could enhance the concentration of ionic defects responsible for transport (e.g. by doping tetravalent Ce into Cr_2O_3) and thereby increase the oxidation rate, as observed for CeO_2 coated Ni at 1100-1200°C.[23] Alternatively, the incorporated active element oxide particles within a scale might hinder lattice diffusion by physically blocking the cross-sectional area available for cation transport or by acting as cation vacancy sinks.[2]

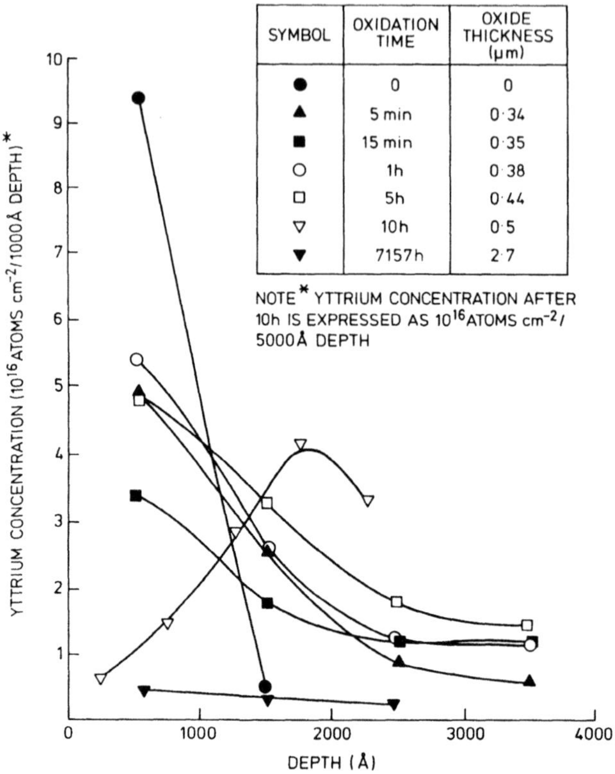

Fig. 7 RBS yttrium depth profiles during oxidation of Fe-20Cr-25Ni stainless steel implanted with 9.0 x 10^{16} yttrium ions cm^{-2}.

4. THE EFFECT OF ACTIVE ELEMENTS UPON Cr_2O_3 SCALE FAILURE

Loss of surface protection of Cr_2O_3 forming alloys can be an undesirable consequence of scale failure by cracking and spallation. Failure can occur during isothermal oxidation but is more probable on subsequent cooling to lower temperatures. The modes of failure of the duplex spinel/Cr_2O_3 scale formed on the Fe-20Cr-25Ni stainless steel during oxidation in carbon dioxide at temperatures $\lesssim$ 950°C have been examined using thin layer activation[29] and acoustic emission[30] techniques concurrently with microbalance weight change measurements. No cracking or spallation was apparent during the actual growth of the scale provided it was bonded to the steel. A limited incidence of cracking during oxidation was associated with failure along the apex of convolutions in regions of buckled scale which had become detached from the steel, but this did not result in spallation at temperature. In contrast, provided the scale was above a threshold thickness, cooling from the oxidation temperature resulted in extensive cracking and spallation from both the predominant bonded and the less common buckled scale regions. The principal parameters governing Cr_2O_3 scale failure, therefore, would appear to be, firstly, scale-substrate interfacial adhesion during oxidation and, secondly, the stresses developed during scale growth and more significantly during cooling as a consequence of the difference between the thermal expansion coefficients of the scale and the substrate metal/alloy. Spallation involves fracture through the scale thickness and along (or within the scale parallel to) the scale-alloy interface, as shown diagrammatically for the Fe-20Cr-25Ni stainless steel in Figure 8. These events have been modelled using stored elastic strain energy considerations for the cases of strong and weak scale-substrate adhesion under compressive stress loading.[31]

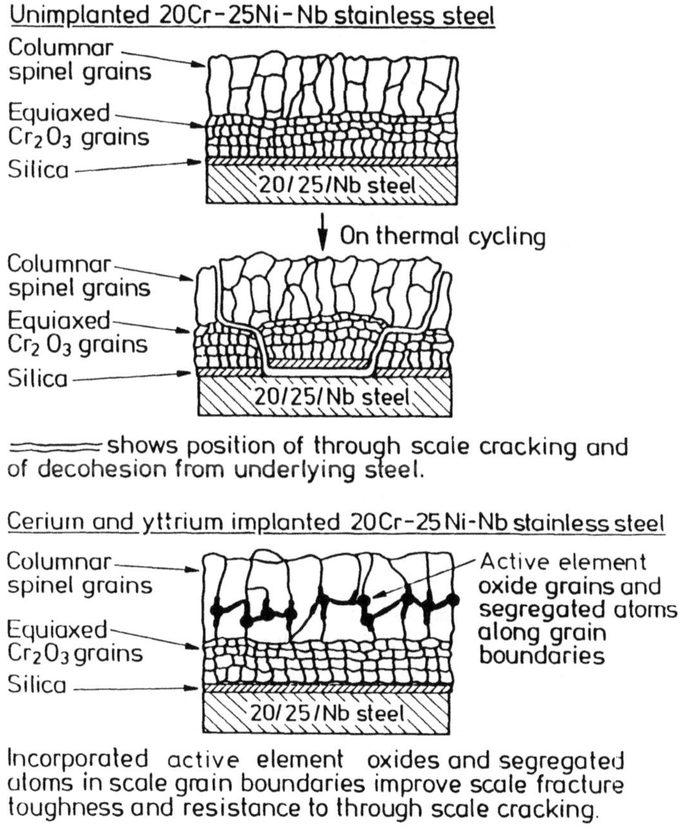

Fig. 8 Schematic representation of spallation on cooling to 25°C of the scale formed on Fe-20Cr-25Ni stainless steel at elevated temperatures and of its inhibition by cerium and yttrium ion implantation.

The third, and in some instances the most dramatic rôle of the active elements in alloy protection is inhibition of Cr_2O_3 scale cracking and spallation. The current understanding of Cr_2O_3 scale failure would imply that the influence of the active elements could derive from a reduction in in-scale compressive strains, an improvement in scale fracture toughness and an enhancement in scale-substrate interfacial adhesion. The microstructural characterisation of the location and form of the active elements within Cr_2O_3 scales, described in Section 3, has clearly eliminated the viability of many previously proposed mechanisms, such as chemical bonding, impurity gettering, oxide keying/pegging and oxide plasticity.[2,32]

Considering first a possible reduction in in-scale compressive strains, no reliable in-situ experimental measurements appear to exist on growth strains in either active element modified or unmodified Cr_2O_3 scales. Indirect evidence from the Fe-20Cr-25Ni stainless steel would imply that cerium ion implantation increased oxidation growth induced strains, which were relieved at temperature by creep deformation of the substrate steel.[33] However, as has been mentioned, a more significant compressive strain within the scale is that which develops on cooling and this will increase with scale thickness. Therefore, as a direct consequence of the inhibition of scale growth by active elements, cooling an active element modified thin scale after a given oxidation time may produce insufficient strain energy to initiate the observed spallation of the corresponding thicker unmodified Cr_2O_3 scale.

Most previous experiments reported in the literature have evaluated the active element effect by comparing spallation behaviour after the same oxidation time. However, this can be misleading. A detailed appreciation of the total influence of reactive elements on Cr_2O_3 scale failure can be ascertained only from a comparison of the behaviour of scales of similar thickness. Cerium or yttrium ion implantation into the Fe-20Cr-25Ni stainless steel increased the threshold thickness for the initiation of spallation of scales formed at 750 to 950°C from 0.7 μm to 2.8 μm, i.e. by a factor of four.[32] It is then possible using the parabolic rate constants for oxidation (Figure 2) to calculate the temperature dependence of the time to initiate spallation from cerium and yttrium implanted Fe-20Cr-25Ni stainless steel. This is compared in Figure 9 with the corresponding relationship for unimplanted Fe-20Cr-25Ni stainless steel[34] and demonstrates that four orders of magnitude improvement in protection time is afforded by cerium and yttrium ion implantation over the temperature range studied.

The scale incorporating an active element can sustain therefore a greater strain on cooling before failure occurs. In

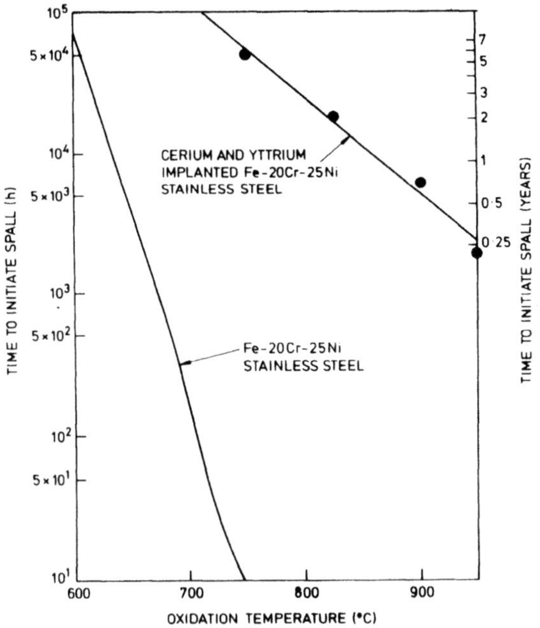

Fig. 9 Influence of cerium and yttrium ion implantation upon the initiation on cooling to 25°C of spallation of oxide formed on Fe-20Cr-25Ni stainless steel.

the decohesion of the duplex scale from the Fe-20Cr-25Ni stainless steel, a plane of weakness is provided by a thin silica layer, which forms at the scale-steel interface by oxidation of a minor silicon component in the steel.[35] A possible cause for the improved adhesion as a result of ion implantation could be a change in scale morphology at the steel interface, such that the silica layer is not continuous. Alternatively and more probably, as shown schematically in Figure 8, the active element oxide grains and grain boundary segregants within the scale inhibit failure by increasing the energy required for through-scale crack initiation and propagation. This could originate from enhanced mechanical strength and fracture toughness of the scale or from the reactive element oxide particles arresting any cracks developed during strain relief. A recent analysis has suggested that the main effect of active element incorporation is to reduce the degree of scale imperfection and thus the propensity to crack, while there is little effect on scale fracture toughness.[36]

Modification of scale growth processes by active elements, with the suppression of cation diffusion and increased contribution of inward oxygen transport, could also result in almost total contact between scale and alloy through the elimination of interfacial porosity. It is possible that this change could be of significance in preventing the formation and failure at temperature of the non-bonded buckled Cr_2O_3 scales found on the unmodified metal/alloy. However, with respect to behaviour on cooling, both bonded and non-bonded scales on the Fe-20Cr-25Ni stainless steel crack/spall, the main difference being that for comparable scale thickness a greater temperature drop is required to initiate spallation of non-bonded scales.[29] Overall, therefore, more conclusive evidence is required to establish the nature of the active element effect on scale adhesion.

5. FACTORS MAXIMISING ACTIVE ELEMENT EFFECTIVENESS FOR Cr_2O_3-FORMERS

Several common features emerge from the discussion of the active element effects on Cr_2O_3 scale establishment, growth and failure in Sections 2, 3 and 4. The effects derive from the provision of active element oxide particles to act as heterogeneous nucleation sites at the alloy surface, as sources of active element ionic segregation to block fast diffusion paths in the scale, as inhibitors of grain growth, as promoters of improved scale mechanical properties, and as sites for crack arresting. Several factors can be identified as maximising their influence. The first is the desirability of oxide particle formation from the onset of oxidation. The second is that a smaller particle size provides a high grain boundary segregant concentration and avoids the direct initiation of scale failure possible with large particles. A third factor is that a dense uniform particle distribution provides many sites for particle related mechanisms and close overlap between regions of high segregant concentration around particles. These characteristics will be influenced by the identity and quantity of the active element and by the form of its distribution (alloying addition, oxide dispersion, surface oxide coating or implant) prior to oxidation.

With regard to the choice of the active element, it is worth repeating several basic principles. The formation of stable active element oxide particles within the Cr_2O_3 scale requires that any active element addition must have a higher oxygen affinity than chromium. An oxide addition must not dissociate or dissolve appreciably into the scale matrix. The active elements increase in effectiveness from Group V to II across the Periodic Table and with increasing atomic number down each Group,[1] which indicates a correlation with increasing ionic radius.[1] This trend is is consistent with boundary diffusion blocking by segregated active elements since large ions experience a high driving force for segregation to grain boundaries (relaxation of misfit strain energy). For the oxidation at 1000-1200°C of Ni-20Cr alloys containing a range of active element additions, rate reductions increased with increasingly negative values of the standard free energy of formation of the dispersed active element oxide.[23,37]

This could be associated with an increase in matrix oxide/active element oxide interfacial energy, increasing active element oxide particle solubility and thereby enhancing segregation, or an increase in the blocking ability of each active element on a boundary due to changes in ion:vacancy association energy. However, irrespective of these chemical implications, other factors could govern the choice of the active element addition. For example, cerium containing alloys may exhibit incipient melting at lower temperatures than the corresponding yttrium-containing alloys, which would affect fabrication procedures.[38]

The various sources of active elements have both advantages and disadvantages in providing optimised active element distributions. Ion implants, oxide dispersions and coatings can provide a uniform active element distribution across a surface. In contrast alloy additions in the form of intermetallic precipitates could generate clusters of active element oxide particles and, as a consequence, local variations in effectiveness. The average separation of active element oxide particles within a scale is a function of the peak concentration in the original active element distribution. Ion implantation and oxide coatings are the most practical procedures for ensuring locally high active element concentrations within the scale. For example, 10^{17} cerium ions cm^{-2} reduced scale growth and spallation on Fe-20Cr-25Ni stainless steel at 900-950°C, whereas the less peaked distribution with a lower maximum concentration from a 0.13% cerium alloy addition only improved spallation resistance.[39]

The total quantity of active element available could also determine particle separation. At lower levels effectiveness increases with reactive element concentration (e.g. Refs. 11,25,40), while at higher concentrations active element effects saturate.[10,11] This may reflect a limitation on particle spacings imposed by microstructural growth processes so that adding further active element increases the particle size but not the amount of segregation. Alternatively there could be a threshold of segregant concentration which gives a maximum blocking of boundary diffusion so that reducing the particle spacing merely increases the segregant level further above this threshold. This behaviour is shown schematically in Figure 10. Excessive active element additions can be detrimental to oxidation performance.[2]

Incorporation of a active element into a scale should be ensured by its prior location in the substrate as immobile oxide particulates. In some instances[5,6] in-service oxidation of implants or alloying additions of an active element has appeared to be too slow or to produce an insufficiently fine oxide dispersion for dense heterogeneous scale nucleation to occur, but this was achieved by prior internal oxidation. Oxide coatings and dispersions would ensure particulate availability at the surface prior to service.

The surface treatments do have several potential disadvantages. Ion implantation is a line of sight technique and therefore re-entrant surfaces are not readily accessible, while adherence and uniformity of coverage[27] have to be assured for oxide coatings. Also, these would not provide a reservoir of active element to renew protection in the event of failure of a modified scale. Nevertheless, protection of the Fe-20Cr-25Ni stainless steel by both surface treatments has been achieved over extended periods, often defined by the test limits e.g. > 20 kh at 750°C, 6 kh at 825 and 900°C. Active element alloying additions within the solubility range could offer some benefits, but higher active element concentrations with resultant intermetallic formation have undesirable side effects. Provision of active element oxide dispersoids is used commercially but necessitates fabrication by powder metallurgical procedures.

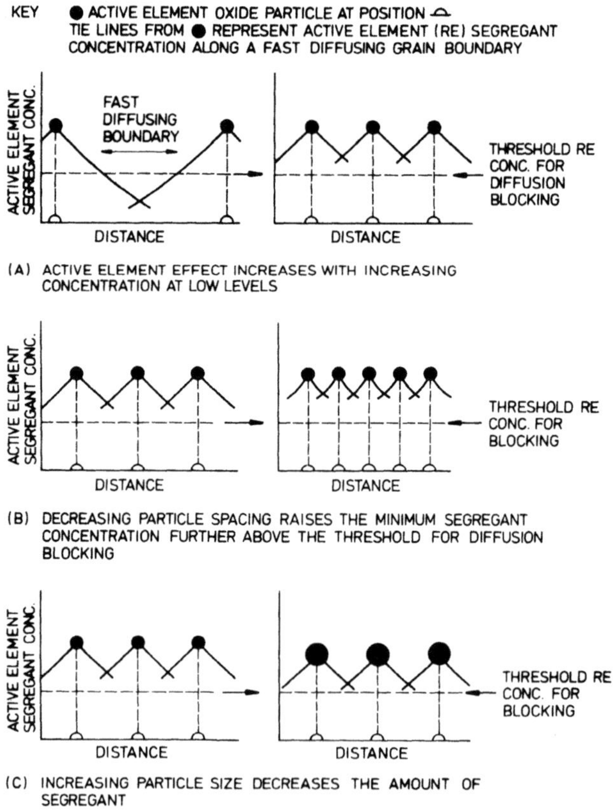

Fig. 10 Schematic representation of the influence of active element oxide particle spacing and size upon the effectiveness for grain boundary segregation.

6. CONCLUSIONS

A reasonable understanding has now emerged as to how active elements improve the oxidation behaviour of Cr_2O_3 formers. This involves contributions from the following three principal mechanisms:

(i) heterogeneous scale nucleation on active element oxide particles at the substrate surface promotes the development of a protective Cr_2O_3 scale;

(ii) inhibition of transport along the dominant grain boundary pathways by active element segregants limits oxide growth by cation diffusion. The active element oxide particles act as sources for the segregants within the scale;

(iii) failure of a Cr_2O_3 scale is inhibited through modifications to its microstructure and transport properties, affecting the stress state and mechanical behaviour of the scale and its adhesive bonding to the substrate.

Of these three the least well understood is the rôle of active elements in improving scale failure. Further progress will necessitate the development of improved techniques for measuring scale growth stresses, mechanical properties and adhesive bonding to the substrate.

ACKNOWLEDGEMENTS

We wish to thank all those whose work made this review possible and apologise for any omissions.

Figures 5 and 6 were kindly supplied respectively by Dr. C.M. Cotell (AT and T) and Dr. R.J. Hussey (NRC, Canada) for which we are most grateful.

This work is part of the longer term research carried out within the Underlying Programme of the UK Atomic Energy Authority.

REFERENCES

1. L.B. PFEIL: UK Patent No. 459848, 1937.
2. D.P. MOON: To be published in Mater. Sci. Technol., 1989.
3. D.P. WHITTLE and J. STRINGER: *Phil. Trans. R. Soc. Lond. A.*, 1980, **295**, 309-329.
4. T.N. RHYS-JONES, H.J. GRABKE and H. KUDIELKA: *Werkst. Korros.*, 1987, **38**, 65-72.
5. D.P. WHITTLE, M.E. EL-DAHSHAN and J. STRINGER: *Corros. Sci.*, 1977, **17**, 879-891.
6. P.Y. HOU and J. STRINGER: *Oxid. Met.*, 1988, **29**, 45-73.
7. J. STRINGER, B.A. WILCOX and R.I. JAFFEE: *Oxid. Met.*, 1972, **5**, 11-47.
8. P.Y. HOU and J. STRINGER: *Mater. Sci. Eng.*, 1987, **87**, 295-302.
9. M.J. BENNETT, B.A. BELLAMY, C.F. KNIGHTS, N. MEADOWS and N.J. EYRE: *Mater. Sci. Eng.*, 1985, **69**, 359.
10. T.N. RHYS-JONES, H.J. GRABKE and H. KUDIELKA: *Corros. Sci.*, 1987, **27**, 49-73.
11. M.J. BENNETT, G. DEARNALEY, M.R. HOULTON and R.W.M. HAWES: in 'Ion implantation into metals', (Ed. V. Ashworth et al), 264-276; 1982, Oxford, Pergamon Press.
12. A. ATKINSON and R.I. TAYLOR: in 'Transport in non-stoichiometric compounds', (ed. G. Simkovich et al), 285-295; 1985, New York, Plenum Press.
13. J.H. PARK, W.E. KING, N.L. PETERSON and S.J. ROTHMAN: in 'Oxidation of metals and associated mass transport', (ed. M.A. Dayananda et al), 103-107; 1987, Warrendale, Penn., The Metallurgical Society Inc.
14. A. ATKINSON: *Rev. Mod. Phys.*, 1985, **57**, 437-470.
15. M.J. BENNETT, H.E. BISHOP, P.R. CHALKER and A.T. TUSON: *Mater. Sci. Eng.*, 1987, **90**, 177-190.
16. C.M. COTELL, G.J. YUREK, R.J. HUSSEY, D.F. MITCHELL and M.J. GRAHAM: *Proc. Electrochem. Soc.*, 1988, **88-5**, 268-277.
17. J.C. PIVIN, C. ROQUES-CARMES, J. CHAUMONT and H. BERNAS: *Corros. Sci.*, 1980, **20**, 947.
18. K. PRZYBYLSKI and G.J. YUREK: *J. Electrochem. Soc.*, 1988, **135**, 517-523.
19. C.H. YANG, G.E. WELSCH and T.E. MITCHELL: *Mater. Sci. Eng.*, 1985, **69**, 351-357.
20. M.J. BENNETT, J.A. DESPORT, M.R. HOULTON, P.A. LABUN and J.M. TITCHMARSH: Mater. Sci. Technol., 1988, **4**, 1107-1113.
21. G.J. YUREK, K. PRZYBYLSKI and A.J. GARRATT-REED: *J. Electrochem. Soc.*, 1987, **134**, 2643-44.
22. C.H. YANG, P.A. LABUN, G.E. WELSCH, T.E. MITCHELL and M.J. BENNETT: *J. Mater. Sci.*, 1987, **22**, 449-458.

23. A.T. CHADWICK and R.I. TAYLOR: in Proc. 9th Int. Congr. on Metall. Corr., Toronto, 1984, **3**, 381-387; Ottawa, National Research Council of Canada.
24. H. NAGAI and M. OKABAYASHI: *Trans. Japan. Inst. Metals.*, 1981, **22**,101-108.
25. D.M. DUFFY and P.W. TASKER: *Phil. Mag. A.*, 1986, **54**, 759-771.
26. D.P. MOON: D. Phil. thesis, University of Oxford, 1987.
27. D.P. MOON and M.J. BENNETT: UKAEA Report AERE-R12757, 1987, Harwell Laboratory; To be published in Mater. Sci. Forum.
28. R.J. HUSSEY, P. PAPAIACOVOU, J. SHEN, D.F. MITCHELL and M.J. GRAHAM: Unpublished results.
29. J. ASHER, S. SUGDEN, M.J. BENNETT, R.W.M. HAWES, D.J. SAVAGE and J.B. PRICE: *Werkst. Korros.*, 1987, **38**, 506-516.
30. M.J. BENNETT, J.B. PRICE, C. PLYMPTON and P. COLLEDGE: Unpublished results.
31. H.E. EVANS and R.C. LOBB: in Proc. 9th Int. Congr. on Metall. Corr., Toronto, 1984, **2**, 46-53; Ottawa, National Research Council of Canada.
32. M.J. BENNETT and A.T. TUSON: UKAEA Report AERE-R13309, 1988, Harwell Laboratory; To be published in Mater. Sci. Eng.
33. M.J. BENNETT: in Proc. Conf. 'High Temperature Corrosion', (Ed. R.A.Rapp), NACE-6, 1983, 145-154.
34. P.W.G. SIMPSON and H.E. EVANS: in 'Nuclear Fuel Performance', 1985, **1**, 265-272; London, British Nuclear Energy Society.
35. M.J. BENNETT: in Proc. 10th Int. Congr. Metall. Corr., Madras, 1987, **4**, 3761-3770; New Delhi, Oxford and IBH Publishing Co.
36. J.R. NICHOLLS and P. HANCOCK: To be published in these Proceedings.
37. H. NAGAI, M. OKABAYASHI and H. MITANI: *Trans. Japan Inst. Metals*, 1980, **21**, 341-348.
38. J.E. ANTILL and K.A. PEAKALL: *J. Iron Steel Inst.*, 1967, **205**, 1136-1142.
39. M.J. BENNETT, J.B. PRICE and D.J. SAVAGE: UKAEA Report AERE-R11228, 1984, Harwell Laboratory.
40. G.M. ECER and G.H. MEIER: *Oxid. Met.*, 1979, **13**, 159.

THE INFLUENCE OF ACTIVE ELEMENTS ON THE STRESS-RELAXATION IN THE SCALE AND SUBSTRATE OF ALUMINA-FORMING ALLOYS

J. Jedlinski

Institute of Materials Science, Academy of Mining and Metallurgy

al.Mickiewicza 30, 30-059 Krakow, Poland

ABSTRACT

The influence of active elements on the extent and relaxation of different stresses generated in the scale and substrate of alumina-forming alloys during oxidation are discussed on the basis of observations of the scale and scale-substrate interface morphology.

The intermetallic compound beta-NiAl has been selected as a model of an alumina-former. It is suggested that active elements reduce the growth stresses and change the way of stress-relief by favouring less detrimental mechanisms such as substrate and scale deformation. In the absence of these additions, scale cracking and spalling occur under less severe oxidation conditions. This change in behaviour is attributed to an enhancement of the adhesive forces between the scale and the substrate atoms and also of the cohesive forces between the oxide grains by active elements.

1. INTRODUCTION

Although the beneficial influence of active elements on the oxidation behaviour of alumina-forming alloys, consisting mainly in a considerable improvement of the scale adherence to the substrate, is well-known, no generally accepted explanation of this phenomenon has been yet emerged.[1-4] Recent results suggest that active elements prevent the detrimental effect of impurities, in particular sulphur, which segregate to the scale-substrate interface[5] and thereby adversely affect the bonding strength between the scale and the substrate atoms.[6] In addition, active elements influence the morphology and growth mechanism of alumina scales.[7-9] These effects have been suggested to affect also the extent and/or the way of relaxation of stresses.[7]

The stresses generated during thermal growth of the oxide on the surface of a substrate may be classified into three main groups:[10 11]

- structural, related to the epitaxial growth of the oxide layer as well as to phase transformations in the growing scale,
- growth, dependent on the scale growth mechanism,
- thermal, arising from temperature changes.

All these stresses are relieved in the overall scale-substrate system in the following ways:[10-14]

- deformation of the substrate,
- deformation of the scale,
- detachment at the scale-substrate interface,
- cracking of the scale,
- spalling of the scale.

The contribution of the various relaxation mechanisms depends on the following factors:

- plastic properties of the substrate,
- plastic properties of the scale,
- cohesive forces between the scale grains,
- adhesive forces between the scale and the substrate atoms.

In view of recent developments, it is evident that, at least, some of the factors mentioned should be affected by active elements. Unfortunately no direct method for determination of stress-levels or of the stress-relief in dynamic scale-substrate systems has been found, although some attempts to apply dedicated techniques have been made.[15-17] Therefore, indirect methods are still frequently used. One of them is the observation of the morphology and the microstructure of the scale.

In the present paper some results are reported of the application of the latter method to formulate some suggestions concerning the influence of implanted yttrium and lanthanum on the extent of stresses, as well as on their relaxation during the oxidation of beta-NiAl intermetallic compound, which is a model of alumina-forming binary alloys. The results of previous studies on this material[7-9] have indicated that implanted yttrium improves both the protective properties of the alumina scale and its adherence to the substrate. It also modifies the oxidation mechanism by suppressing the inward diffusion of oxygen through the scale. On the other hand, ion-implanted lanthanum improves the protective properties of the scale but not its adherence to the substrate. Both additions modify the scale morphology, but do not affect its composition.

2. MATERIALS AND EXPERIMENTAL PROCEDURES

The preparation of flat disc samples of beta-NiAl, without and with ion-implanted yttrium or lanthanum, has been described elsewhere.[7-9] In order to check if the active elements affect the scale morphology, only a part of the surface of the samples was implanted, the rest of it being masked during the implantation process. The samples were oxidized under isothermal and thermal cycling conditions at temperatures in the range 1373-1573 K, in both oxygen and air. Details of the oxidation procedure have also been reported previously.[7-9] The morphology and the microstructure of the scale and substrate have been studied using Scanning Electron Microscopy (SEM) and various X-ray techniques. Both the outer and inner surface of the scale as well as the fracture cross-sections were examined. The inner surfaces of adherent scales were revealed following chemical removal of the substrate. Poorly adherent scales were collected in small crucibles during their spalling.

3. RESULTS

3.1. Outer Surface of the Scale

The surfaces of the samples of all materials oxidized under thermal cycling conditions are shown in Fig. 1.[7-9] Large regions, where spalling had occurred during oxidation, are visible on the active-element-free and lanthanum-implanted

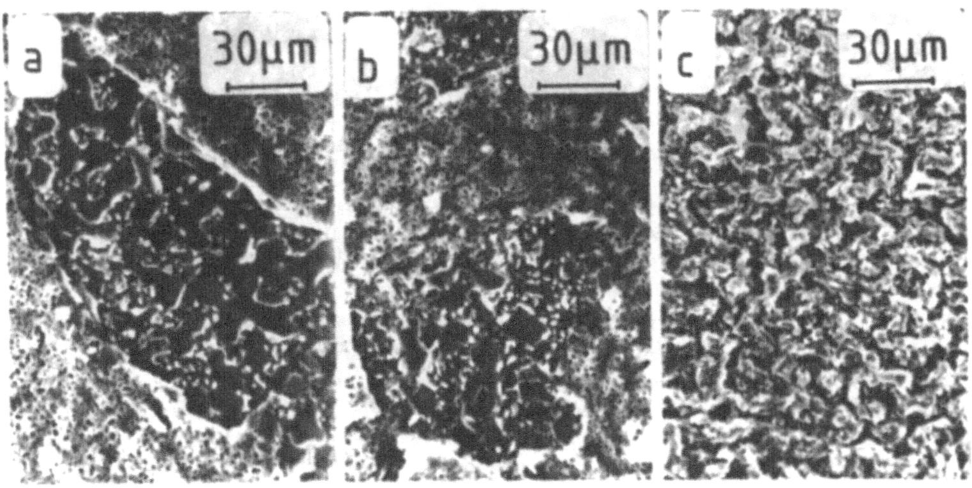

Fig. 1: Surfaces of the scales formed on unimplanted (a), lanthanum- (b) and yttrium- (c) implanted beta-NiAl under thermal cycling conditions (1473 K, air, 1 cycle = 2 h, a & b - 20 cycles, c - 100 cycles).[9]

samples. In contrast, no spalling has been observed on the yttrium-implanted material. Moreover, the poorly adherent (Fig. la & b) and the well-adherent (Fig. lc) scales exhibit different morphologies.

The influence of both active elements on the scale morphology is more clearly visible on samples, on which only part of the surface had been implanted with Y or La. As can be seen in

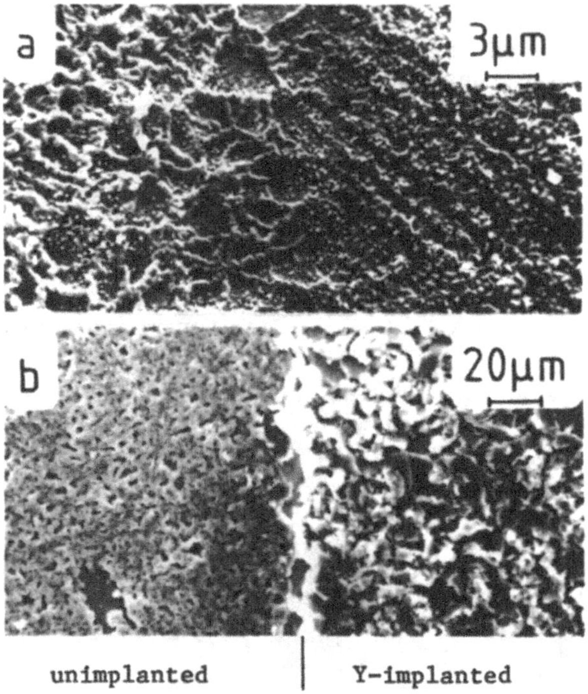

Fig. 2: Surfaces of the scales formed on unimplanted (left) and yttrium-implanted (right) parts of beta-NiAl-samples: (a) 1373 K - 240 h, oxygen, (b) 1473 K - 200 h, oxygen.

Fig. 2, the ridges which have grown extensively on the unimplanted material do not appear on the yttrium-implanted samples. In the case of lanthanum-implanted material, the formation of ridges was retarded, but not eliminated (Fig. 3). It is less pronounced and requires relatively higher temperatures to be initiated. The differences in surface morphology are even better demonstrated in Figs. 4-7. The growth of ridges on unimplanted beta-NiAl is preceeded by the formation of blade-like grains (Fig. 4a). After their "nucleation" (Fig. 4b), the ridges gradually cover the whole surface of the scale (Fig. 4c-e). Occasionally, some alumina whiskers grow on the scale surface (Fig. 4e).

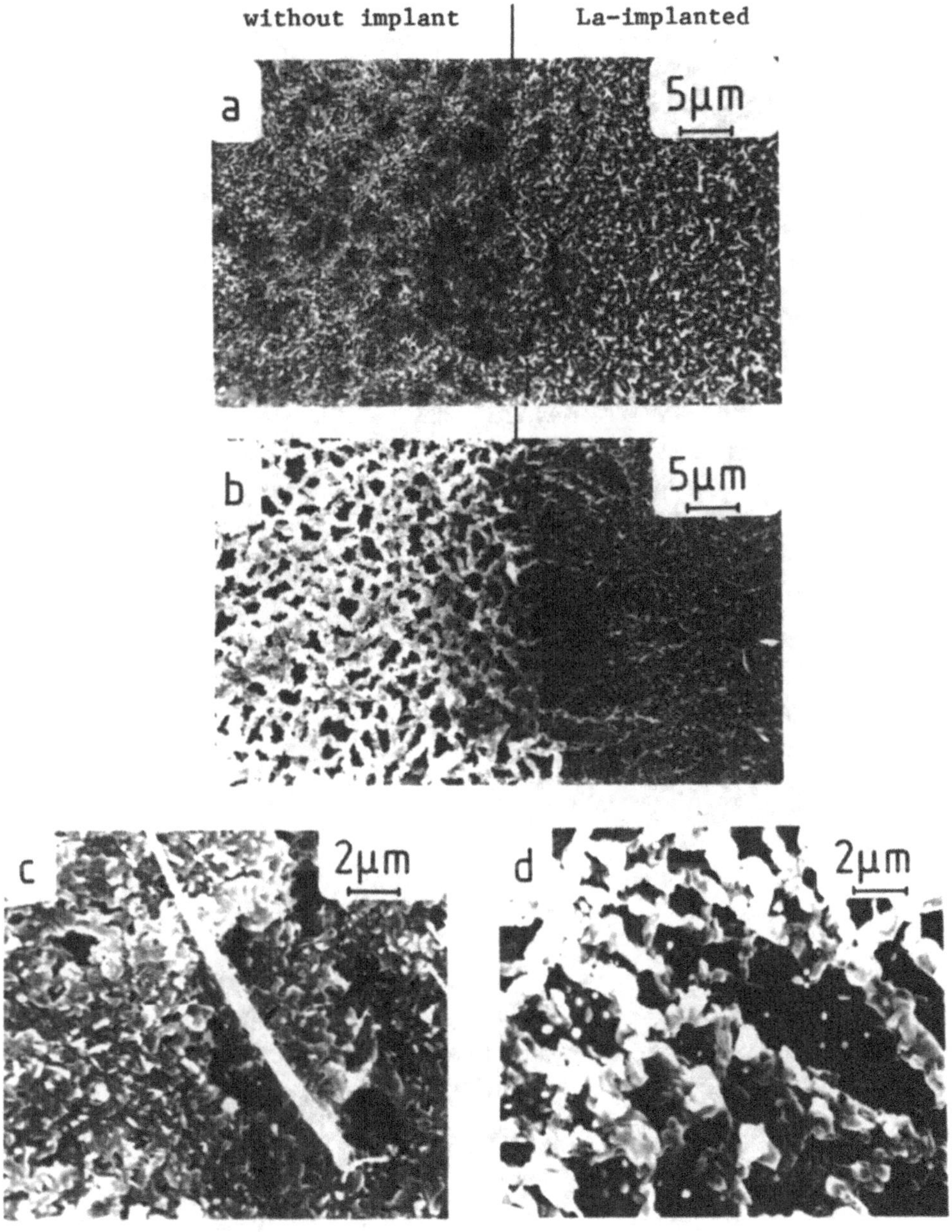

Fig. 3: Surfaces of the scales formed in air on unimplanted and lanthanum-implanted beta-NiAl samples: (a) 1273 K - 72 h, (b) 1373 K - 72 h, (c) & (d) 1473 K - 72 h.

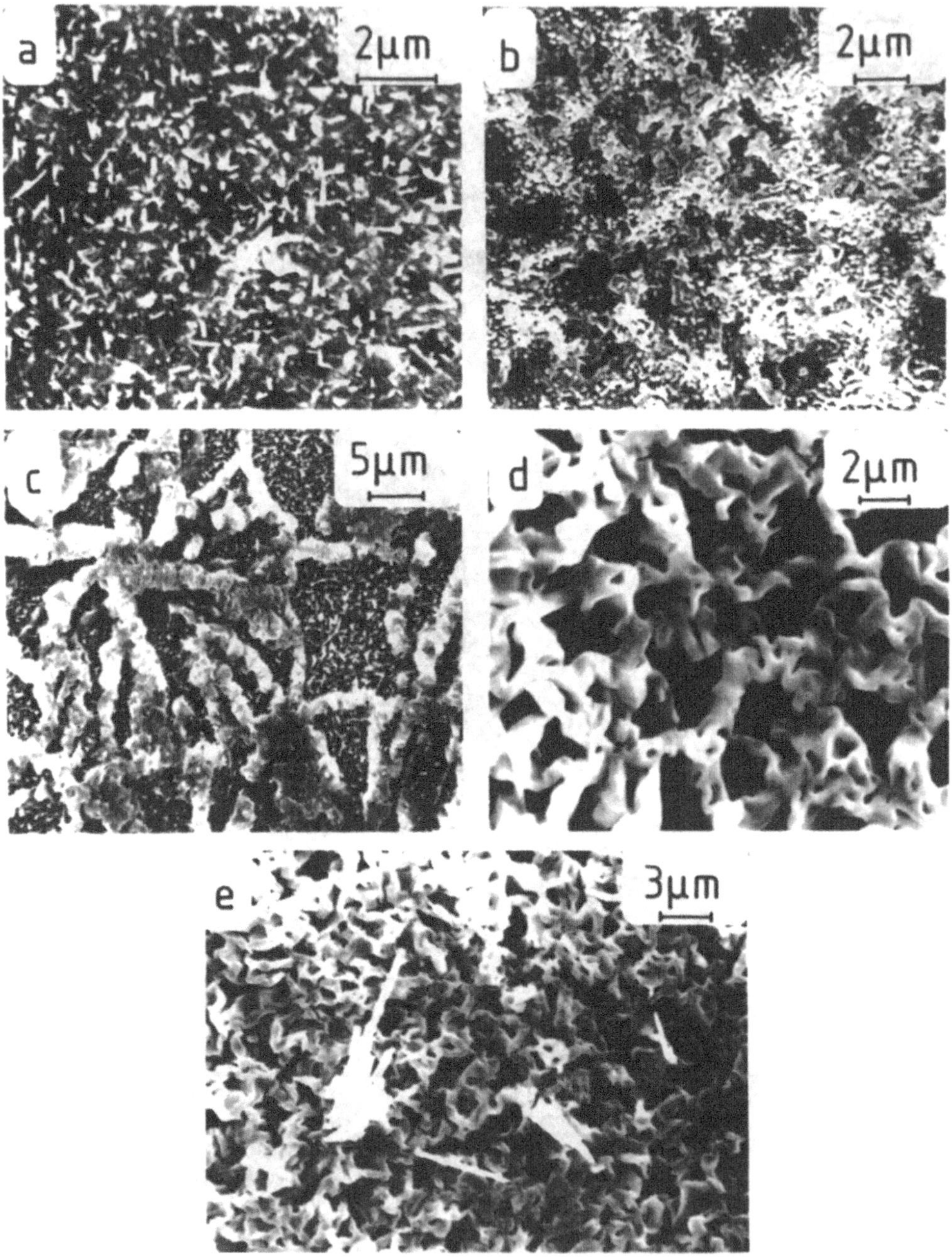

Fig. 4: Surfaces of the scales formed on unimplanted beta-NiAl during the consecutive stages of oxidation, showing the progressive change from blade-like grains to ridges covering the surface.
a) 1373 K - 3 h, b) 1323 K - 24 h, c) 1373 K - 24 h, d) 1473 K - 24 h, e) 1523 K - 72 h; 1 atm. O_2.

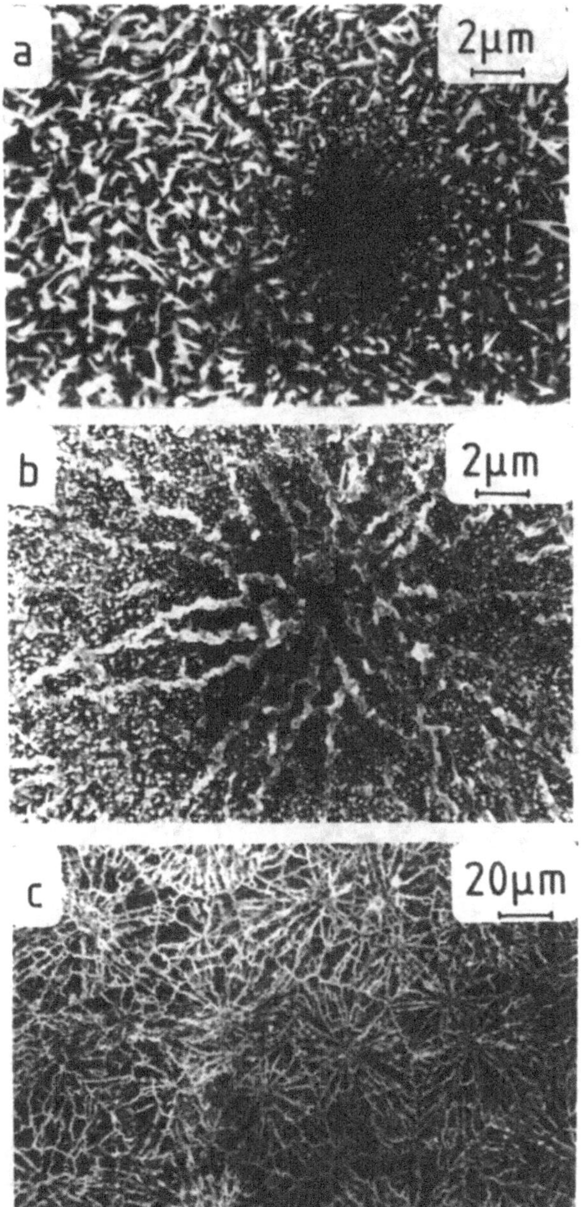

Fig. 5: Surfaces of the scales formed on lantanum-implanted beta-NiAl during the consecutive stages of oxidation (in air):
a) 1273 K - 72 h, b) 1373 K - 72 h, c) 1473 K - 72 h.

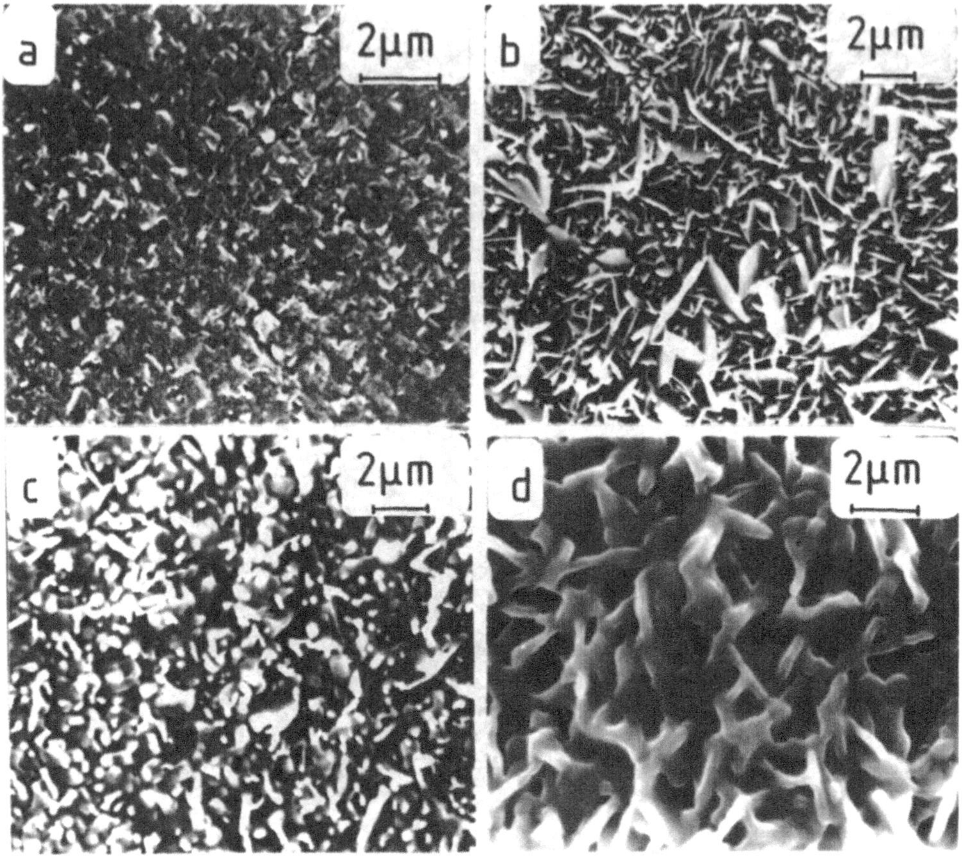

Fig. 6: Surfaces of the scales formed on yttrium-implanted beta-NiAl during early and intermediate stages of oxidation (1 atm. O_2):
a) 1373 K - 3 h, b) 1323 K - 24 h, c) 1423 K - 120 h, d) 1473 K - 24 h.

On lanthanum-implanted NiAl, the ridges grow in the previously formed radial cracks in the scale (Fig. 5). This radial symmetry of the ridges network (the so-called "lace structure"[18-20]) has not been observed on unimplanted samples.

The evolution of the morphology of the scale surface developing on yttrium-implanted beta-NiAl proceeds through different stages (Figs. 6 and 7), no ridges being observed at all. Instead, the blade-like grains grow and, subsequently, loose their sharp contours (Fig. 6b-d). The resulting surface morphology (Fig. 7a) is similar to that observed on sintered

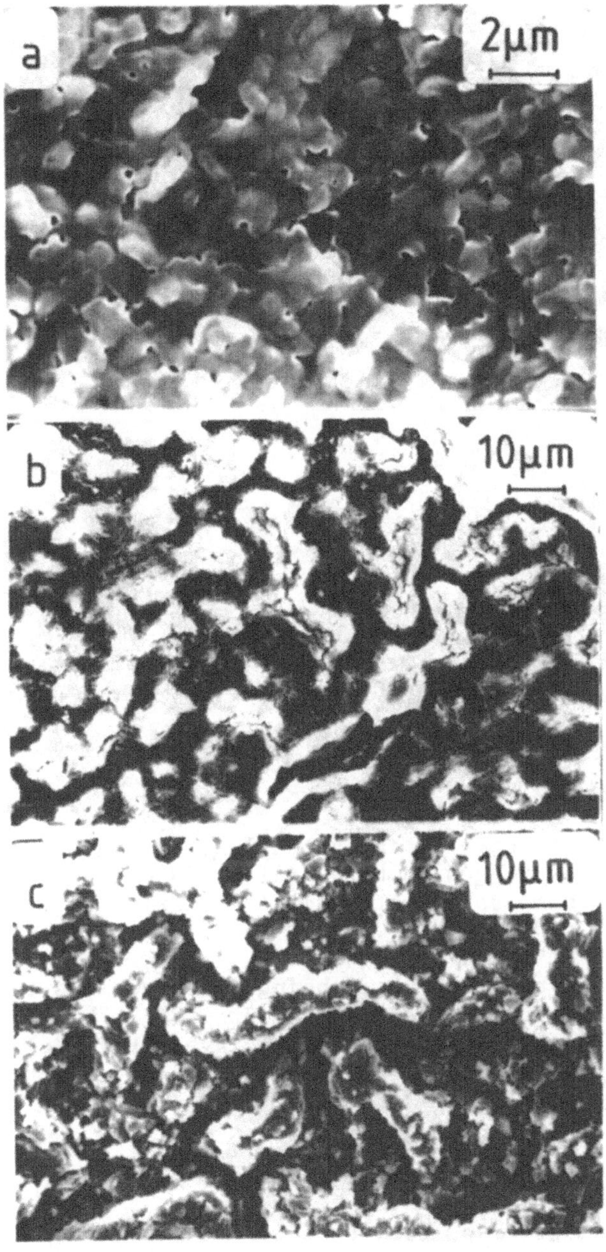

Fig. 7: Surfaces of the scales formed on yttrium-implanted beta-NiAl at later stages of oxidation (1 atm. O_2)[7,9]: a) 1473 K - 120 h, b) 1473 K - 200 h, c) 1523 K - 72 h.

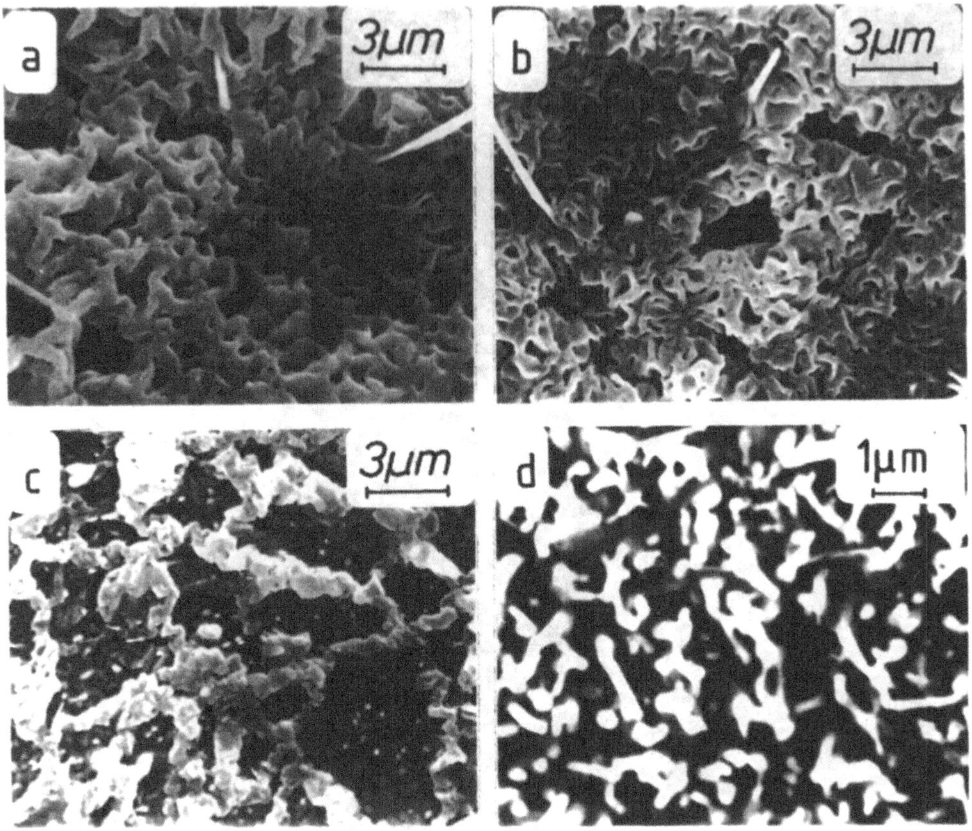

Fig. 8: The influence of implanted lanthanum dose on the formation of ridges on beta-NiAl (1473 K - 24 h, oxygen) : (a) unimplanted material, (b) - $2x10^{14}$, (c) - $2x10^{15}$ and (d) $2x10^{16}$ La^{+}/cm^{2}.[9]

ceramics, with numerous voids being observed. As the reaction proceeds, cracking of the scale occurs (Fig. 7b), leading to the growth of a highly convoluted oxide layer (Fig. 7c). Blocking of ridge formation is dependent upon the amount of active element in substrate (Fig. 8).[9]

3.2. The Scale Cross-Sections

The cross-sections of the scales formed on unimplanted and yttrium-implanted beta-NiAl at different stages of reaction are shown in Fig. 9. The scale growing on the unimplanted material exhibits a typical columnar structure, while on the Y-implanted material an equiaxed structure has developed. Although these scales have been grown in slightly different conditions, no

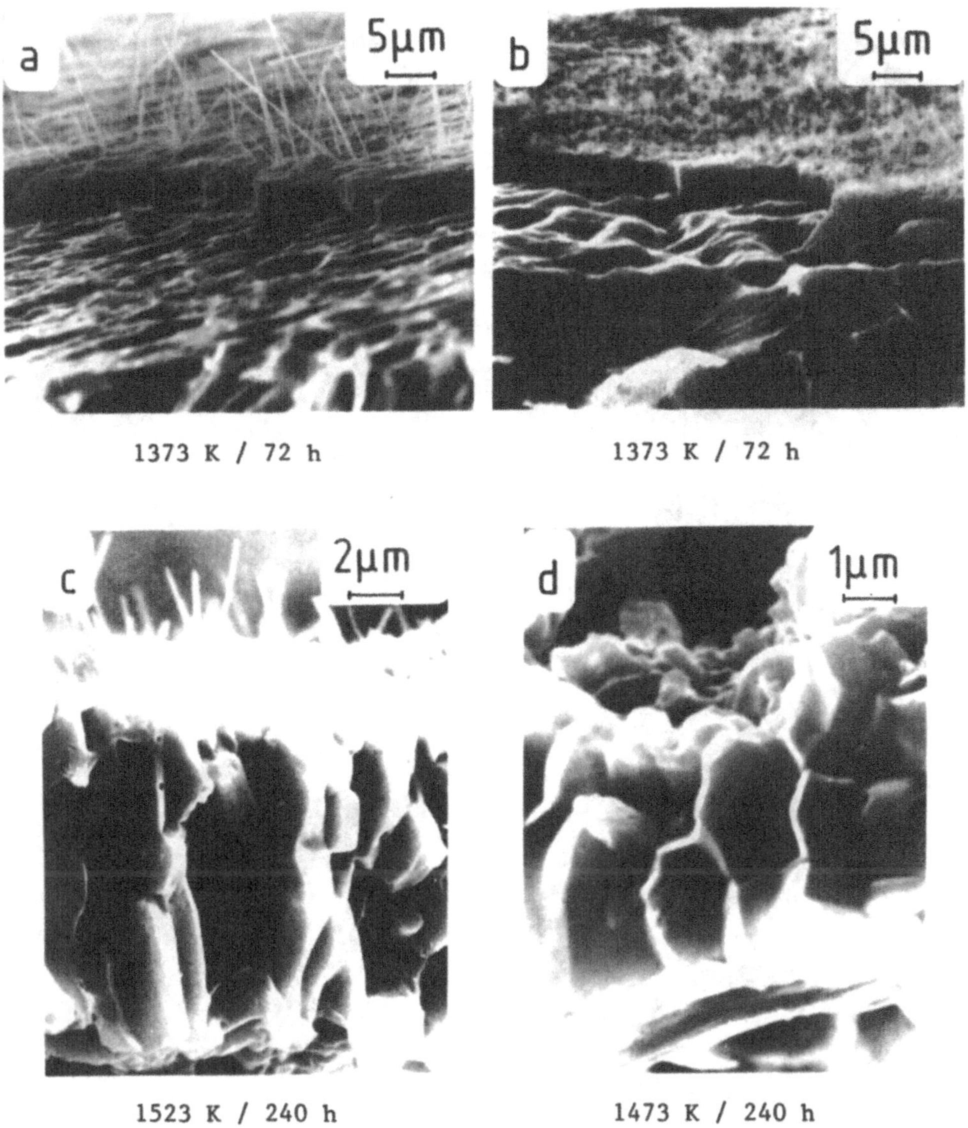

Fig. 9: The fractured cross-sections of the samples, (a & b), and the scales, (c & d), formed in oxygen on unimplanted, (a & c), and yttrium-implanted, (b & d) beta-NiAl.

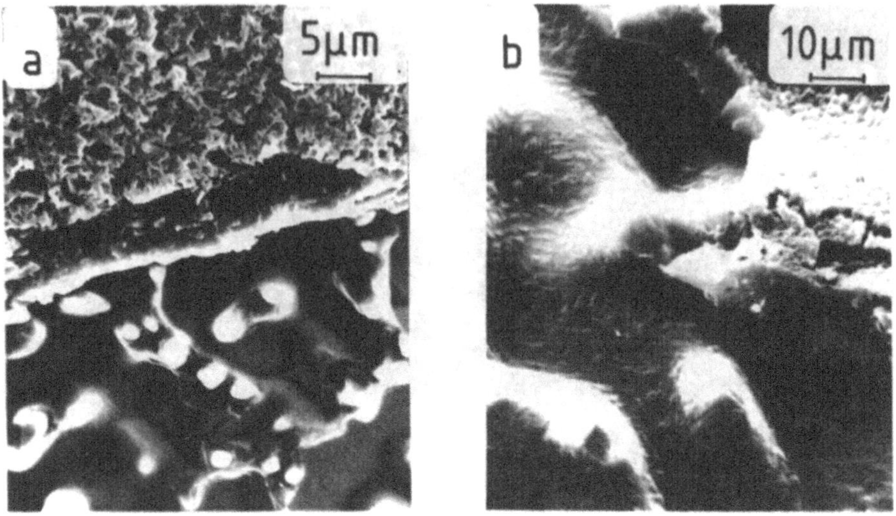

Fig. 10: The scales and scale-substrate interfaces on (a) unimplanted (1523 K - 24 h, oxygen) and (b) yttrium-implanted (1473 K - 240 h, oxygen) beta-NiAl.[9]

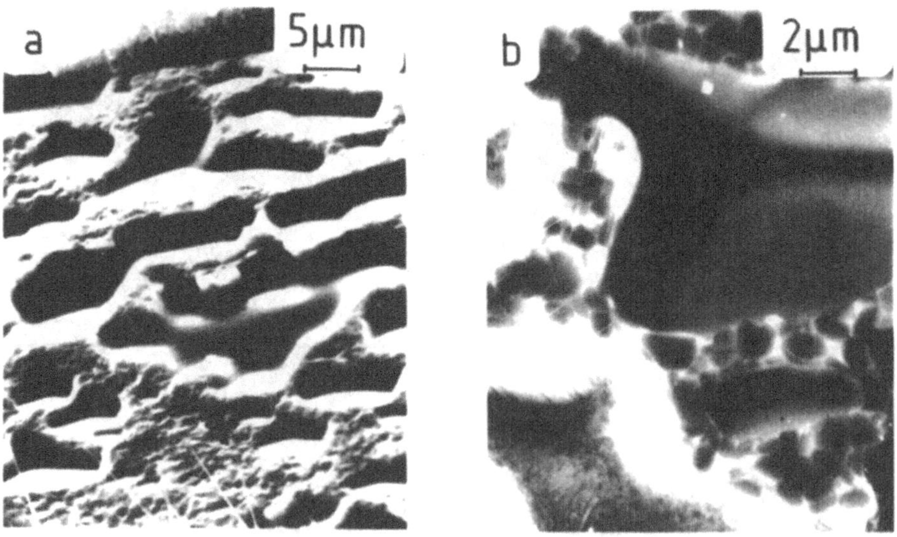

72 h 240 h

Fig. 11: The cavities at the scale-substrate interface on unimplanted beta-NiAl, 1473 K, oxygen.

spectacular variations in their grain size are observed. Other features of the scale on the unimplanted material are the whiskers as mentioned already and some occasional voids.

3.3. The Substrate Surface

From Figs. 9a & b it can be deduced that the oxidation affects in different ways the substrate morphology of unimplanted and of yttrium-implanted material. Indeed, further results presented in Figs. 10-12 strongly support this conclusion. The gradual growth of interfacial porosity has been observed in the case of addition-free beta-NiAl, with two types of morphologies being observed (Figs. 10 and 11): initially flat cavities and, on their edges, the imprints of the spalled oxide. The depth of the initially flat cavities increases as the reaction proceeds, due to thermal etching of the substrate (Fig. 11). The surface of the substrate is, apart from the presence of cavities, flat with no corrugations being observed. In contrast, the surface of the yttrium-implanted substrate is highly corrugated (Figs. 9b and 10b).

In the case of lanthanum-implanted material, the situation is more complex (Fig. 12b). Two kinds of morphologies co-exist at the stage of reaction presented: in some regions corrugations are observed, while in other areas cavities and imprints of the spalled oxide are seen.

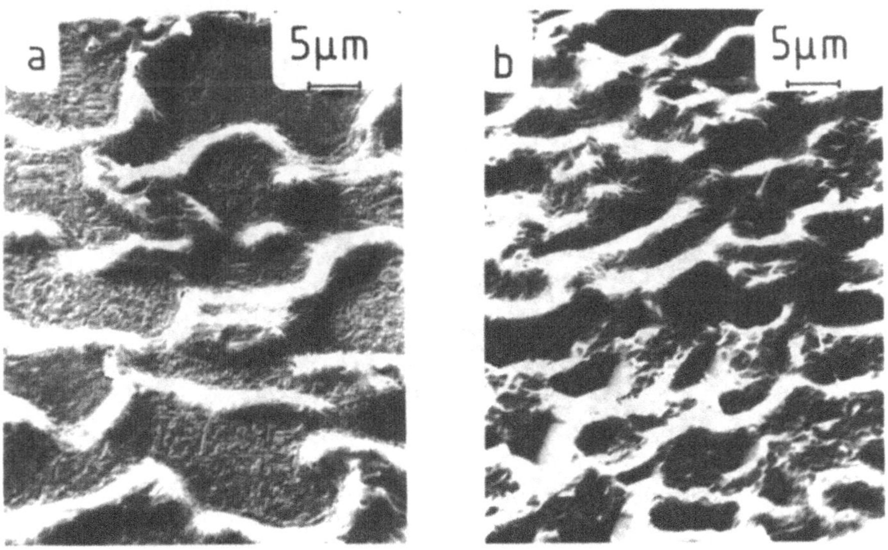

Fig. 12: The substrate surface of (a) unimplanted, and (b) lanthanum-implanted beta-NiAl (1473 K - 24 h, oxygen).

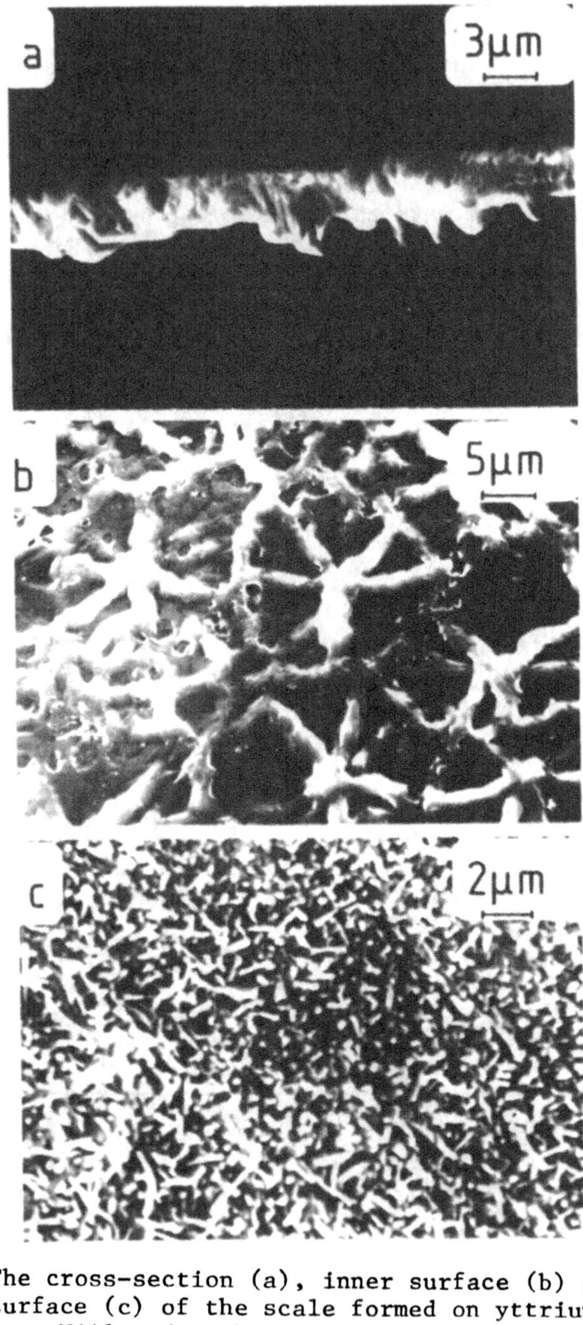

Fig. 13: The cross-section (a), inner surface (b) and outer surface (c) of the scale formed on yttrium-implanted beta-NiAl under thermal cycling conditions (1473 K, air, 10 cycles, 1 cycle = 5 h).

3.4. Inner Surface of the Scale

The formation of wavy scale-substrate interfaces during the oxidation of yttrium-containing beta-NiAl was confirmed by the observations of the inner surface and the cross-section of the scales (Figs. 13 and 14). It should be noted that oxide intrusions are formed before the convoluted scale appears (Fig. 13a and b). In addition, numerous voids develop at the inner scale surface (Fig. 13b), while cracks, hardly visible due to the irregular morphology, are present at the outer surface of the scale (Fig. 13c). On the other hand, a rather flat scale-substrate interface is formed during oxidation of unimplanted material (Fig. 15). A gradual loss of the regions, in which contact between the scale and the substrate is maintained, was observed.

4. DISCUSSION

The present results strongly suggest that a close correlation between the adherence of the scale and its morphology exists. The poorly adherent scales exhibit a highly ridged morphology, whereas no formation of ridges has been observed in the case of well adherent scales. Moreover, the limited effect of lanthanum in improving the oxidation behaviour of beta-NiAl is also consistent with insufficient blocking of ridge formation. Apart from the growth of ridges, interfacial porosity occurs also in the case of poor adherence, leading to the detachment of the scale from the substrate. This is not the case for yttrium-implanted beta-NiAl.

The following mechanisms have been evoked for the stress generation associated with the growth of scales on alumina-forming alloys:

- transformation of unstable alumina polymorphs to a stable alpha-phase, which is accompanied by ca. 10% volume contraction, inducing tensile stresses in the oxide layer;[21]
- lateral growth of the scale due to formation of "new" oxide within the bulk oxide layer.[8,22]

In addition, the tendency to maintain contact between an outward growing scale and the substrate may also produce stresses.[14] This is the case for, at least, yttrium-implanted material.[8] Obviously, thermal stresses generated by changes in temperature should also be taken into account.

As far as the formation of ridges is concerned, this has been attributed to the healing of cracks in the scale resulting from the relief of tensile stresses generated by the phase transformation.[21] The preferential growth of ridges in these cracks has been confirmed (Fig. 5). However, the formation of ridges may be attributed also to the scale deformation related to its

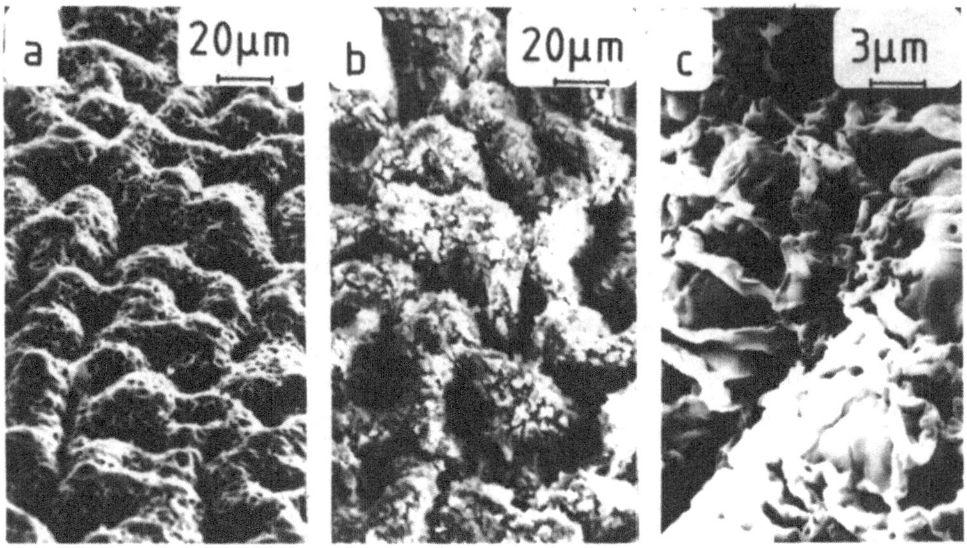

Fig. 14: Inner (a, c) and outer (b) surface of the scale formed on yttrium-implanted beta-NiAl (1473 K - 240 h, oxygen).

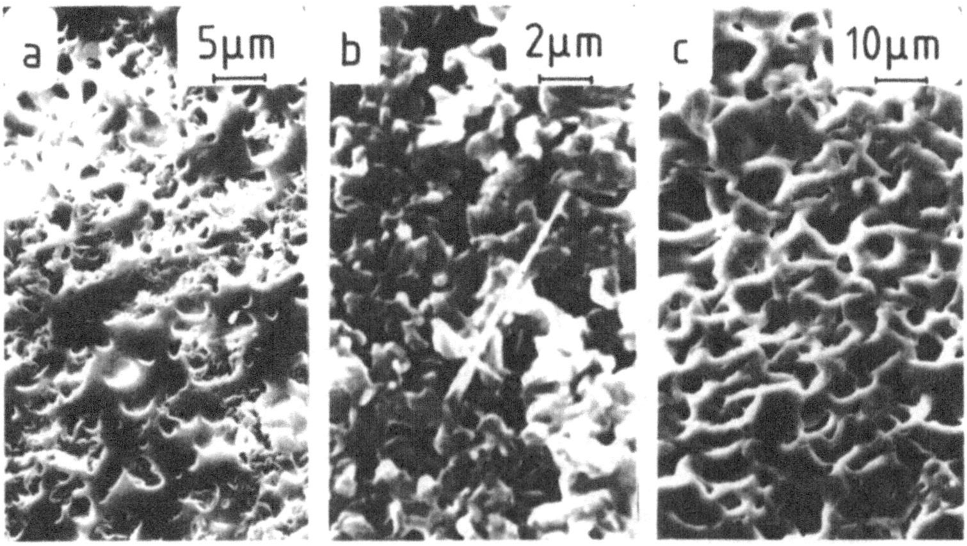

Fig. 15: Inner (a, c) and outer (b) surface of the scale formed on unimplanted beta-NiAl under thermal cycling (a & b - 1473 K, air, 10 cycles, 1 cycle = 5h) and isothermal (c - 1573 K - 24 h, oxygen) conditions.

lateral growth.[8] Both of the postulated mechanisms for the growth of ridges are perfectly acceptable. Moreover, they may be operating simultaneously. The occurance of cracks indicates poor cohesion between the oxide grains. The rôle of active elements is, thus, to facilitate during early stages of the oxidation another mode of stress relief than cracking of the scale. In fact, no cracks have been found in scales grown on yttrium-implanted substrates until a relatively late stage of oxidation is reached. The generation of these cracks in the scale will be discussed later.

The different symmetry of ridges on unimplanted and lanthanum-implanted material may be related to the formation of cavities with non-radial symmetry at the scale-substrate interface in the former case, and to the growth of asperities at this interface during the earlier stages of oxidation in the latter one. The subsequent cracking of the scale occurs preferentially at the asperities, leading to the radial symmetry of the cracks. This is consistent with theoretical considerations concerning the conditions of scale cracking initiation.[23,24]

It follows from current and previous results[9] that ion-implanted yttrium, being much more effective in improvement of the scale adherence than lanthanum, causes a reduction of the stress by decreasing the scale growth rate and by suppressing the inward diffusion of oxygen, by which the formation of "new" oxide within the scale is avoided. Further the mode of stress-relaxation is modified. The stresses generated during oxidation of unimplanted material cause scale deformation to occur at rather early stages of oxidation leading to cracking and spalling. The local detachment at the scale-substrate interface is believed to play a positive role in the stress accommodation, enabling the scale to deform further.[14]

In the case of yttrium-containing material, during the later stages of the scale growth, "new" oxide is formed also at the scale-substrate interface, not only at the scale-gas interface (Figs. 13b, 14a & c). This effect is preceded by the formation of voids and/or cracks in the scale (Figs. 7, 13a). These features are believed to be the paths for inward diffusion of oxygen.[2,7,9] Simultaneously, a highly corrugated substrate surface is observed (Figs. 9b, 10b), which has to be attributed to a deformation of the substrate. This behaviour can be explained by yttrium strongly enhancing the bonding between the scale and the substrate atoms, if they involve nickel and aluminum.[6] The tendency to maintain contact between the growing scale and the substrate leads to the development of such an interface.[14] The generation of cracks in the scale should be attributed to the relief of thermal stresses occuring under thermal cycling conditions (Fig. 13b) by the scale-substrate interface achieving such an unsuitable distribution and height of asperities and by the scale developing such a thickness that the growth stresses can not be accommodated in any other way. The presence of voids in the scale is probably related to condensation of vacancies resulting from aluminium atoms diffusing from the substrate to the scale-gas interface.

Yttrium, favouring aluminum diffusion in the scale,[8] should also facilitate the diffusion-controlled deformation and/or stress-free sintering of the oxide layer.[10,13] The scale formed during the intermediate stage of the oxidation (Fig. 7a) seems to correspond well with this suggestion. The numerous voids formed simultaneously are additional paths for fast inward diffusion of oxygen. This leads to the development of ridges at the inner surface of the scale (Fig. 14a & c). The formation of cracks in this scale at relatively late stages of reaction suggests that the cohesion between the oxide grains is improved by yttrium.

The above considerations also apply to the situation under thermal cycling conditions. In this case, however, much higher stresses are expected. Nevertheless, there is no reason to believe that other modes of stress relief should be invoked. Moreover, it seems that the corrugated scale-substrate interface under some conditions can facilitate the relaxation of the stresses.[10] This takes place when the asperities are fine and uniformly distributed and the scale thickness does not exceed a critical value. At the top of the asperities, due to the probable presence of yttrium, the contact between the scale and the substrate is maintained. According to the high strength of such bonding, the adhesion of the scale to the substrate is improved.

Both conclusions related to the effect of yttrium on the cohesion of grains and the adhesion of the oxide layer to the underlying substrate are reasonable, although the overall amount of yttrium present in the substrate as well as in the scale is not high. This is due to its strong tendency to segregate to energetically favoured regions, being primarily the scale-substrate interface and the scale as well as the substrate grain boundaries. As far as the scale-substrate interface is concerned, because of its irregular shape, there is no reason to try to describe the overall stress distribution, but one has to consider the local situation.

Finally yttrium favours the growth of a scale with equiaxed grains, while the scale formed on unimplanted material is columnar (Fig. 9). Thus, deformation of the scale can occur much easier in the case of Y-doped material.

5. CONCLUDING REMARKS

The present results suggest that completely different modes of stress-relaxation are operative in the case of poorly adherent scales as compared to well adherent ones. In the former case formation of cracks in the scale is observed. Simultaneously, interfacial porosity grows continuously leading to extended detachment at the scale-substrate interface. In addition, the growth of ridges indicates that scale deformation occurs to some extent. However, the latter process does not seem to be easy because of the columnar structure of the scale. These modes of stress-relief are not sufficiently effective so that extensive spalling of the scale is observed.

The addition of implanted yttrium leads to the elimination of scale cracking during early stages of oxidation, suggesting an enhancement of the cohesion between the oxide grains and/or a decrease of the stress level. The stresses are lower due to the thinner scale thickness (yttrium decreases the rate of oxidation) and also due to the suppression of lateral growth of the scale. Moreover, strong bonding between the scale and the substrate causes substrate deformation, which relaxes also the stresses. Simultaneously, diffusion-controlled processes, such as creep and sintering of oxide, which are facilitated by yttrium, decrease the residual stresses existing in the scale- substrate system. Cracking of the scale is also retarded, which is probably due to its better ability to deform.

Due to the change of the oxidation mechanism, the enhancement of the cohesive forces between the oxide grains and of the adhesive forces between the scale and substrate and by favouring easier deformable scale structures, yttrium reduces the growth stresses and favours less-detrimental modes of stress-relief, such as substrate and scale deformation. Therefore, spalling of the scale was not observed under the present experimental conditions (more than one hundred cycles and two weeks of isothermal oxidation at 1573 K).

By contrast implanted lanthanum has a negligible effect upon the adherence of the scale to the substrate, although it significantly affects the scale growth rate. At present, this discrepancy can not be explained, but it could be attributed to the insufficient effectiveness of La in enhancing the bonding forces in the scale and at the scale-substrate interface. Possibly, the distribution of lanthanum in the substrate or in the scale is different to that of yttrium.

ACKNOWLEDGEMENTS

The author is indebted to Professor S. Mrowec (Academy of Mining and Metallurgy, Krakow) for his encouragement to carry out this study and for many helpful discussions. The Alexander von Humboldt Foundation is acknowledged for their financial support to the author's participation at this conference.

REFERENCES

1. WHITTLE, D.P. and STRINGER, J.: Phil. Trans. R. Soc. London, 1980, A295, 309-329.

2. JEDLINSKI, J.: Proc. Conf. "German - Polish Seminar on High Temperature Corrosion of Metals and Alloys", Jülich, 1987, KFA-Report, in press.

3. HUNTZ, A.M.: Mater. Sci. Eng., 1987, 87, 251-260.

4. STOTT, F.H. and WOOD, G.C.: ibid, 1987, 87, 267-274.

5. SMEGGIL, J.G., ibid, 87, 1987, 261-265.

6. ANDERSON, A.B., MEHANDRU, S.P. and SMIALEK, J.L.: J. Electrochem. Soc. 1985, 132, 1695-1701.

7. JEDLINSKI, J.: Proc. 10th Int. Cong. on "Metallic Corrosion", Vol.4, 3709-3718, 1987, Madras.

8. JEDLINSKI, J. and MROWEC, S.: Mater. Sci. Eng., 1987, 87, 281-287.

9. MROWEC, S. and JEDLINSKI, J.: Trans. TMS-AIME, 1988, 247, 1099-1108.

10. STRINGER, J.: Corros. Sci. 1970, 10, 513-543.

11. APPLEBY, W.K. and TYLECOTE, R.F.: ibid, 1970, 10, 325-341.

12. HOLMES, D.R. and PASCOE, R.T.: Werkstoffe und Korrosion, 1972, 23, 855-859.

13. KOFSTAD, P.: in Proc. 6th Conf. "High Temperature Corrosion" (ed. R.A. Rapp), 123-138, 1983, Houston, Texas.

14. KOFSTAD, P.: Oxid. Met. 1985, 24, 265-276.

15. DELAUNAY, D., HUNTZ, A.M. and LACOMBE, P.: Corros. Sci., 1980, 20, 1109-1117.

16. ASHARY, A., MEIER, G.H. and PETTIT, F.S.: Proc. Conf. "High Temperature Protective Coatings", 105-119, 1983, Atlanta, Georgia, 112th Annual AIME Meeting.

17. HUNTZ, A.M., ZHAO, J.G., BOUMAZA, A. and MOULIN, G.: Mater. Sci. Tech., 1988, 4, 470-478.

18. SMIALEK, J.L.: Metall. Trans. A, 1978, 9A, 309-320.

19. HINDAM, H.M. and SMELTZER, W.W.: J. Electrochem. Soc., 1980, 127, 1630-1635.

20. HOMMA, T., HINDAM, H.M., PYUN, Y. and SMELTZER, W.W.: Oxid. Met., 1980, 17, 223-233.

21. DOYCHAK, J.: Ph. D. thesis, Case Western Reserve University, 1986, Cleveland, Ohio.

22. GOLIGHTLY, F.A., STOTT and F.H., WOOD, G.C.: Oxid. Met., 1976, 10, 163-187.

23. EVANS, A.G., CRUMLEY, C.B. and DEMARAY, R.E.: ibid, 1983, 20, 193-216.

24. EVANS, H.E. and LOBB, R.C.: Corros. Sci. 1984, 24, 209-222.

THE INFLUENCE OF FINE STRUCTURE, MORPHOLOGY AND COMPOSITION OF ALLOY AND OXIDE ON THE GROWTH OF Cr_2O_3 SCALES

M. Hirbod

R&D ABB Stal,

Finspong 612 20, Sweden

ABSTRACT

The kinetics of morphological and structural development of Cr_2O_3 scales on Fe-Cr alloys, with and without yttrium additions, under isothermal conditions at 1000°C have been studied in an attempt to elucidate the growth processes. The steady state scale on the yttrium-free alloy is uniformly continuous and highly convoluted, with a porous, equiaxed grain structure. It is considered that such a configuration results from transport processes in the scale and the development of high compressive stresses, leading to lateral growth of the oxide. In particular, both inward movement of oxygen and outward movement of chromium occur, enabling new oxide to form within the bulk scale with accompanying oxide dilatation.

The presence of active elements, such as yttrium, in the alloy has a marked effect on the oxidation process. In particular, there are considerable changes in the microstructure and the adherence of the scale. This results in a large decrease of the amount of scale spalling upon cooling or thermal cycling. The influence of the active element is accounted for in terms of its effect on transport processes in the scale, i.e. the suppression of chromium diffusion. This enables oxygen to penetrate to the scale/metal interface where new oxide is formed without significant stress development.

THE EFFECT OF YTTRIA DISPERSIONS ON THE GROWTH MECHANISMS AND MORPHOLOGY OF CHROMIA AND ALUMINA SCALES

W.J. Quadakkers[1], H.Holzbrecher[2], K.G.Briefs[1], H.Beske[2]

Kernforschungsanlage Jülich

1 Institut für Reaktorwerkstoffe

2 Zentralabteilung für Chemische Analysen

Postfach 1913, 5170 Jülich, FRG

ABSTRACT

The oxidation behaviour of a chromia and an alumina forming ODS alloy has been compared with that of conventional wrought model alloys with similar compositions. Scale structure and morphology, as well as scale growth mechanisms, were studied using metallography, X-ray diffraction, scanning electron microscopy, energy dispersive X-ray analysis and an ^{18}O tracer technique.The main effects of the yttria dispersions on the properties of the chromia and alumina scales were found to be improved adherence, decreased growth rates and enhanced selective oxidation. The results indicate that the main reason for these changes in scale properties is a change in transport mechanism in the scales. The oxides on the conventional alloys grow by both metal and oxygen transport, the addition of oxide dispersions suppresses the outward scale growth. It is shown that the change in properties of the oxide scales on the ODS alloys can be explained by assuming that the reduced outward growth is the dominating effect caused by the addition of yttria dispersions. The results indicate that the effectiveness of the dispersions depends on the amount and the distribution. Because of their faster scale growth rates, chromia forming alloys seem to require larger amounts of yttria than alumina formers in order to obtain the same beneficial effect.

1. INTRODUCTION

It has been known for a long time that addition of oxide dispersions significantly affects the oxidation behaviour of high temperature alloys; oxide scales on Oxide Dispersion Strengthened (ODS) alloys possess better spallation resistance than scales formed on conventional cast or wrought alloys of the same compositions. The reason for this improved scale adherence has been studied in several investigations, but a generally accepted theory has not yet been put forward.

A frequently proposed mechanism is the so called "pegging" effect.[1,2] It is supposed that protrusions which form around rare-earth-metal containing particles extend from the oxide scale into the metal and pin the scale to the underlying alloy. Luthra and Briant[3] proposed that yttrium containing particles are concentrated at the scale-metal interface and improve the mechanical integrity of this interface. Tien and Pettit[4] suggested that the oxide dispersions in the alloy act as prefered sites for the nucleation of vacancies which are therefore prevented from condensing at the scale-metal interface. Recently, Funkenbusch et al.[5] claimed that poor scale adherence is caused by alloy impurities such as sulphur, which diffuse to the scale-alloy interface and weaken the bond between alloy and scale. Oxide dispersions act as sulphur segregation sites, thereby preventing the sulphur diffusion to the scale-metal interface. Nagai et al.[6,7] studied the effectiveness of different oxide dispersions on the oxidation rate of Fe-20Cr. The results indicated that the effectiveness of the dispersion increased with its thermodynamic stability.

It has been established that improved scale adherence can also be achieved by small additions of reactive elements in the metallic form rather than as oxide[8], although it is not sure that the effects are completely identical. Ramanarayanan et al.[9] observed that, in alumina-forming alloys, yttrium leads to maintenance of a fine-grained oxide by retarding grain-boundary migration. In these fine-grained scales, growth stresses can be more effectively relieved. Recently the same authors reported that yttria dispersions in Ni-Cr alloys strongly decreased the growth rate of chromia scales.[10] As the critical strain energy for oxide scale spallation is inversely proportional to the scale thickness, thinner scales generally have a better adherence. In the case of the alumina-forming alloy Fe-20Cr-5Al, yttria had no significant influence on the growth rate of the scale and the improved adherence was attributed to a change in oxide grain structure.[10]

Several authors have tried to relate the changes in scale properties by reactive elements or reactive element oxides to changes in the growth mechanism of the scales. The most widely used methods for such experiments are those in which inert metal markers are placed on the metal surface prior to oxidation.[10,11] The location of the marker after oxidation is then determined from scale cross sections. Because the size of the marker is generally quite large compared to the thickness of chromia and especially alumina scales, these experiments can only give qualitative information of the transport phenomena in these scales. Young and de Wit[12] showed that inert marker experiments can even lead to erroneous conclusions on the occurring growth mechanism. More accurate information can be obtained from two stage oxidation experiments in which the atmosphere in the first or second oxidation period is enriched by ^{18}O.[12,13,14] The location of the ^{18}O in the scale can be determined by nuclear reaction analysis or by depth profiling using secondary ion mass spectroscopy (SIMS). This method has been used to study the influence of metallic reactive elements on high temperature oxidation, but has not been employed, as far as the authors know, to detailed investigations of the scale growth mechanisms on ODS alloys.

In the present investigation the influence of yttria dispersions on the growth and adherence of chromia and alumina scales has been investigated. For this purpose the oxidation behaviour of two ODS alloys, a chromia-former and an alumina-former was compared with that of two conventional wrought model alloys Ni-25Cr and Fe-20Cr-5Al. The different transport phenomena in the oxide scale and the differences in scale morphology are correlated with the observed scale growth and adherence.

2. EXPERIMENTAL

The model alloys Ni-25Cr and Fe-20Cr-5Al, supplied as forged bars, were solution treated at 1150°C for 4 h to obtain a homogeneous grain structure (Fig.1). The two ODS alloys MA 754 (Ni-20Cr-0.5Y_2O_3) and MA 956 (Fe-20Cr-4.5Al-0.5Y_2O_3), were used in the as-received condition.(The main minor alloying element in both alloys was 0.3% Ti). All materials were cut into rectangular specimens 30x10x2 mm and the surfaces were polished to a 600 grit finish and cleaned in acetone before use. The oxidation experiments were carried out in a muffle furnace at 1000 and 1100°C using synthetic dry air as the test environment. Weight changes were measured at regular intervals of time by interrupting the experiments and cooling the specimens to ambient temperature. Scale analysis was carried out using optical microscopy, scanning electron microscopy (SEM), energy dispersive X-ray analysis (EDX), and X-ray diffraction.

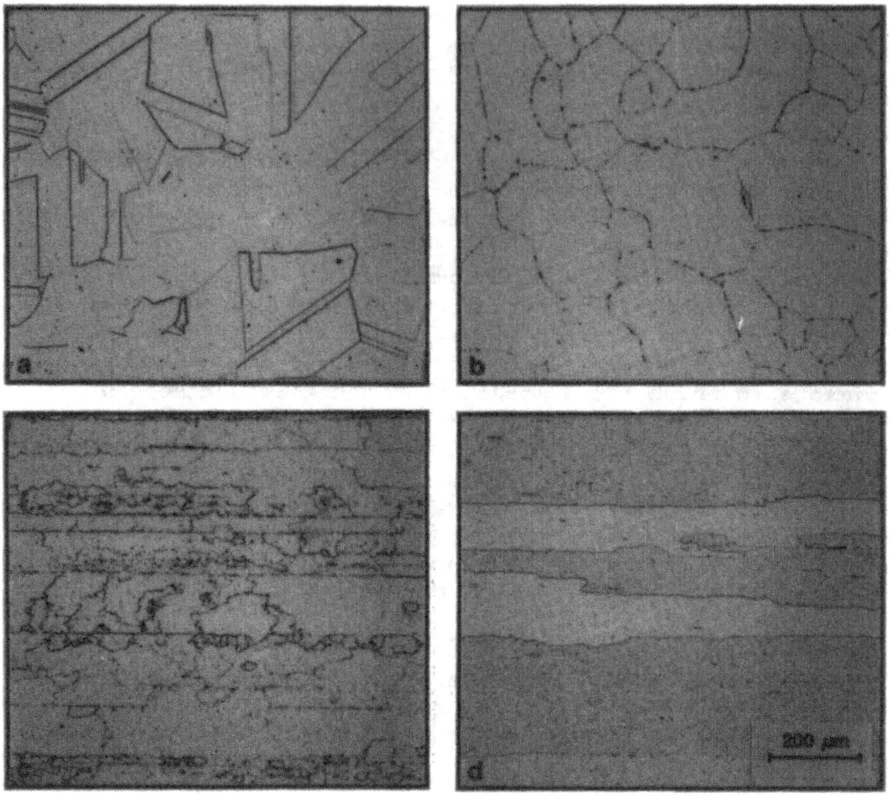

Figure 1: Microstructures of the investigated alloys.
(a) Ni-25Cr (b) Fe-20Cr-5Al (c) MA 754 (d) MA 956.

For studying the scale growth mechanisms, two-stage oxidation experiments were carried out at 1000°C. The specimens were placed in a gas tight quartz tube and moved into the furnace after the operating temperature had been reached. In the first stage, the specimens were oxidized for 3 h in synthetic air which was enriched with ^{18}O (enrichment of 10 %). Then the quartz tube was evacuated (around one minute) and non-enriched air was introduced for the second oxidation period (9 h). The isotope distribution in the scale formed during this two-stage oxidation was determined by depth profiling using secondary ion mass spectroscopy (SIMS). The equipment used was a cameca IMSMF with a Cs ion source.

3. OXIDATION KINETICS AND SCALE STRUCTURE OF THE CHROMIA FORMERS

Fig.2 shows the oxidation kinetics of the chromia-forming alloys Ni-25Cr and MA 754. At 1100°C the two alloys showed a qualitatively similar behaviour: spallation (indicated in the

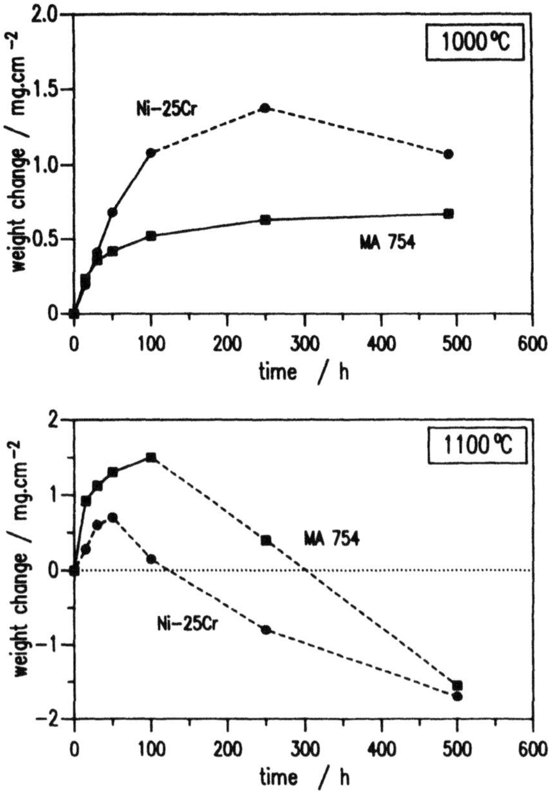

Figure 2: Weight changes of the chromia-formers during oxidation in synthetic air at 1000 and 1100°C. Dotted lines indicate that the measured weight changes are influenced by scale spallation.

diagrams by a dotted line) occurred after an initial weight increase. Due to the very early spallation of the scale on Ni-25Cr, the weight increase in the initial stage was smaller than that for MA 754, although in reality the oxidation rate of the first mentioned alloy was higher. This can clearly be seen from the results at 1000°C where spallation of the scale on Ni-25Cr did not occur up to exposure times of about 100 hours. The heavy spallation of the scale on Ni-25Cr at 1100°C led to a significant chromium depletion in the surface-near region of the alloy. Consequently the re-formed scale contained larger amounts of the spinel phase, $NiCr_2O_4$, in the outer part of the scale. After 500 h at 1100 °C the scale on MA 754 still only consisted of chromia. The spinel phase formation on Ni-25Cr was not only observed after spallation but also under conditions where scale spalling did not

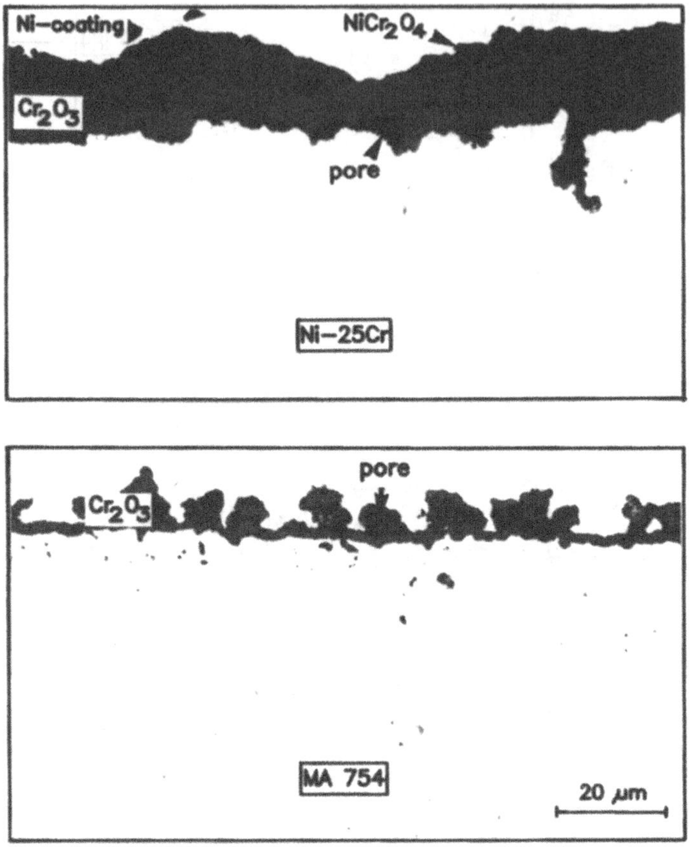

Figure 3: Cross sections of the oxide scales on the chromia-formers after 300 h uninterrupted oxidation at 1000°C.

occur. This is shown in Fig. 3 illustrating the differences in the composition and thickness of the scales on the two chromia-forming alloys after 300 h non-interrupted exposure at 1000°C. The thick scale on Ni-25Cr clearly contained small amounts of $NiCr_2O_4$ in addition to chromia. The X-ray diffraction results even indicated the formation of very small amounts of NiO (Fig. 4). Under the given conditions, MA 754 formed a pure chromia layer which shows that the ODS alloy oxidized more selectively in spite of its lower chromium content.

The scale morphology after different exposure times at 1100°C is shown in Fig 5. The scale on Ni-25Cr exhibited a convoluted morphology already after short oxidation periods. This structure coarsened on further exposure. The scale on MA 754 was initially flat and only small oxide nodules were growing locally at the

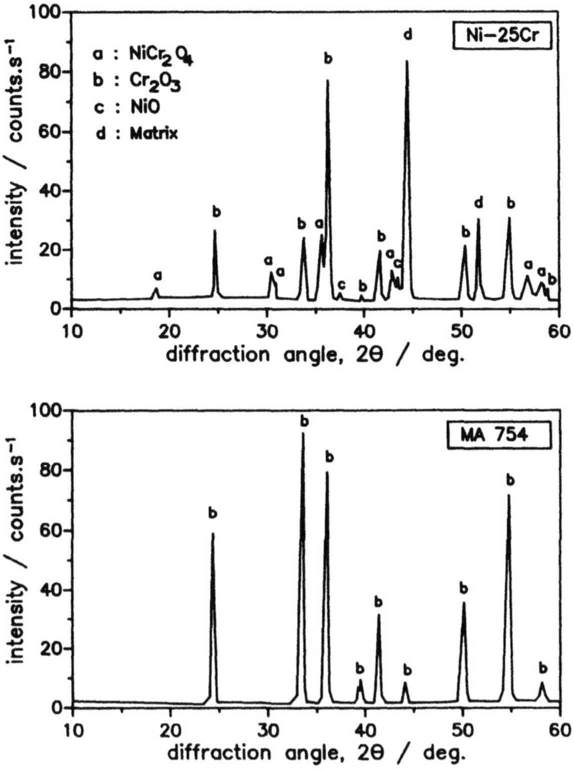

Figure 4: X-ray diffraction analysis of the scales shown in Fig.3.

outside of the scale. The nodules, which coarsened on subsequent exposure, seemed to be no compact oxide particles, at least after longer exposure times. This can be seen from Fig. 3 where large pores are clearly visible in the scale of MA 754.

4. OXIDATION KINETICS AND SCALE STRUCTURE OF THE ALUMINA-FORMERS

The observed weight changes during exposure of the alumina-formers at 1000 and 1100°C are shown in Fig. 6. The scale formed on the ODS alloy was far more better adherent than that on the conventional alloy. Contrary to the chromia-forming alloy (Fig.2), the alumina-former MA 956 even showed a perfect scale adherence up to 500 h at 1100°C.

At 1100°C alloy Fe-20Cr-5Al started to form iron-rich oxides after long-term exposure due to aluminium loss after repeated scale spalling. X-ray diffraction revealed that under conditions where no scale spalling occurred, both alloys formed a single phase α-alumina layer. The difference in scale thickness under these

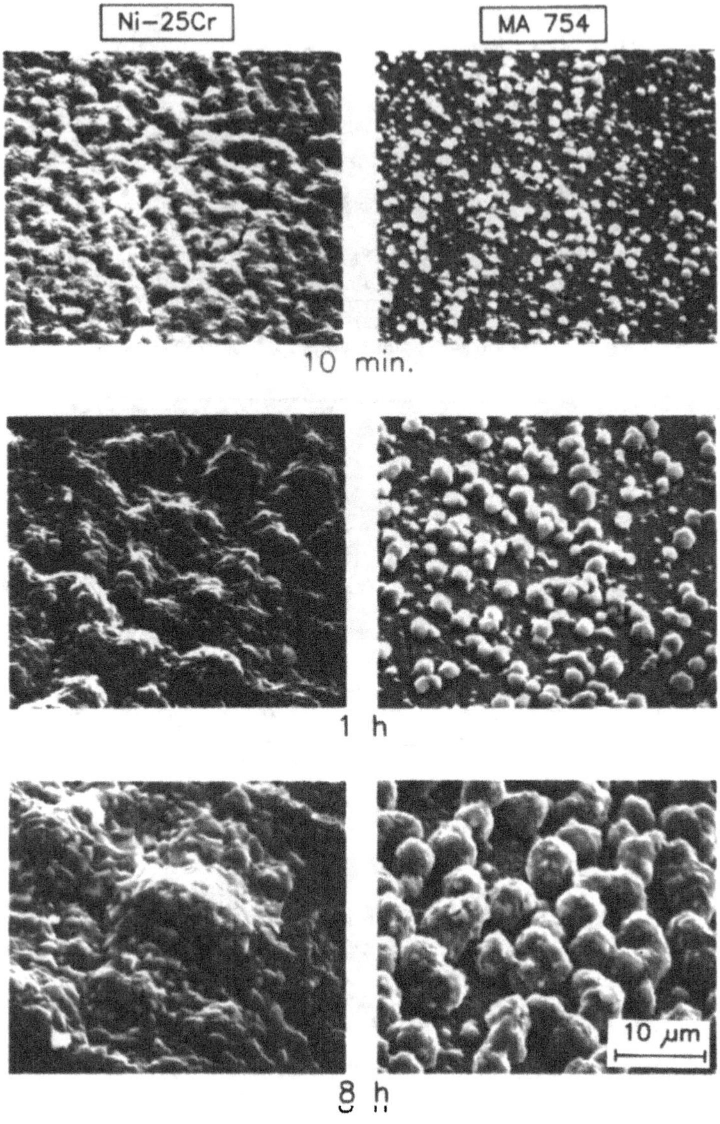

Figure 5: Scale morphology of the chromia-formers after different times of oxidation at 1100°C.

conditions is shown in the metallographic cross-sections after 300 h uninterrupted exposure at 1000°C (Fig. 7). EDX-analysis revealed, that the alumina on Fe-20Cr-5Al contained small amounts of chromium and, especially in the outer scale, iron (Fig. 8). The alumina on MA 956 contained small amounts of titanium; no chromium or iron was detected.

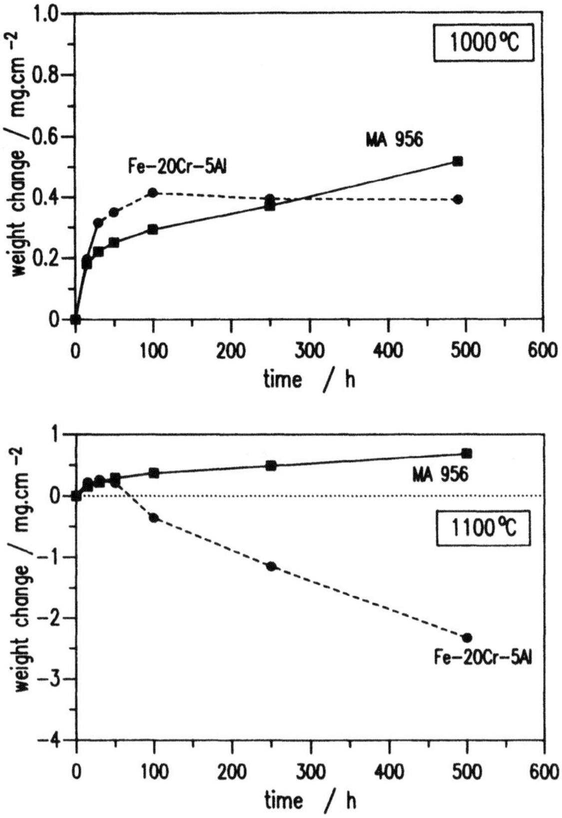

Figure 6: Weight changes of the alumina-formers during oxidation in synthetic air at 1000 and 1100°C. Dotted lines indicate that the measured weight changes are influenced by scale spallation.

The scale morphology after different exposure times of the two alumina-formers is shown in Fig. 9. The scale on alloy Fe-20Cr-5Al shows a convoluted morphology even after very short oxidation periods. At least after long term oxidation, large gaps seemed to be formed beneath the convolutions (Figs. 7, 8). In contrast,the scale on MA 956 was extremely flat and only very small nodules were seen growing on top of the scale; the number of nodules increased with increasing oxidation time and only slight coarsening was noticed.

5. TRANSPORT MECHANISMS IN THE CHROMIA AND ALUMINA SCALES

In order to obtain more accurate information on the scale transport phenomena which are responsible for the observed scale growth, morphology and adherence, two-stage oxidation experiments

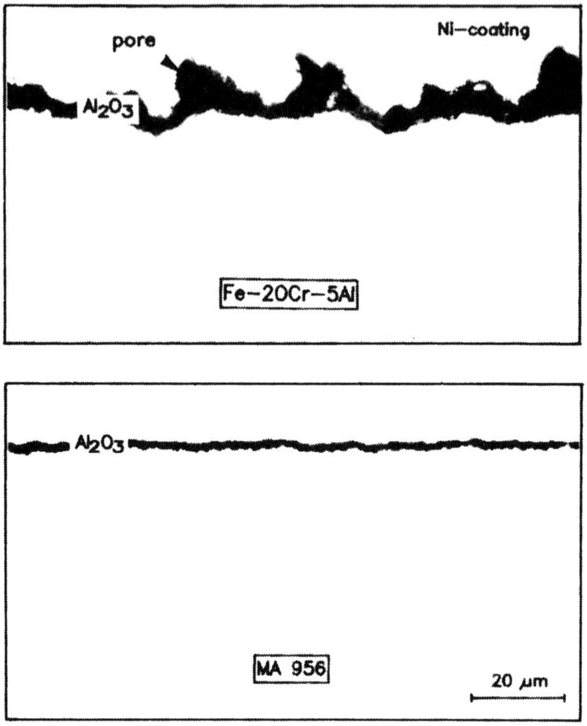

Figure 7: Scale cross-sections of the oxide scales on the alumina-formers after 300 h uninterrupted oxidation in synthetic air at 1000°C.

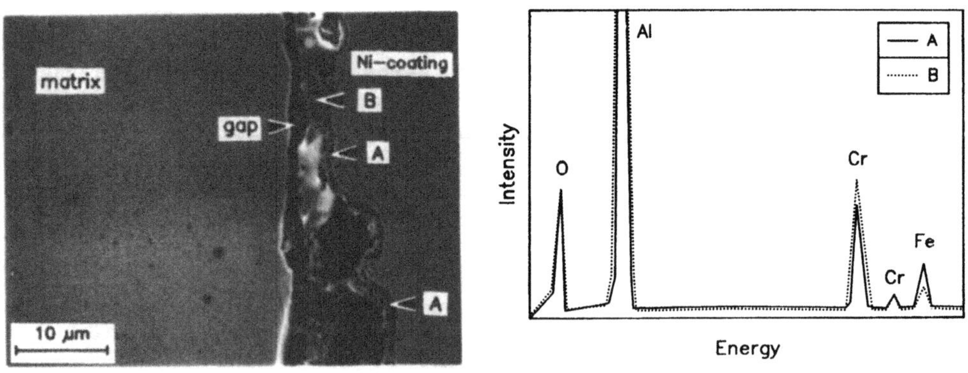

Figure 8: EDX-analysis of the oxide scale formed on Fe-20Cr-5Al after 300 h uninterrupted oxidation in synthetic air at 1000°C.

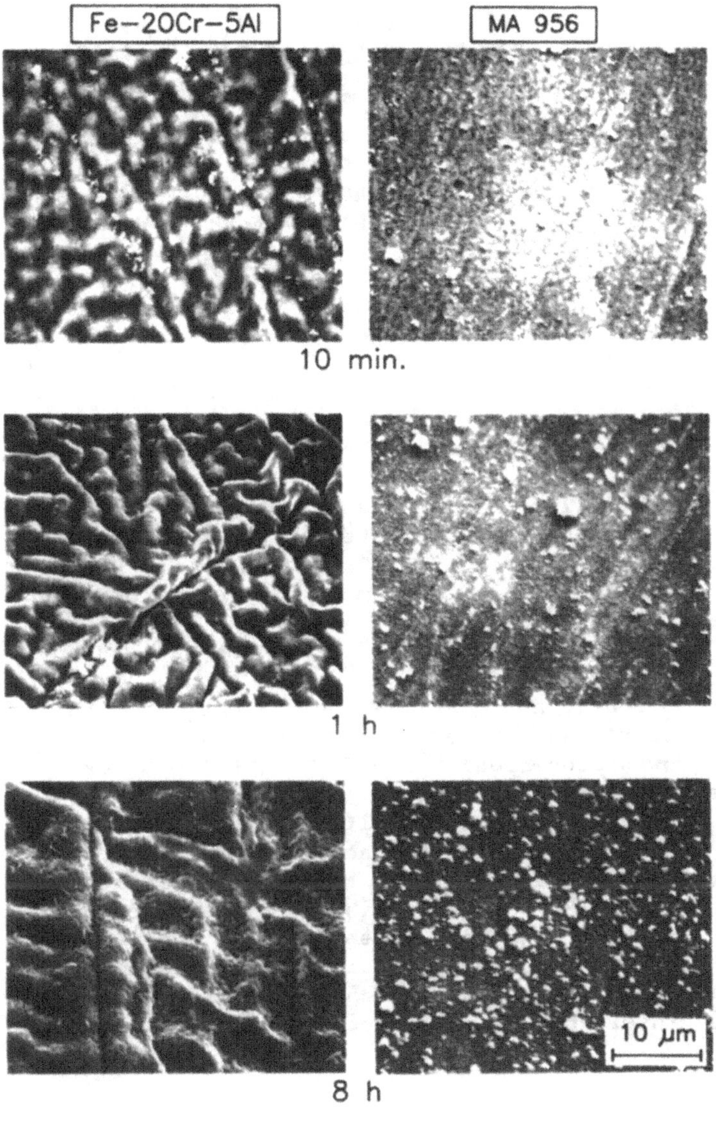

Figure 9: Scale morphology of the alumina-formers after different times of oxidation in synthetic air at 1100°C.

were carried out as described in the second section. The synthetic air used for the first oxidation period (3 h) contained oxygen which was 10 %-enriched with O^{18}; the second period in non-enriched air was 9h. The experiments were only conducted for very short exposure times at 1000°C. At 1100°C the dispersion-free model alloys showed scale spalling and a convoluted scale morphology even after relatively short test durations which obviously would have made the results of the two-stage oxidation experiments completely unreliable.

The distibution of the oxygen isotopes in the scales formed during two-stage oxidation of the different alloys is shown in Figs. 10 and 11. As the ^{18}O-enrichment was only 10 %, the measured curves had to be recalculated to profiles which would have been obtained if the total amount of oxygen in the synthetic air during the first oxidation period would have been ^{18}O. The results clearly indicate that the oxide scales on the two conventional alloys, Ni-25Cr and Fe-20Cr-5Al, grew by both inward and outward transport processes. For the ODS alloys the growth mainly occurred near the scale-metal interface: only a sharp ^{16}O-peak was visible near the scale-gas interface. Especially in the case of the alumina-former MA 956, the amount of outward growth was very small.

The reproducibility of the results obtained by the two-stage oxidation experiments is illustrated in Fig. 12. It shows oxygen isotope distributions in the alumina scale formed on MA 956 after two-stage oxidation with experimental parameters different from those in Fig. 11. The oxidation times in the two oxidation periods were only 1 and 2 hours and the ^{18}O-enriched gas was applied in the second instead of the first oxidation stage. Although the profile is obviously not quantitatively identical with that shown in Fig. 11 for MA 956, the experiment clearly verifies the occurring growth processes: the oxide grows almost exclusively by inward transport of oxygen and only a very limited outward aluminium transport occurs. The different slopes of the ^{18}O-profile in the outer part of the scale in Figure 12 nicely demonstrate the dominating transport phenomena.[14]
Whereas the outer ^{18}O-enrichment is associated with outward aluminium transport, the less steep slope of the ^{18}O-curve nearer to the middle of the scale is an almost ideal example of an isotope distribution caused by exchange of ^{18}O diffusing to the scale-metal interface via grain boundaries with ^{16}O in the lattice.

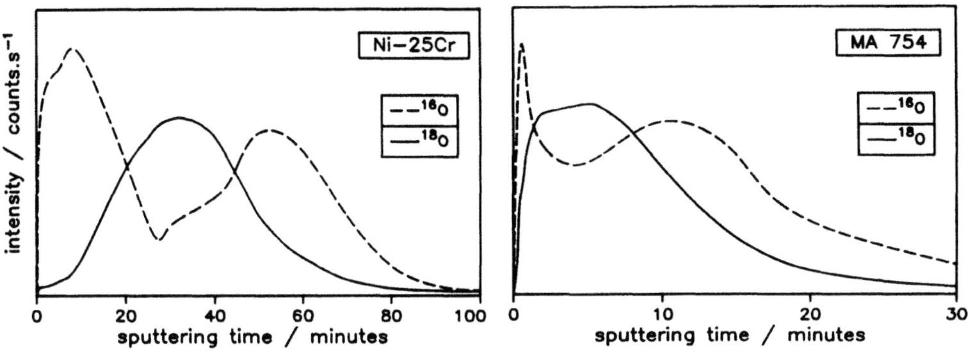

Figure 10: Oxygen isotope distributions in the oxide scales on the chromia-formers after three hours oxidation in ^{18}O-enriched air and subsequently nine hours in non-enriched air at 1000°C.

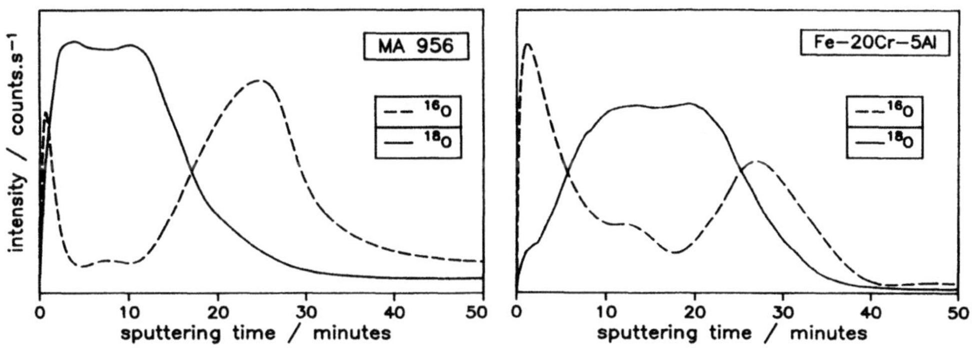

Figure 11: Oxygen isotope distributions in the oxide scales on the alumina-formers after three hours oxidation in ^{18}O-enriched air and subsequently nine hours in non-enriched air at 1000°C.

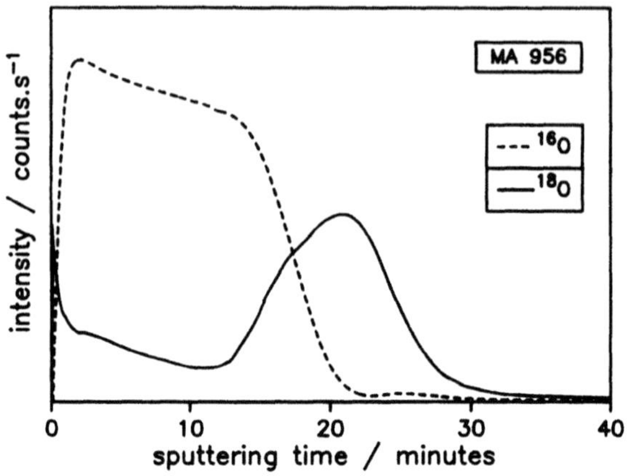

Figure 12: Oxygen isotope distributions in the oxide scale on MA 956 after one hour oxidation in non-enriched air and subsequently two hours in ^{18}O-enriched air at 1000°C.

6. DISCUSSION

The oxidation experiments with alumina- and chromia-forming high temperature alloys have verified that addition of yttria decreases the oxide growth rate and improves scale adherence. Consequently, the addition of yttria dispersions has a very similar effect on oxidation as described in literature for the addition of yttrium or other reactive elements in the metallic form.[8] The similarity in the effects is probably caused by the strong tendency of the reactive elements to oxidize below a growing surface scale, such that they will quite rapidly be transformed into internal oxide precipitates.

Concerning the influence of reactive element oxides on the growth rates of alumina scales, some authors claim that, although adherence is improved, no change in growth rate occurs.[10] The present results, however, show that under the given conditions the growth rate of alumina is affected by reactive element (oxide) addition as has also been observed by Reddy et al.[13]

The results from Figs. 3 and 4 also illustrate that in the case of the chromia-forming alloys the minimum concentration of chromium for obtaining selective oxidation is decreased by addition of yttria. Under conditions where no spalling occurred, on both investigated alumina-forming alloys, single phase α-alumina layers were found (Figs. 7 and 8). Contrary to MA 956,

Fe-20Cr-5Al formed an alumina scale which contained small amounts of iron and chromium showing a tendency to form additional oxide phases. This indicates that also in the case of the alumina-forming alloys, the minimum concentration of the alloying element responsible for obtaining selective oxidation is decreased by yttria dispersions. This is in agreement with results from several investigators[15,16] and has also been found in a previous study by one of the present authors.[17]

The oxidation experiments revealed that the improvement of the scale properties due to the yttria addition is less significant for the chromia-former than for the alumina-forming alloy. The time for the onset of spalling is around four to eight times longer for MA 754 than for Ni-25Cr (Fig. 2). The alumina-forming ODS alloy showed perfect scale adherence up to 500 h even at 1100°C whereas the scale on Fe-20Cr-5Al spalls after extremely short durations (Fig. 6). From these results one should not conclude that oxide dispersions are always more effective in alumina-formers than in chromia-formers; it is likely that the exact amount and distribution of the oxide dispersion play a significant role in the effectiveness for scale property improvement, as will be discussed below.

The results of the two-stage oxidation experiments demonstrate that growth of both the alumina and the chromia scale on the conventional wrought alloys occurs by oxgen as well as metal transport. A comparison of the oxygen isotope distributions in the scales of the different alloys indicates that oxide dispersions do not alter the transport processes from anion to cation diffusion (or vice versa). Addition of yttria only leads to a suppression of the metal ion transport through the scale.

As discussed in the introduction, several mechanisms have been proposed to explain the influence of reactive element oxide additions on the oxidation behaviour of high temperature alloys. Some authors have suggested that different mechanisms are operating simultaneously. Occasionally it has been proposed that the mechanism is different for different types of oxides.

The present investigations indicate that the mechanism by which yttria influences the oxide scale properties of the investigated materials is similar for both envisaged oxide types and is related to the transport phenomena in the scale:

- For both the chromia- and alumina-formers yttria additions reduced oxide growth rate and the tendency for spalling; for both oxide types, outward scale growth is suppressed.
- The effect of yttria on the scaling behaviour was more significant for the alumina-former than for the chromia-former; also the suppression of outward scale growth is more significant for the alumina-former, than for the chromia-forming alloy.

These results indicate that the change in scale growth rate and adherence due to yttria addition is correlated with the suppression of outward scale growth..

Using the scale formation on the chromia-formers as an example, Fig. 13 shows schematically how the differences in transport phenomena in scales on the alloys with and without yttria, lead to the observed differences in oxide structure and properties. If both alumina and chromia grow by cation and anion diffusion, it is obvious that the suppression of one of the two fluxes by yttria leads to a reduced oxide growth rate as observed in the present investigation. The smaller growth rates will certainly affect scale adherence because growth stresses are smaller in thinner scales.[18] Additionally cation transport from the scale-alloy to the scale-gas interface causes vacancy generation in the underlying alloy. Preferential condensation of vacancies at the scale-alloy interface results in pore formation and consequently in local scale detatchment which obviously has a detrimental effect on scale adherence. Local scale detatchment might also lead to the convoluted scale morphology observed on the yttria-free alloys. Cathcart[19] observed void formation beneath alumina scales on NiAl alloys and concluded from the morphological features of those voids that aluminium vapour phase transport occurred across the voids.It appears reasonable to assume that the rate of transfer of aluminium atoms to the scale is different if it occurs via the vapour phase rather than via the solid phase. This would lead to local differences in scale growth rates resulting in a convoluted morphology. It is also possible that the convolutions are related to deformation of the scale due to oxide formation within the scale[20,21] which occurs because of parallel transport of oxygen and aluminium in the yttria-free alloys.

The transport of cations from the alloy to the scale-gas interface probably occurs via short circuit diffusion paths, e.g. grain boundaries. This will be the case not only for the cations of the main oxide forming element, but also for the more noble elements, such as nickel and iron. If the yttria dispersions block the diffusion paths for chromium cations through chromia and for aluminium cations through alumina, it is obvious that the transport of the other cations is also reduced (Fig. 13). Blocking the diffusion of these elements (e.g. nickel) will prevent spinel oxide formation at the scale-gas interface. In other words, the suppression of outward growth leads to enhanced selective oxidation (Figs. 3, 4 and 8).

It is not yet completely understood in which way yttria dispersions might suppress outward cation movement. Little information is available on the interaction of reactive element oxides with the lattice and especially the grain boundaries of the oxide. Detailed scale analysis using transmission electron

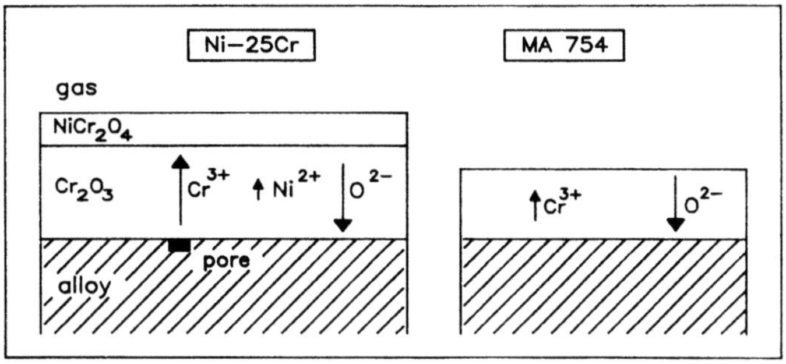

Figure 13: Schematic illustration of the transport processes in the oxide scales of the ODS alloy MA 754 and the wrought model alloy Ni-25Cr. The reduced cation transport in the scale on the ODS alloy leads to a decreased growth rate, less pore formation by vacancy condensation and more selective oxidation. The phenomena in the alumina scales were found to be similar, except for the spinel formation.

microscopy indicates that oxide dispersions are incorporated in the growing scales and inhibit diffusion along grain boundary paths.[22] High concentrations of oxide dispersions in the scale are required for effective cation diffusion blocking.[22]

If the yttria concentration is too small or if the yttria particles are not evenly distributed, cations can diffuse via paths which are not effectively blocked, leading to local external growth. The small nodules seen on the scales formed on the ODS alloys (Figs. 5 and 9), especially MA 754, could be explained by such a mechanism. If effective diffusion blocking is only possible by an even distribution of yttrium or yttrium oxide throughout the scale, a faster growing oxide such as chromia would require a higher concentration of oxide dispersion in the alloy than an alumina-forming alloy in order to obtain the same beneficial effect. This would explain why in the present investigation the yttria addition was more beneficial in the case of the alumina-former than for the chromia-former, although both ODS alloys contained the same amount of oxide dispersion.

Considering the initial stages of oxidation, it is obvious that the presence of oxide dispersions in the alloy will influence the morphology of the transient oxide layer, by affecting either the oxide nucleation or the initial oxide grain growth. This will certainly be reflected in a smaller grain size of the oxide scales on the ODS alloys compared to those on the yttria-free alloys after longer oxidation times, as observed by Ramanarayanan et al.[10] If we consider grain boundaries as possible cation diffusion paths, it is, however, difficult to understand how a smaller oxide grain size, and thus a larger number of grain boundaries, would lead to a decreased outward growth of the scales on the ODS alloys. A smaller grain size might be beneficial for the mechanical scale properties[10], however, it is not clear from the present study whether this is an important factor for the observed changes in the mechanistic and macroscopical scale properties by oxide dispersions.

7. CONCLUSIONS

The oxidation experiments with alumina-forming and chromia-forming high temperature alloys have demonstrated that the addition of yttria decreases oxide growth rate, improves scale adherence and enhances selective oxidation. The present studies on scale formation and morphology combined with ^{18}O-tracer analysis of scale growth mechanisms indicate that these changes in the macroscopic properties of both types of oxide due to oxide dispersions can be explained by a change in the transport processes within the oxide scale. Chromia and alumina scales grow by both cation and anion diffusion, probably along rapid diffusion paths. Presence of yttria dispersions in the alloy leads to a suppression of cation transport. It is obvious that the suppression of one of the two fluxes responsible for oxide growth, leads to a decreased scale growth rate. As growth stresses are smaller in thinner scales, the decrease in growth rate will lead to improved scale adherence. Additionally a reduction of cation transport reduces pore formation at the scale-metal interface originating from vacancy condensation, and thereby prevents the formation of a convoluted, poorly adherent scale. Decreased transport of nickel- and/or iron-cations reduces spinel type oxide formation at the oxide-gas-interface and consequently enhances selective oxidation. Indications have been found that a limited amount of outward growth of the oxide scales on ODS alloys occurs locally, at places where the blocking of cation diffusion was not effective due to low amounts and/or uneven distributions of the oxide dispersions. Effective cation blocking in the chromia scales seems to require higher amounts of yttria than in the alumina scales.

ACKNOWLEDGEMENTS

The authors gratefully acknowledge Dr. Wallura and Mr. Els for assistance in carrying out the SEM/EDAX investigations, Mr. Gutzeit for the optical metallography and Dr. Schulze and Mr. Lersch for X-ray diffraction analysis.

REFERENCES

1. B. LUSTMAN, Trans. TMS-AIME, 1950, 188, 995-998
2. C.W. PRICE, J.G. WRIGHT and G.R. WALLWORK, Met. Trans. 1973, 4, 2423-2427
3. K.L. LUTHRA and C.L. BRIANT, Oxid. Met. 1986, 26, 397-416
4. J.K. TIEN and F.S. PETTIT, Met. Trans. 1972, 3, 1587-1599
5. A.W. FUNKENBUSCH, J.G. SMEGGIL and N.S. BORNSTEIN, Met. Trans. 1985, 16A, 1164-1166
6. H. NAGAI, T. MURAI and H. MITANI, Trans. JIM, 1979, 20, 299-310
7. H. NAGAI, Y. TAKEBAYASHI and H. MITANI, Met. Trans., 1981, 12A, 435-442
8. D.P. WHITTLE and J. STRINGER, Philos. Trans. R. Soc. London, 1980, A295, 309-329
9. T.A. RAMANARAYANAN, M. RAGHAVAN and R. PETKOVIC-LUTON, J. Electrochem. Soc. 1984, 131, 923-931
10. T.A. RAMANARAYANAN, R. AYER, R. PETKOVIC-LUTON and D.P. LETA, Oxid. Met., 1988, 29, 445-472
11. C.S. GIGGINS and F.S. PETTIT, Met. Trans., 1971, 2, 1071-1078
12. E.W.A. YOUNG and J.H.W. DE WIT, Surface Interface Anal. (1987), ESCASIA conference, Veldhoven, Netherlands, 1985
13. K.P.R. REDDY, J.L. SMIALEK and A.R. COOPER, Oxid. Met. 1982, 17, 429-449
14. S.N. BASU and J.W. HALLOVAN, Oxid. Met. 1987, 27, 143-155
15. T.N. RHYS-JONES, H.J. GRABKE and H. KUDIELKA, Corr. Sci. 1987, 27, 49-73
16. P.Y. HOU and J. STRINGER, Mat. Sci. Engng. 1987, 87, 295-302
17. W.J. QUADAKKERS, C. WASSERFUHR, A.S. KHANNA and H. NICKEL, Conference "Scale Growth", London, March 23, 1988, Mat. Sci. and Technol., 1988, 4, 1119.
18. M.J. BENNETT and A.S. TUSON, Mat. Devel. Div., Harwell Lab., Oxfordshire, U.K., Report AERE R 13309
19. J. CATHCART, Conf. on High Temp. Ordered Intermetallic Alloys, Boston, USA, Nov. 1984, Mat. Res. Soc., Pittsburgh, 1985, p. 445 - 459, Acession nr. 85 (10): 72 - 559
20. F.A. GOLIGHTLY, F.H. STOTT and G.C. WOOD, Oxid. of Met., 1976, 10, 163-187
21. C.H. YANG, G.E. WELSCH and T.E. MITCHELL, Mat. Sci. and Engng. 1985, 69, 351-357
22. D.P. MOON, Mat. Devel. Div., Harwell Lab., Oxfordshire, U.K., Report AERE R 12870, 1987

IMPROVEMENTS IN THE ADHESION AND PROTECTIVENESS OF SCALES FORMED ON IRON-CHROMIUM STEELS BY THE APPLICATION OF BORON COMPOUNDS

S R J Saunders* and J A Little+

* National Physical Laboratory, Teddington, UK
\+ Metallurgy and Material Sciences Department, Cambridge University, Cambridge, UK

ABSTRACT

The effects of boron-containing compounds, present in the vapour phase or directly applied, on the oxidation behaviour of iron-chromium alloys are reviewed. Generally long-term beneficial effects are observed in oxidising, carburising and sulphidising environments in reducing ingress of oxidants and in promoting adherent scales. The mechanisms of these effects are discussed and it is suggested that transport processes during scale growth are modified by the presence of boron reducing growth stresses, and/or facilitating stress relief processes by the development of a fine grained scale allowing grain boundary sliding.

1. INTRODUCTION

High temperature corrosion resistance of metallic materials results from the formation of a protective oxide layer on the surface. The properties of this oxide can be optimised to produce a mechanically and chemically stable diffusion barrier by modification to the alloy composition, but frequently requirements for protective film formation and bulk mechanical properties of the alloy conflict so that other means of enchancing protective scale formation must be devised. In this connection surface coatings have been developed and a well-established technology exists to provide coatings tailored for particular applications.[1] In addition the oxide layer that forms on uncoated materials may be modified to enhance performance by selective oxidation methods,[2] by the application of simple surface treatments or by additions of trace amounts of gaseous species to the working environment.[1] This latter process is analogous to inhibition in aqueous corrosion although the technique has not been as extensively developed.

Modification to oxide growth processes has been a major research topic for many years and there exists a vast body of information on the effects of rare-earth elements or compounds and, to a lesser extent, on the effects of some noble metals. The literature on the former topic has reached almost obsessive proportions, so much so that other possibilities have been largely ignored. Some other avenues of research have, however, been pursued notably at the National Physical Laboratory and more recently at the University of Cambridge, into the effects of boron-compounds on the nature of protective film formation on iron-chromium alloys. The results obtained have generally been beneficial with evidence of increasing oxide spallation resistance, decreasing scale growth rates and improvements in the character of oxide layers as diffusion barriers in the presence of boron. This paper attempts to review the significant effects of boron-compound additions and to draw together the available evidence about the controlling mechanisms. It is widely accepted that rare-earth compounds influence grain boundary diffusion processes because of segregation effects associated with the large ionic radius of the rare-earths. The effects observed with boron compounds may well be explained on the basis of the small ionic radius of boron, the possibility of covalency, and of second phase compound formation or segregation effects at oxide grain boundaries.

2. METHODS OF APPLICATION

Two main methods of applying boron-compounds have been devised. Firstly, simple dip treatments in alcoholic solutions of boric acid or its salts followed by evaporation of the solvent by slight heating in air can be used and this will leave a thin deposit on the alloy surface.[3] The amount of deposit can be controlled by varying the concentration of the solution; other variants have included suspending the sample above the solution and controlling the deposit thickness by the duration of exposure to the vapour.[4] Secondly, volatile boron compounds can be introduced into the atmosphere either during the early stages of the exposure or be present continuously.[5] Typical sources for this type of application have been boric oxide and boric acid salts, but even pyrex glass[3] has been shown to be effective in this regard. Some workers have also studied the effects of boron alloy additions,[6] ion implantation[7] and boron diffusion coatings[8] and while alloying has often had detrimental effects on oxidation, recent work with borided surfaces and ion implantation showed significant benefits. A summary of these various methods has been compiled by Simms[9] and is listed in Table I. From quoted deposit thickness data Simms calculated that the boron doping at the surface is in the range 10^{16}-10^{19} atoms cm^{-2}, although for pyrex contamination and possibly some other vapour phase treatments it is likely that boron additions are effective at lower levels.

Table I
Boron content of various coatings

Formula	Mol Wt	Doping level	Boron content of coating (atoms cm^{-2})	Ref
B_2O_3	69.62g	0.01 μg $cm^{-2}h^{-1}$ ☆	< $8.65x10^{13}$ (atoms $cm^{-2}h^{-1}$)	19
$Na_2B_4O_7$	201.22g	3 ±1 mg cm^{-2}★	$3.59x10^{19}$ ±$.20x10^{19}$	10
		< 500 μg cm^{-2}★ (10 μg cm^{-2})	< $5.99x10^{18}$ ($2.99x10^{16}$)	3
CaB_2O_4	125.70g	3 mg cm^{-2}±1 mg cm^{-2}★	$2.87x10^{19}$ ±$9.58x10^{18}$	10
H_3BO_3	61.83g	< 0.4 mg cm^{-2}★	< $3.90x10^{18}$	4,26
H_3BO_3	61.83g	0.01 g $cm^{-2}h^{-1}$ ☆	< $9x10^{19}$ (atoms $cm^{-2}h^{-1}$)	16

☆ continuous additions ★ pre-deposition

3. RESULTS

3.1 Oxidising Environments

a) Air

Early work using superficial boron-compound additions was reported by Rolls and Shaw[10] in a study of the effects of $Na_2B_4O_7$ and CaB_2O_4 on iron in an attempt to reduce loss of metal by scaling during hot-rolling processes in the manufacture of sheet steel. This work developed from use of proprietary powder mixtures containing boron compounds developed some ten years previously.[11,12] At temperatures of about 1000 °C in oxygen the oxidation rate of carbon steel was reduced by up to about 50% during 6 h exposure periods. Analysis of the scales suggested that an iron-boro-ferrite compound (vonsenite - Fe_3BO_5) formed as an incomplete layer at the scale/metal interface. The activation energy for growth of the oxide was 39 kJ mol^{-1} which was the same as that for the untreated steel, but the relative thicknesses of the higher oxides to wustite were increased for the borate treated steel.

Much more dramatic reductions in scaling rate were reported by Hendy *et al*[3] during an investigation of the effects of superficially applied boron-compounds to iron-chromium alloys with chromium contents up to 20%. An archetypal breakaway oxidation phenomenon is found with an Fe-10Cr binary alloy upon oxidation in air at 600-800 °C. Initially a thin protective duplex film of $(FeCr)_3O_4$ and Fe_2O_3 forms and there then follows a transition to much higher rates believed to be associated with mechanical disruption of the scale caused initially by strain in the outer Fe_2O_3 layer.[13] Application of 'borax' ($Na_2B_4O_7$) stabilised the protective period of oxidation such that for the more heavily doped samples the transition to higher rates of attack did not occur in 30 kh exposure at 800 °C in air (Figure 1). In some cases

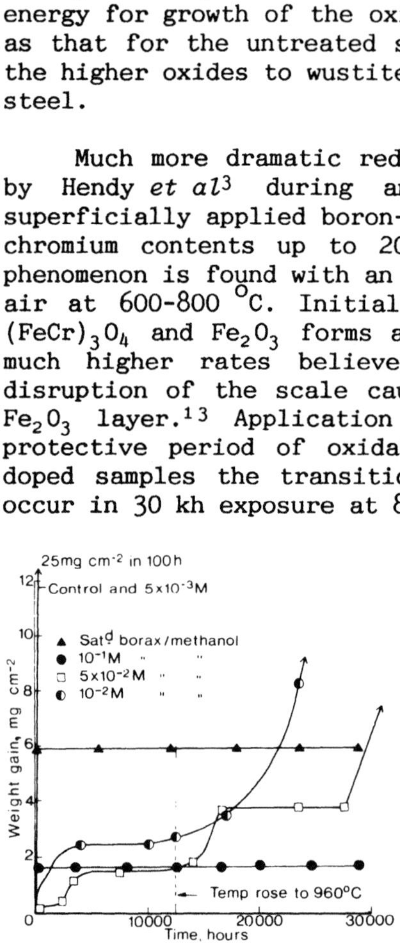

Figure 1

Weight gain as a function of time for Fe-10 Cr steel superficially treated with boron-containing compounds and exposed to air at 800 °C.

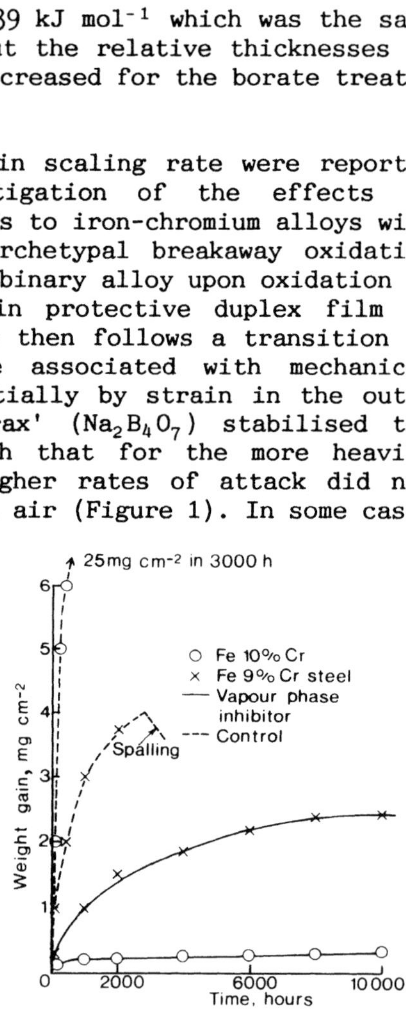

Figure 2

Weight gain as a function of time for chromium steels exposed in air, and air + B_2O_3 vapour at 600 °C.

where a high initial deposit of borax was used an increase in oxidation rate was observed before the stable film became established. Beneficial effects were also reported on a range of steels for a number of other boron compounds in alcoholic solutions such as lithium borate, zinc borate, boric acid, boric oxide and trimethyl or triethyl borate.[9,14] In cases where thick scales were produced, for example with the low (2¼) chromium steels, an additional layer was found in the scale which was shown to contain iron, boron and oxygen but its phase could not be identified.[4] Fracture sections showed that major alteration to the oxide structure occurred when B_2O_3 was present causing the equiaxed/columnar oxide grain structure of the untreated steel to become entirely equiaxed.

Historically the work on superficial application of borates at the National Physical Laboratory arose from the observation that after a furnace tube had become contaminated with pyrex glass breakaway oxidation of iron-chromium alloys was prevented[3] and this suggested that vapour phase inhibition processes would also be possible. Controlled experiments were subsequently carried out, therefore, using B_2O_3 vapour as a source which was placed upstream of specimens in flowing air; Figure 2 shows that long-term protection of iron-chromium steels could also be achieved by this method,[3] although it was apparent that more effective protection was achieved with a simple binary alloy than with a commercial steel.

b) Steam

Steam turbine blades can suffer severe erosion from impact with entrained oxide spalled from heat exchangers in steam power generating systems. Clearly, therefore, control of oxide spallation is a critical factor in minimising this problem, and the effect of vapour phase and superficial boron-compound addition has been investigated.[15,16] Small reductions of about 30% in oxidation rate of 2¼ Cr 1Mo steel at 600 °C were observed when B_2O_3 vapour was present in the steam. However, more significantly, perhaps, was that while this steel was very susceptible to oxide spallation, this was greatly reduced when B_2O_3 vapour was present.[15] A much greater effect on the oxidation rate was observed with a higher chromium steel (9 Cr 1Mo) treated with either B_2O_3 vapour present in the steam or superficially prior to exposure and again spallation resistance was improved.[16] A boron-containing phase was found in the scales formed on the 2¼ Cr 1Mo steel which was identified by X-ray diffraction as vonsenite and, as with scales formed in air, fracture sections revealed a change in the oxide grain structure from duplex equiaxed/columnar for untreated samples to entirely equiaxed when B_2O_3 vapour was present in the steam, and the extensive microporosity of the scales produced in pure steam was greatly reduced by exposure to the boron-containing environments, (Figure 3)[15].

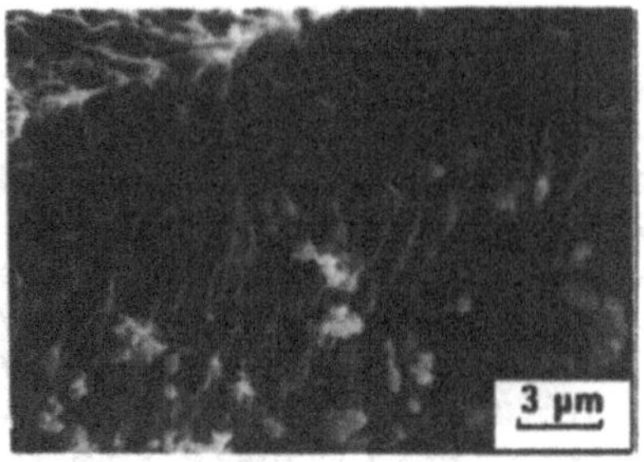

a)

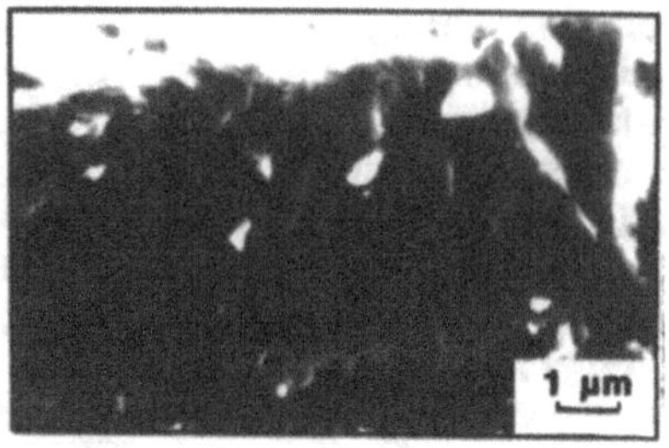

b)

Figure 3

Scanning electron micrographs of fracture of scales formed on 2¼ Cr 1 Mo steel after 100 h exposure to steam at 600 °C, a) control, b) exposed to boric acid vapour during oxidation.

3.2 Carburising Environments

9 Cr ferritic steels exposed to CO_2-1%CO gas mixtures such as are used in the advanced gas-cooled reactor (AGR) are susceptible to carburisation. It has been proposed that when the steel is fully carburised, carbon formed in the Boudouard reaction precipitates in the oxide because the effective sink for carbon (the steel) is no longer available. This carbon precipitation process in the oxide causes a transition from protective to non-protective oxidation kinetics (breakaway), particularly at temperatures greater than about 600 °C.[17] Although operating temperatures of steel for AGR use were generally lower than 600 °C and breakaway oxidation was not considered to be a problem, it was prudent to develop measures to lower the risk of occurrence of this type of attack. Accordingly, as part of a wider programme of coating development the potential of both superficially applied and vapour phase boron-containing inhibitor was examined in simulated AGR atmospheres.

Long-term oxidation testing of samples of 9 Cr 1Mo steel dipped in aqueous solutions of sodium and potassium tetraborate prior to exposure to the AGR environment was carried out at 580 °C, and in the best cases reductions in oxidation rate of over an order of magnitude were obtained.[18] However, the behaviour was somewhat variable with non-protective oxide growth appearing at localised sites on the surface. It is not clear, however, whether there was continued oxidation at high rates at these sites since this may have only occurred during an earlier stage of oxidation before the protective film became established. An investigation of the effect of varying the alkali metal:boron ratio was carried out and less protection was observed when solutions containing non-stoichiometric ratios were used, but no explanation was given to account for this result. X-ray diffraction analysis of the scale yielded a pattern consistent with vonsenite but potassium was also shown to be present, and evidently a complete description of the scale had not been possible.

An evaluation of the effect of B_2O_3 vapour was also made in the AGR atmosphere, and specimens already in breakaway oxidation as well as protectively oxidised samples and unoxidised samples were exposed for 30 kh at temperatures up to 580 °C.[19] It was found that the specimens already in breakaway were not affected by exposure to B_2O_3 vapour, but the other samples had extrapolated weight gains to 250 kh reduced by a factor of 2-3 times compared with untreated steels. It was also noted that the rate of carbon uptake in the alloy was reduced to an even greater extent (Figure 4) which was particularly significant in view of the proposed mechanism of breakaway oxidation in this alloy.

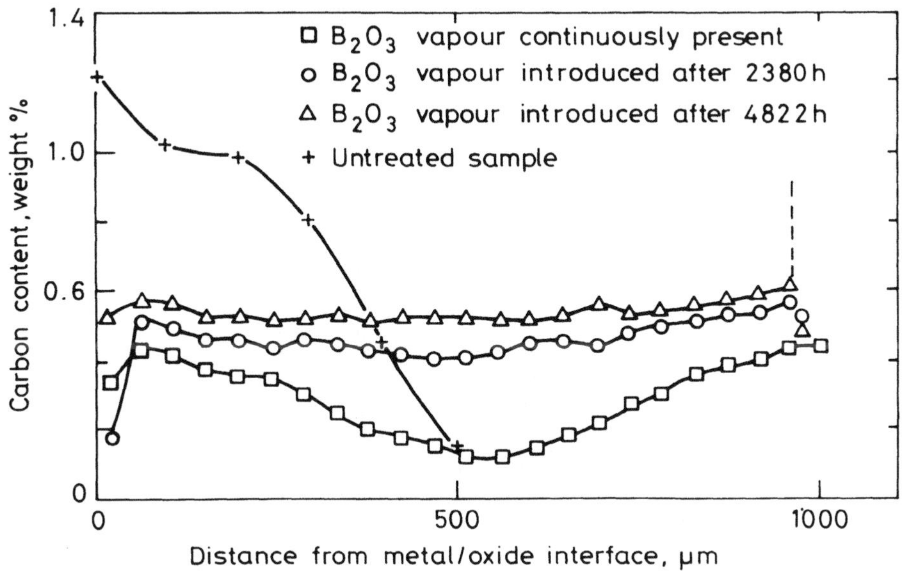

Figure 4

Carbon profiles through various samples of 9 Cr 1 Mo steel exposed to CO_2-1 CO at 580 °C.

3.3 Sulphidising Environments

While very little work in sulphidising atmospheres has been carried out, the few reported[20] results are quite intriguing. Fe-10 Cr, Fe-20 Cr and AISI 310 stainless steel samples were

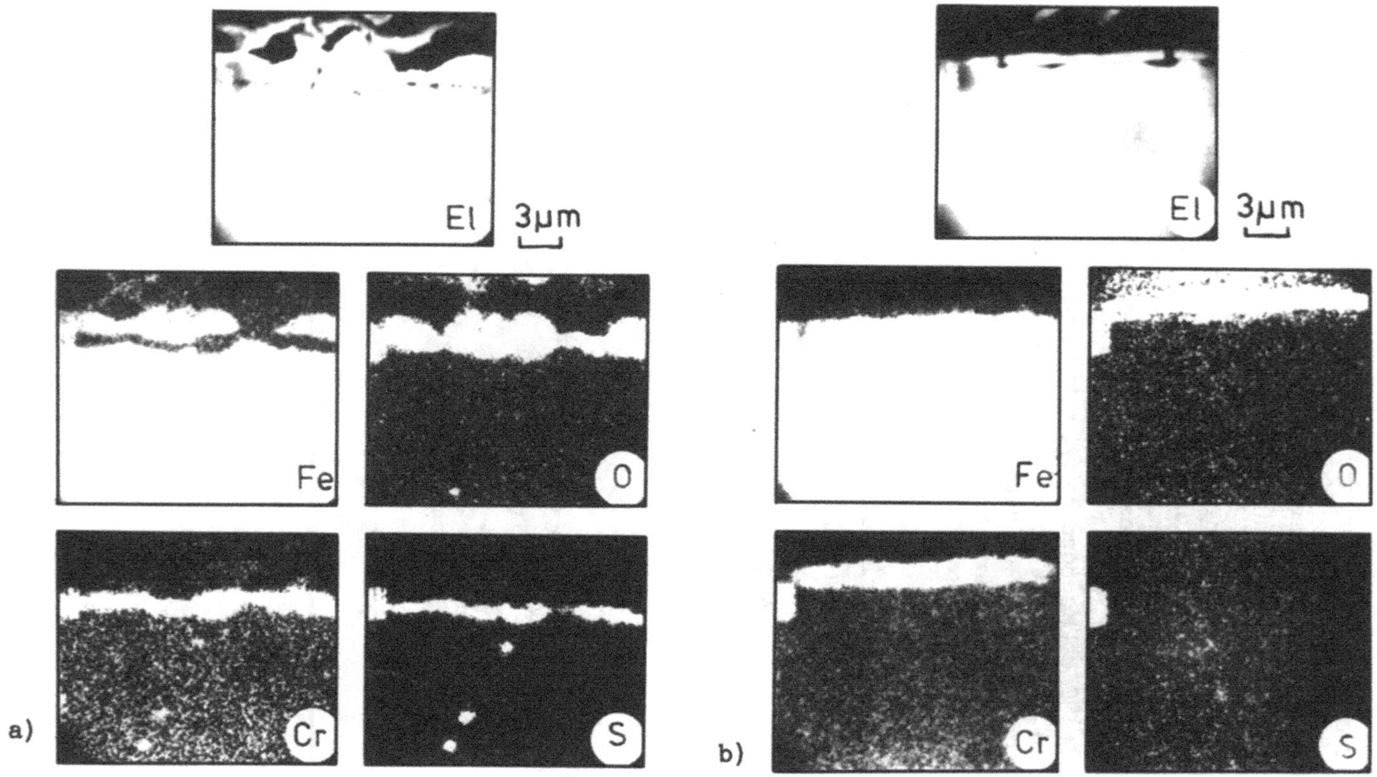

Figure 5

Microprobe analysis of preoxidised Fe-10 Cr after 200 h exposure to Ar-10%SO_2 at 800 °C.
a) 1 min preoxidation in air, b) 1 min preoxidation in air + B_2O_3 vapour.

exposed to Ar-10% SO_2 at 800 °C, with and without various preoxidation treatments in air and in air containing B_2O_3 vapour. Fe-20 Cr was resistant to attack by the Ar/SO_2 gas mixture so that there was little effect of the preoxidation treatment. However, both Fe-10 Cr and AISI 310 stainless steel were severely attacked by the sulphidising gas. Preoxidation in both types of atmosphere greatly reduced rates of scaling in 200 h tests, but slightly thicker scales resulted when B_2O_3 vapour was present. Of particular note, however, was the fact that the presence of B_2O_3 vapour completely prevented sulphur pickup in the scale formed on the Fe-10 Cr alloy (Figure 5), and greatly reduced it on AISI 310 stainless steel.

4. MECHANISTIC STUDIES AND DISCUSSION

It is evident that two types of protective regime can be provided by exposure to boron-containing compounds; thin films, less than about 100 nm in thickness, can be stabilised by minute traces of boron, and much thicker films, often tens of microns in thickness, can have their rate of growth significantly reduced, and in both cases resistance to spalling has been increased. We will review the evidence currently available on the detailed microstructure and composition of these oxide scales, and critically consider the various hypotheses that have been proposed to account for these very remarkable effects.

4.1 Thin Films

Possibly the most definitive study on the effect of boron on the stabilisation of thin films formed on Fe-10 Cr at 600 °C was carried out by Lea.[21] The research was particularly noteworthy because of the difficulty of analysing for boron which in this case was further masked by traces of chlorine. Auger electron spectroscopy combined with argon ion sputtering was used to obtain depth concentration profiles through oxide layers 60 nm in thickness. Figure 6 shows a profile obtained after sputtering through the naturally formed oxide layer and one through a layer stabilised by the presence of boron. The presence of boron in the film caused the normally iron-rich outer layer to become greatly enriched in chromium. Boron was detected only in the outer 5-10 nm with a maximum of 6 at % at the surface. Chromium enrichment of thin films formed during the early stages in steam of a 9 Cr 1Mo steel treated with boric acid has recently been observed by Rowley and Little[16] during an examination of stripped oxide films in the analytical transmission electron microscope, but analysis for boron using characteristic X-rays was not possible. In very recent work by Brydson *et al*[22] using energy loss spectroscopy, however, the presence of boron has been confirmed. The fine structure of the boron edge yielded additional information about boron-oxygen bonding which suggested that boron was located at the trigonal interstices in an oxygen close-packed lattice.

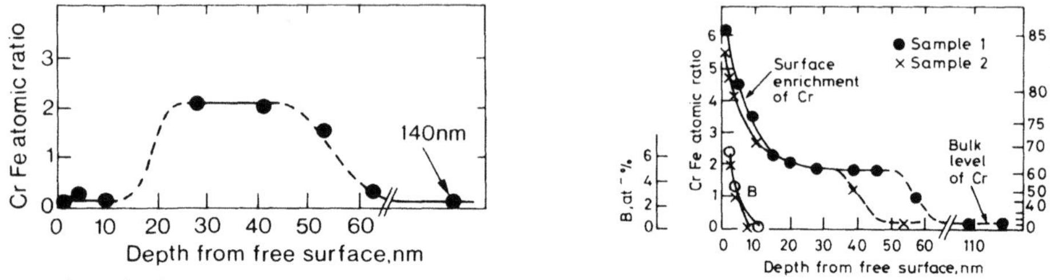

a) control

b) Air + Pyrex contamination

Figure 6

Concentration profiles through oxide formed on Fe-10 Cr after exposure at 600 ^{o}C to a) 5h in clean air and b) 28 days in air plus contamination from 'pyrex' glass.

The oxide formed on untreated Fe-10 Cr is made up of an outer layer of Fe_2O_3 and an inner chromium-containing spinel (M_3O_4). Lloyd *et al* have suggested that failure of the protective oxide results from the development of growth stresses associated with formation of the outer Fe_2O_3 layer by anion diffusion.[13] Wagner-Hauffe considerations suggest that substitution of one trivalent ion (Cr^{3+} or Fe^{3+}) by another (B^{3+}) would not be expected to alter the defect structure. However, Saunders *et al*[20] pointed out that if B^{3+}-ions were located at trigonal interstices, as would be expected from the small size of the ion, then, to maintain electrical neutrality, an equal number of trivalent iron or chromium ions sites would need to be vacated. If boron doping occurred in the outer Fe_2O_3 layer, as was apparently the case in the pyrex-contaminated oxide, then the increase in the number of iron vacancies might cause a change in the growth process from one controlled predominantly by anion diffusion to either a mixed mode or one entirely controlled by cation diffusion. In either case a reduction in oxide growth stresses would be expected, which would tend to stabilise the film. If, on the other hand, boron also doped the inner spinel layer then, as Azaroff[23] pointed out, voidal diffusion, which has a lower activation energy than the usual vacancy or interstitialcy mechanisms, would be possible. Voidal diffusion in a close-packed structure of the spinel type occurs when less than half the octahedral sites are occupied and the diffusion path is through unoccupied octahedral–tetrahedral–octahedral sites; when more than half the sites are normally occupied, vacancy diffusion occurs. In the M_3O_4 spinel structure half the octahedral sites are normally vacant, so that an increase in the octahedral site vacancy concentration caused by boron doping could lead to enhanced diffusion by the voidal mechanism.

The chromium enrichment observed in the outer layers of the pyrex-contaminated oxide may have arisen by a contribution from enhanced diffusion in the spinel layer by the voidal mechanism but since no boron was found in the spinel it seems most likely that the chromium enrichment was simply the result of more time being available to allow normal diffusion processes to occur. However, in other cases where larger amounts of boron were present, doping of the inner layer may have occurred, so that the voidal mechanism may well have been an important factor in enhancing chromium diffusion. For example, in the discussion of thick films in the next section it will be shown that a boron-containing layer occurred at the interface between an M_3O_4 and Fe_3O_4 layer.

Various other suggestions have been proposed. Hendy *et al*[3] suggested that the effect of boron was to introduce O-B-O covalent bonds into the oxide structure and thereby impede oxygen ion diffusion in Fe_2O_3 again resulting in a reduction in oxide growth stress. Some recent support for this suggestion was provided by the work of Simms[9] using XPS. He investigated the chemical shift associated with the boron peak produced in spectra obtained from thin films treated with boron-compounds. A binding energy of 192 eV was found which is in good agreement with that from a borate anion which is claimed to have a degree of covalency. Lea[21] suggested that chromium enrichment stabilised the film and that the rôle of boron was to 'extract' chromium from the inner layers to form the thermodynamically stable double oxide $CrO \cdot 2B_2O_3$. Crystal field calculations of Dunitz and Orgel[24] show that chromium in octahedral sites has a large stabilisation energy whereas iron receives no such stabilisation from occupancy of that site. Thus, according to Cox *et al*[25] chromium usually remains immobile and this accounts for the commonly observed partition of iron and chromium oxide into duplex structures. Another possibility, therefore, is that incorporation of boron into the oxide lattice might alter the crystal field stabilisation energy to allow easier movement of chromium.

4.2 Thick Films

In the context of the following discussion thick films are defined as scales consisting of an inner fine-grain equiaxed chromium-containing spinel, an intermediate pure iron oxide with a spinel structure consisting of large columnar grains and often containing pores, magnetite (Fe_3O_4), and for oxidising environments such as air or steam an outer haematite layer (Fe_2O_3). The inner chromium-containing spinel usually forms approximately within the original dimensions of the alloy with the pure iron oxides forming externally.

Formation of thick films in the presence of boron-containing compounds is characterised by four distinguishing effects:

i) a discrete-boron containing layer is frequently found at the interface between the chromium-containing spinel and the magnetite layer,

ii) chromium is present in the outer magnetite layer,

iii) a fine-grained, less porous oxide grain structure equiaxed throughout the entire scale thickness,

iv) reduced spalling of the haematite layer.

The analytical and microstructural evidence for these characteristics is given below.

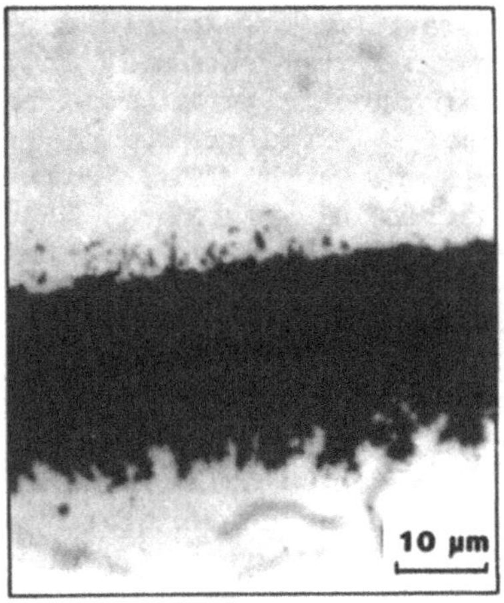

Figure 7

Optical micrograph of 2¼ Cr 1 Mo steel after oxidation in steam + B_2O_3 vapour at 600 °C for 100 h.

Optical micrographs of cross-sections through thick scales formed in air, steam or CO_2 with either vapour phase B_2O_3 present or superficially applied treatments usually show the presence of an additional dark-coloured phase at the M_3O_4/Fe_3O_4 interface. In the case of doping by B_2O_3 vapour this phase sometimes extends into the Fe_3O_4 layer as illustrated in Figure 7. X-ray diffraction analysis of scales containing this layer have produced evidence for vonsenite formation,[15,18] but Simms[9] showed that the diffraction patterns can be very complex with many extra reflections being present which suggested other boro-ferrite compounds. Simms and Little[26] also showed that the development of the triplex scale structure followed from initial formation of the boro-ferrite, and after longer oxidation time haematite was observed on top of the boron-containing layer. With continued growth magnetite nucleated in or above the boro-ferrite (and below the haematite) whilst the M_3O_4 layer nucleated below. Both the magnetite and M_3O_4 were then observed to spread laterally through the scale structure.

In addition to the somewhat indirect evidence from X-ray diffraction, the intermediate dark-coloured layer was found to contain boron using the nuclear microprobe analyser.[15] This result was confirmed by Simms and Little[26] using laser microprobe mass spectrometry (LAMMS) where a lateral resolution of about 5 μm was achieved, and Figure 8 shows an optical micrograph of the laser-treated scale and three spectra taken from regions of the scale, which identify the boron-containing oxide. Kent *et al*[19], using X-ray mapping, were able to show the presence of boron in the magnetite layer formed on 9 Cr 1Mo steel after exposure to simulated AGR atmospheres containing B_2O_3 vapour.

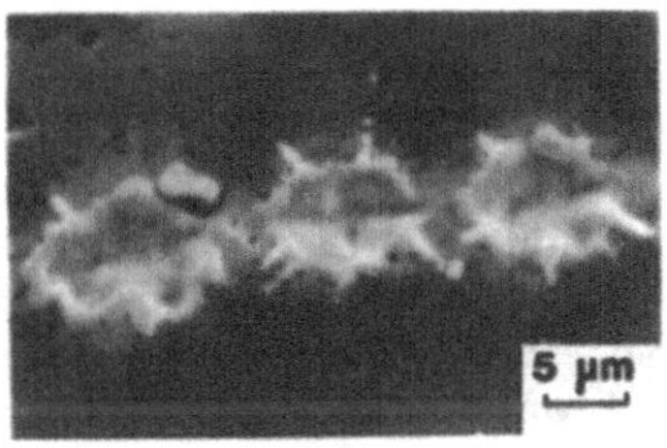

Figure 8

Scanning electron micrograph of scale formed on 2¼ Cr 1 Mo steel after superficial B_2O_3 treatment and exposure to oxygen at 600 °C for 100 h, subsequently analysed by LAMMS; and 3 associated LAMMS spectra.

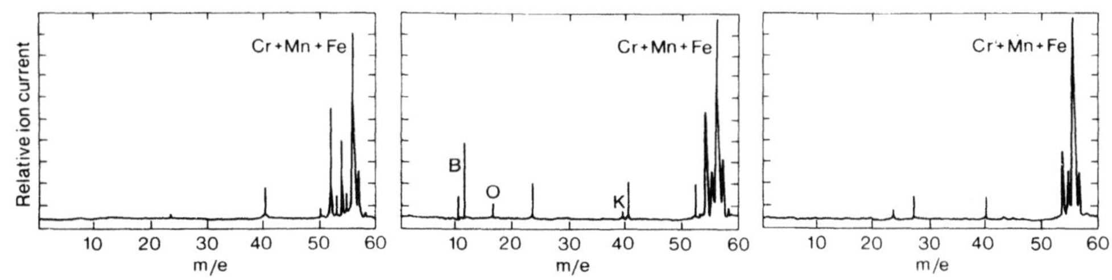

a) M_3O_4 b) Fe-B-O c) Fe_3O_4

The evidence for chromium in the outer magnetite layer is less well documented. However, Hall and Saunders[15] reported chromium in the magnetite layer, and recent work by Rowley and Little[16] showed that the chromium content of the magnetite layer could be as high as 14% in the case of a 9 Cr 1Mo steel exposed to B_2O_3-doped steam at 600 °C.

The fine-grained structure of scales grown in the presence of boron-containing compounds has already been illustrated in Figure 3 which contrasts that structure with the large columnar grains normally found in the outer magnetite layer from an untreated steel. Figure 9 shows the typical pore structure of such a layer. Oxide spallation is frequently observed with these thicker scales especially when a haematite layer is formed, and it has been reported that incidence of spallation is greatly reduced by the presence of boron compounds.[15]

Figure 9

Scanning electron micrograph of a plan view of a fracture section of the Fe_3O_4 layer on 2¼ Cr 1 Mo steel[3] after oxidation in pure steam at 600 °C for 100 h.

In order to discuss the effect of boron additions to these scales it is important to understand the development of these duplex oxide structures in boron-free conditions. Atkinson and Smart[27] have recently discussed the development of scales of this type and concluded that the inner layer forms in the interfacial voids, created by the coalescence of cation vacancies, by oxidant transport through pores in the outer columnar layer growing by outward transport of cations. The problem has been often posed as to how the pores remain open since one would expect that oxidation reaction with the incoming oxidant would seal the pores. Although much evidence for pores in the outer layer exists, e.g. Figure 9, no direct evidence of a continuous network from the gas/scale interface to the scale/alloy interface has been reported. Atkinson and Smart[27] concluded, therefore, that the pores do indeed seal by the formation of new oxide but new pores form elsewhere. The pores are believed to be the result of localised stresses in the oxide arising from oxide formation within the scale which, although generally compressive, could result in local tensile regions in neighbouring grain boundaries. Kofstad[28] suggested that since grain growth and deformation are both diffusion processes, the large columnar grains are evidence of higher growth rates and deformation of the outer layer of the scale.

Thus, the effect of boron-compounds in preventing the development of the columnar grain structure is evidence of either a reduction in growth stress in the system or the presence of impurities at grain boundaries preventing grain growth. The absence of pores in the outer layer also would indicate a less stressed oxide, and this is consistent with the general observation of a reduction in oxide spallation. It would also explain the greatly reduced carbon-uptake in CO_2-1%CO atmospheres since presumably direct access of oxidant to the alloy/scale interface is much more difficult, so that the Boudouard reaction cannot occur. A reduction in grain size of the oxide by the action of segregants or second phase formation would also allow stress relief by grain boundary sliding, so that the pores developed in the more highly stressed columnar scales would not develop. However, Kofstad[28] points out that if the relative rates of diffusional transport and grain boundary sliding varied through the thickness of the scale, porosity would be expected to develop. It has also been suggested that a glassy phase may form at the grain boundaries and further enhance stress relief by viscoelastic deformation processes.

There appears to be little direct evidence to indicate which of these two processes might operate - a reduction in stress *per se* or more efficient stress relief by grain boundary sliding. Some transmission electron microscopy on transverse sections through boron-containing scales was carried out by Simms[9] but he did not detect the presence of any segregation or second phase at the oxide grain boundaries. However, because of the difficulty of preparing suitable thin samples in cross-section, further similar work is required. Figure 10 is a micrograph of such a transverse section of oxide formed on 2¼ Cr 1Mo steel after boric acid pre-doping.

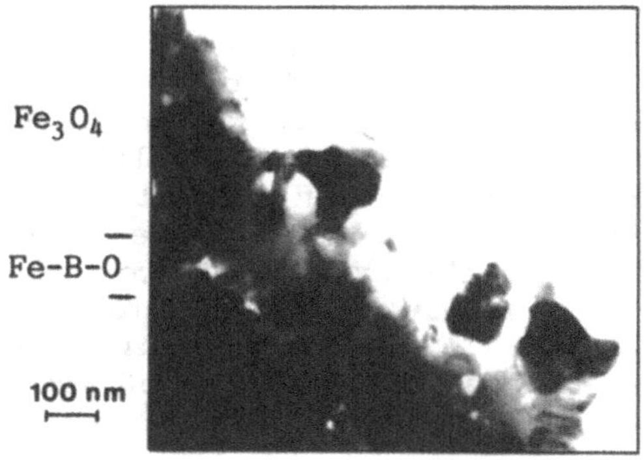

Figure 10

Transmission electron micrograph of a transverse section of oxide on 2¼Cr 1 Mo steel after superficial treatment with boric acid vapour and exposure to O_2 at 600 °C - showing the thin B-containing layer and the fine-grained Fe_3O_4.

It has also been suggested that boron compounds may operate by a pore blocking process due to formation of boro-ferrite compounds.[19] Pore blocking would also have the effect of reducing inward transport of oxidant. Kent *et al*[19] pointed out that, with a vapour phase process, the flux of boron-containing material arriving at the sample surface must be sufficient to form a boro-ferrite compound, and that this might only occur after some slowing of the rate of oxide formation had occurred.

The observation of incorporation of chromium in the outer magnetite layer may be related to a change in the transport processes in the spinel layer caused, as previously discussed, by the boron ions occupying trigonal sites in the lattice.

In the case of the sulphidising environments the oxide scale formed was predominantly Cr_2O_3, so that much of the above discussion might not apply. However, in general terms it is believed that the same considerations are relevant. There is much dispute in the literature about the ingress of sulphur through chromium oxide, but the general view is that gaseous transport occurs via pores or cracks although some workers suggest that lattice diffusion processes may also be important.[29] Certainly pores are observed in Cr_2O_3, and it is suggested that these are also the result of growth stresses.[28] Thus, if boron acted to reduce stress in some way or acted to block any pores generated, a reduction in sulphur ingress would be expected, as was observed. The details of the mechanism, however, remain unclear.

5. SUMMARY

The effects of boron doping on oxide formation processes have been shown to be very significant and long lasting for a variety of oxidising, carburising and sulphidising atmospheres. Progress in understanding the mechanism of boron inhibition is being made, but as yet no complete explanation exists. It is clear, however, that stresses in the boron-treated oxide are greatly reduced, and this may be related to an alteration in the transport processes whereby outward diffusion of chromium and iron is enhanced thus reducing the stresses resulting from new oxide formation within the existing layer generated by inward oxygen diffusion. This change in transport may be caused by boron occupying trigonal sites in the oxygen lattice which results from its small ionic radius causing an equal number of vacancies to arise in the normally occupied octahedral sites. In the case of spinels this could result in a change of mechanism from vacancy diffusion to the easier voidal diffusion process. Alternatively, stress relief processes may be facilitated by the formation of a fine-grained scale which would allow greater plastic deformation by grain boundary sliding processes. The grain refinement may result from the formation of a second phase or segregation effects at oxide grain boundaries. It has also been suggested that pore blocking processes by the formation of boro-ferrite complexes could explain some of the reported effects of boron treatments in preventing ingress of oxidants.

REFERENCES

1 S.R.J. SAUNDERS and J.R. NICHOLLS, Materials Science and Technology, to be published.

2 F.H. STOTT, F.M.F. CHONG and C.A. STIRLING, in "High Temperature Corrosion in Energy Systems", Ed. M. Rothman, Published by AIME Warrendale, 1985, 253-267.

3 P. HENDY, B. KENT, G.O. LLOYD, J.E. RHOADES-BROWN and S.R.J. SAUNDERS, Proc. of Sixth European Congress on Metallic Corrosion, Eurocorr '77, Soc. Chem. and Industry, London, 1977, 105-111.

4 N.J. SIMMS and J.A. LITTLE, J Mat. Sci. Letters, 1987, 6, 171-174.

5 G.O. LLOYD, B. KENT, S.R.J. SAUNDERS and C. LEA, Proc. Trans. R. Soc. Lond., 1980, A295, 334-335.

6 E. SCHHEIL and K. KIWIT, Arch. Eisenhüttenw., 1934, 9, 405.

7 A. GALERIE, M. CAILLET, and M. PONS, Mat. Sci. and Engineering, 1985, 69, 329-340.

8 P. ROWLEY and J.A. LITTLE, Proc. of Int. Symposium on Coatings, Newcastle, 1988, in press.

9 N. SIMMS, PhD Thesis, University of Cambridge, 1988.

10 R. ROLLS and R.D. SHAW, Corrosion Science, 1974, 14, 431 - 441 and 443 - 450.

11 M.G. SCHAUMBERG, Fr. Pat. No 1,489,122, June 1967.

12 J.M. LEES, Br. Pat. No 1,179,850, Feb 1970.

13 G.O. LLOYD, S.R.J. SAUNDERS, B. KENT and A. FURSEY, Corrosion Science, 1977, 17, 269-299.

14 J.E. RHOADES-BROWN and S.R.J. SAUNDERS, National Physical Laboratory, unpublished work.

15 D.J. HALL and S.R.J. SAUNDERS, Proc. of Eurocorr '87, Karlsruhe, 1987, Dechema, 279-284.

16 P. ROWLEY and J.A. LITTLE, Proc. of Conference Corrosion 1989, New Orleans, 1989, NACE Houston. Researches in Progress, to be published.

17 P.C. ROWLANDS, R.B. DOOLEY, M.J. HOLT and F.G. HICKS. Proc. of Conf. on Metal-Slag-Gas Reactions and Processes, Eds. Z.A. Foroulis and W.W. Smeltzer, Electrochemical Soc., 1975, 409-411.

18 R.B. DOOLEY and D.B. MEADOWCROFT, "Borate coatings for the protection of 9%Cr AGR boiler steels", CEGB Report RD/D/N199175, 1975.

19 B. KENT, D.B. MEADOWCROFT and S.R.J. SAUNDERS, "The effect of a vapour phase borate inhibitor on the oxidation of 9%Cr steels in high pressure simulated AGR gas", NPL report DMA(D)286, 1982.

20 S.R.J. SAUNDERS, B. KENT and S. WHATELY, Proc. of Eighth Int. Conf. on Metallic Corrosion, Mainz, 1981, DECHEMA, 786-791.

21 C. LEA, Metal Science, 1979, 13, 301-306.

22 R. BRYDSON, P. ROWLEY, J. LITTLE, D. HALL and S.R.J. SAUNDERS, Cambridge University/NPL to be published.

23 L.V. AZAROFF, J. of Applied Physics, 1961, 32, 1658-1662 and 1663-1665.

24 J.D. DUNITZ and L.E. ORGEL, J. Phys. Chem. Solids, 1957, 3, 20-29 and 318-323.

25 M.G. COX, B. McENANEY and V.D. SCOTT, Nature, 1972, 237, 140-142.

26 N. SIMMS and J.A. LITTLE, Proc. of Conf. Corrosion '88, St Louis, 1988, NACE Houston, paper 342.

27 A. ATKINSON and D.W. SMART, J. Electrochem. Soc., 1988, 135, 2886-2893.

28 P. KOFSTAD, Oxidation of Metals, 1985, 24, 265-275.

29 M.F. STROOSNIJDER and W.J. QUADAKKERS, High Temperature Technology, 1986, 4, 83-96.

EFFECTS OF Ta AND Y ADDITIONS ON THE OXIDATION AND HOT CORROSION BEHAVIOUR OF A γ/γ' ALLOY

M. Vilasi, J. Steinmetz, P. Steinmetz

Laboratoire de Chimie du Solide Minéral, Universite de Nancy I,
Unité associée au CNRS - UA 158
BP 239, 54506 Vandoeuvre les Nancy, Cedex, France

ABSTRACT

The results of investigations on the oxidation and hot corrosion of a γ/γ' alloy identified in a NiCoCrAlYTa coating for Nickel-base superalloys, are presented. Its composition corresponds to the formula: $Ni_{37}Co_{27}Cr_{27}Al_8Ta$.

Yttrium-containing alloys (0.1, 0.5, and 1 at%) and a tantalum-free alloy ($Ni_{37}Co_{27}Cr_{27}Al_9$) were tested in isothermal and cyclic oxidation at 850°C and 1100°C in air.

The same alloys were also tested in conditions corresponding to hot corrosion: periodic contamination with Na_2SO_4, then heating in an air + SO_2 gas mixture.

At the two temperatures studied, the γ/γ' alloys are chromia-formers. With yttrium-free alloys, at 850°C, the influence of tantalum on the oxidation kinetics is slightly beneficial while at 1100°C, the situation is inverse (tantalum additions increase the oxidation rate). In this last case, tantalum is converted into a porous and non-adherent oxide, $TaCrO_4$. The presence of Y at elevated temperature, 1100°C, dramatically improves the oxide scale adherence.

Yttrium and tantalum effects on the hot corrosion behaviour of γ/γ' chromia-forming alloys are limited. Nevertheless, yttrium is subjected to acidic fluxing and this reaction has probably a deleterious influence if the alloys are alternatively exposed to hot corrosion and high temperature oxidation.

AN ALTERNATIVE VIEW OF THE EFFECT OF ACTIVE ELEMENTS ON THE MECHANICAL PROPERTIES OF SCALES

J.R. Nicholls and P. Hancock

Cranfield Institute of Technology

ABSTRACT

Much work has been published over recent years on the role active elements play in the growth and adhesion of oxide scales. In these studies it is widely acknowledged that the presence of active elements improves the adhesion of both Cr_2O_3 and Al_2O_3 scales as is clearly evident from the improved cyclic oxidation resistance of these alloys. Many mechanisms have been suggested to account for these improvements but few of them have addressed the role active elements play in modifying the mechanical properties of oxide scales.

This paper considers the role of active elements on scale mechanical properties by assessing the condition necessary to cause scale fracture. Examples to illustrate these effects are taken from the published literature for cerium and yttrium additions to alloys forming chromia scales. Fracture mechanics techniques are used to assess the effect of active elements in modifying the fracture behaviour of the scales.

It is shown, that the distribution of defects (cracks, voids, open boundaries) is modified by the presence of the active element and that this effect is a major contribution to improved resistance to scale fracture. Reactive element additions do not modify the fracture toughness (K_{1c}) of chromia scales, nor alter scale plasticity. They do modify the scales growth mechanism and hence the growth stresses generated within these scales. A similar behaviour should be expected for alumina scales, but the difficulty in measuring defect distributions in these thin scales makes it more difficult to quantify this effect.

1. INTRODUCTION

Over the last two decades the 'active element effect' has been a subject of intense study. A recent search of the literature by the authors highlighted 260 references directly citing the role of active elements on oxide scale formation. Of these, 10 qualitatively discussed the role of active elements on oxide mechanical properties and only four were able to provide quantitative data.[1-4]

In these studies, it is widely accepted that active elements promote the formation of protective oxide scales by limiting the rate of scale growth and inhibiting scale failure mechanisms. However, the precise mechanisms by which active elements improve oxide scale adhesion is still open to discussion.

A number of models have been proposed to explain these reactive element effects. Those most widely accepted include:

1. The enhancement of scale nucleation processes. In this mechanism, heterogeneous nucleation of new oxide on reactive element oxide particles (formed as a result of internal oxidation) promotes the early formation of a stable oxide scale.[5-8]

2. Modification of the oxide growth mechanisms. Blocking of outward cation diffusion along short circuit diffusion paths, through the segregation of active elements at oxide grain boundaries has been suggested as a mechanism for chromia and alumina scales.[9-12] This cannot be considered to be universally applicable as in some studies the converse has been observed.[13-14] However, it is generally agreed that active elements promote one preferential diffusion mechanism and hence scale formation at one preferred interface.[15]

3. Reduction of compressive stresses within the oxide due to reduced lateral oxide growth.[16,17]

4. Pegging of the oxide scale to the alloy.[16-21]

5. Formation of a compound layer (graded seal) between the oxide and alloy substrate. This compound layer may well play a dual role by reducing the transport of cations outward from the substrate[22] and by providing a bonding layer that may improve scale adhesion[23].

6. The provision of vacancy sinks will limit diffusional growth of the oxide and will prevent vacancy condensation at the metal/oxide interface to form interfacial voids,[21,24–26] hence improving the scale crack resistance by minimising the size of available defects.

7. Enhanced oxide plasticity.[2,17]

8. Improved chemical bonding between the alloy and its oxide.[27,28]

9. A decrease in the segregation of tramp elements at the interface, particularly sulphur, so enhancing interface adhesion.[29–32]

For a more detailed discussion of these mechanisms reference is made to the many published reviews on this subject.[9,15,20,22,33–38]

Clearly, the effects of active element additions may be classified into three main groups, namely, those

a) promoting the preferential oxidation of an element that forms a stable oxide of low diffusivity, i.e. Cr_2O_3 or Al_2O_3 on the majority of alloys (mechanism 1),

b) reducing the rate of scale growth (mechanisms 2,5,6), and

c) inhibiting scale failure due to either through thickness cracking or scale/substrate interfacial decohesion (mechanisms 3-9).

The effect of active elements on selective oxidation and scale growth are discussed elsewhere[36–38], hence this paper will concentrate on the role active elements play in limiting scale failure.

A method of quantifying the effect of active elements on mechanical properties is presented. A fracture mechanics approach[39] is used to quantify the defect distribution present in the scales. Using this approach many of the effects cited in the literature can be unified into a common theory reflecting the different roles that active elements play in modifying the conditions necessary to cause oxide scale fracture.

2. A MECHANISM OF OXIDE FAILURE

Oxide scales at temperature are invariably stressed. For isothermal exposure, these stress may be generated internally as a result of the oxidation process[2,3,15,37,39], may originate from the geometry of the component[40,41] or be externally applied[41-43]. Additional stresses may also be induced by changes in temperature[43-47] hence, scale failure is most often observed during cooling from service temperatures.

Scale failure results when these 'in situ' stresses exceed some critical value, which can be evaluated using simple fracture mechanics considerations[39,44-46] whether through thickness cracking (tensile failure) or scale spallation (compressive failure) results. Linear Elastic Fracture Mechanics (LEFM) predicts that the stress to cause failure in tension (σ_c) is given by

$$\sigma_c = K_{1c} \cdot (\pi\bar{a})^{-½} \qquad (1a)$$

where K_{1c} is the 'critical stress intensity factor' (fracture toughness) and $\bar{a}$ is the half length of an 'equivalent through thickness crack' in the oxide scale.[39] Hence scale failure is determined by the scale fracture toughness and the distribution of defects present in the scale.

Alternatively equation (1a) can be rewritten in terms of a critical strain (ε_c) within the scale and the energy to form two new oxide surfaces ($2\gamma_o$), namely,

$$\varepsilon_c = (2\gamma_o/\pi E_o\bar{a}) \qquad (1b)$$

where E_o is the oxide elastic modulus and $K_{1c} = (2E_o\gamma_o)^{½}$. In this form the equation is similar to that cited by Manning.[44] Differences relate to the value of the constant and the parameter taken as a measure of flaw size.

This simple analysis suggests that active element additions will be beneficial if they

(i) reduce the level of stress generated within the scale,
(ii) increase the scale fracture toughness (critical stress intensity factor), or

(iii) reduce the equivalent through thickness crack length in the scale for tensile conditions or the equivalent interfacial (or near interfacial) crack length to improve spalling resistance under compressive loading.

Of these factors, quantification of the 'equivalent crack length' (defect size) is probably the most difficult. Levels of stress within the scale can either be measured directly[2,3,15,37] or calculated.[44–46] Fracture toughness data can similarly be calculated directly for the oxide,[39] or estimated from data for bulk oxides[48]. However, until recently[39] no method existed for quantifying the distribution of defects present in the scale.

It has been acknowledged that both scale cracking and spallation vary with scale thickness, with thinner scales providing more resistance to damage. Measurements of the strain to cause through thickness cracks[44,49–51] or spallation[44,46,52,53] have confirmed this, as has the improved resistance of thin oxide scales to erosion damage.[54,55] Manning and co-workers[44,53] assumed a flaw size of 0.2 d (where d is the scale thickness) to calculate the critical strain level to cause scale cracking under tensile load. Under compressive loading, their equation defining the critical strain for spall formation implies that the flaw size is 0.64 d.

2.1. Estimation of the Composite Defect Size in Oxide Scales

In a recent paper by the authors[39] on the application of fracture mechanics to the failure of surface oxides scales, a method for assessing the composite defect size was proposed.

Consider, by way of example, the tensile failure of an oxide scale under a superimposed stress as this is the simpliest geometry to analyse. Linear elastic fracture mechanics (LEFM) requires that the defect be an ideal through thickness crack, as illustrated schematically in Figure 1a. However in real scales such simple defects do not occur and all inherent defects in the scale are invariably partial defects. Figure 1b illustrates schematically a scale containing partial defects.[39]

To analyse such defects, corrections have to be made both for the defect shape and its position within the oxide scale. These correction parameters are known as the 'shape parameter' and the 'membrane parameter'. They are available as nomograms for assessing defects in welds.[56] The authors have shown that these correction factors can equally be applied to assessing defects in oxide scales.[39]

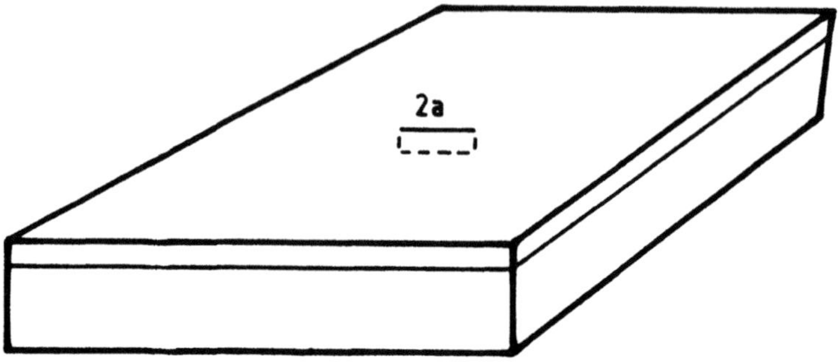

Figure 1a: An idealised surface defect

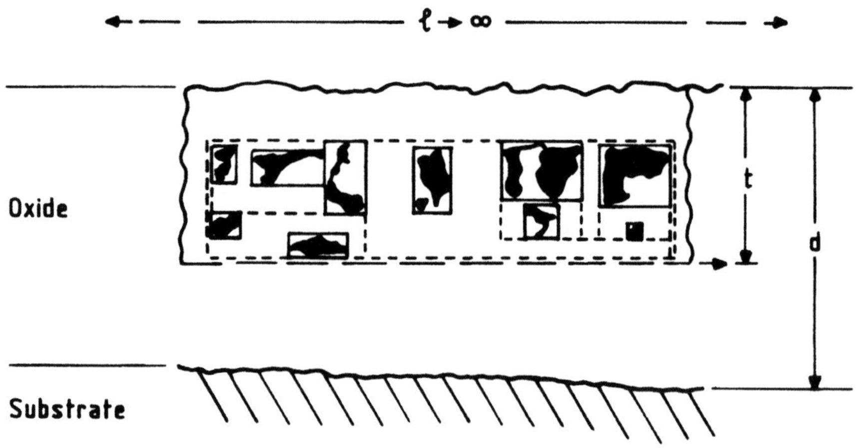

Figure 1b: Method of converting actual defects into an idealised defect.

Analysis proceeds by drawing rectangular boxes around each defect, which in turn defines the length and width of the defect. Interaction rules[39,56] are then applied to neighbouring defects to determine if they should be considered singly or as a composite defect. Broadly these rules state that defects must be separated by a distance greater than the sum of the half lengths or greater than the sum of the half widths if the defects are to be considered in isolation. This interaction analysis proceeds iteratively until all defect interactions have been assessed. Using this procedure a 'composite equivalent defect' of length 'l' and thickness 't' can be defined. For example, in Figure 1b, the defects present result in a composite defect enscribed by the large rectangle.

For many oxide scales, interacting defects exist throughout the scale, hence defects in oxides can be considered to be of finite depth (t) but infinite length.[39]

The last stage in this analysis is to calculate the half length of 'equivalent through thickness crack' $\bar{a}$ (equation 1a) from the calculated thickness of composite defect (t).

For a surface composite defect (i.e. defects within the scale sufficiently close to the oxide/air or metal/oxide interface as to be considered to interact with either surface)

$$\bar{a} = \frac{t \cdot M_m^2}{\pi \quad Q_o^2} \qquad (2)$$

when stressed in tension with no bending component present.[39] In this equation M_m is the membrane correction factor and Q_o is the defect shape parameter. Both parameters vary depending on the length and thickness of the composite defect, and are published as nomograms in PD6493.[56]

Figure 2 illustrates the dependence of $\bar{a}/d$ on t/d for an infinitely long surface defect of thickness t in an oxide scale of thickness d. The plotted points are calculated using PD6493,[56] while the least squares polynominal regression curve is plotted as a solid line. Over the range $0.1 \leq t/d \leq 0.9$ this regression curve has the form,

$$\frac{\bar{a}}{d} = 1.28 \left[\frac{t}{d}\right] + 13 \left[\frac{t}{d}\right]^3 + 360 \left[\frac{t}{d}\right]^7 \qquad (3)$$

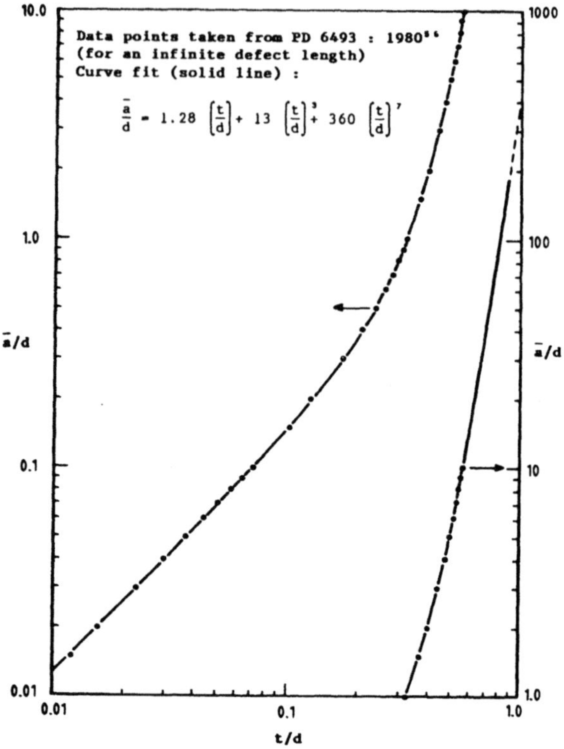

Figure 2: Relationship between idealised defect size $\bar{a}$ and thickness t.

For buried defects the equivalent through thickness crack length is less than for surface defects, which acknowledges the fact that surface defects are more damaging than buried defects. To estimate $\bar{a}$ for buried defects, a surface defect with half the depth of the buried defect should be used. Finally, for stressing conditions involving bending within the oxide, a bending correction factor has to be applied. References 39 and 56 cover the analysis of this more complex stress state.

Following detailed metallography of scales on EN2 steel oxidized at 700°C,[39] on 304 stainless steel at 900°C[61] and on Nimonic 75 at 950°C, $\bar{a}$ is typically 1600 μm for a 200 μm thick FeO/Fe_3O_4 scale on EN2 steel (t/d = 0.55), 66 μm for a 1.8 μm Cr_2O_3 scale on 304 stainless steel (t/d = 0.71) and 132 μm for a 6.6 μm Cr_2O_3 scale on Nimonic 75 (t/d = 0.80).

3. THE EFFECT OF ACTIVE ELEMENTS ON SCALE PROPERTIES

Having derived a method for evaluating the equivalent through thickness crack length, this approach can be applied to examine the influence of active elements on scale properties. From simple Linear Elastic Fracture Mechanics considerations active elements would be beneficial if they a) reduce the equivalent through thickness crack length, b) increase the scale fracture toughness or c) reduce the level of stress generated within the scale.

The influence of active elements on each of these scale properties is discussed below.

3.1. Equivalent Through Thickness Crack Length

To assess the effect of active elements on the equivalent through thickness crack length ($2\bar{a}$) two case studies are examined. Firstly, the effect of small Ce additions on the oxidation of Fe-Cr alloys at 1000°C,[7,57] and secondly the effect of ion implantation of Ce and Y on the scaling behaviour of a Fe-20Cr-25Ni-Nb steel in the temperature range 900-1000°C.[58]

In the first example, the effect of a 0.1wt% Ce addition to Fe-Cr alloys, with chromium contents in the range 10-20wt%, is considered. These alloys have been oxidised for 70 hours,[7,57] at 1000°C in a He - 0.13 bar O_2 environment. Figure 3 illustrates the scale morphologies observed for the Fe-20%Cr alloy with and without 0.1% Ce additions.[7] Clearly, the scale formed on the cerium-containing alloy is far more compact, more uniform and contains fewer gross defects than found on the alloy without the active element addition.

Measured scale thicknesses and maximum depths of defective oxides for alloys containing between 1-20wt% Cr, without or with 0.1wt% Ce additions are summarised in Table 1. The data was measured from micrographs published by Rhys-Jones et al.[7,57] of which Figure 3 is an example.

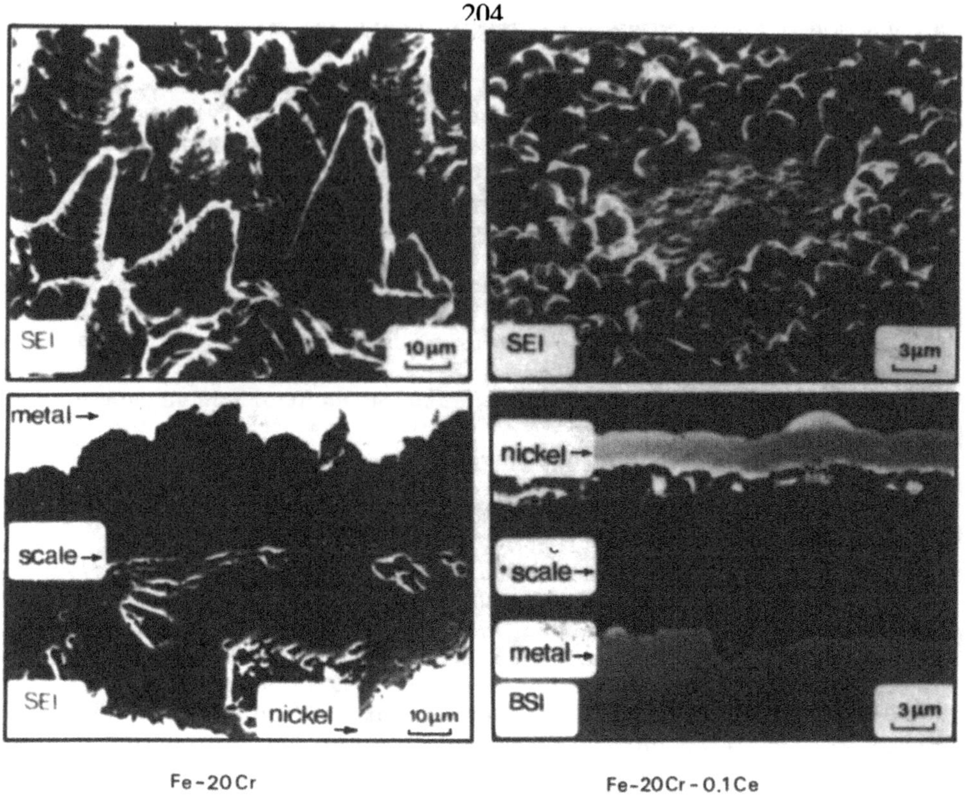

Figure 3: Fe-20Cr oxidised at 1000°C with and without Ce additions (after T. Rhys-Jones et al.[7,57]).

Table 1

Measured oxide thicknesses, composite defect depth and equivalent through thickness defect size $\bar{a}$ for Fe-Cr alloys exposed in a He-0.13 bar O_2 environment at 1000°C

a) 70 hours at 1000°C

Cr %	Ce %	$\bar{d}$ μm	s_d μm	d_{max} μm	t_{max} μm	$\bar{a}$ μm
10	0	146.5	23.6	183	183	∞
	0.1	81.0	19.0	113	113	∞
12	0	237.0	33.2	296	296	∞
	0.1	71.0	10.6	88	88	∞
14	0	130.4	24.5	190	190	∞
	0.1	86.9	22.4	128	128	∞
16	0	53.6	5.2	64	56	9 900
	0.1	11.3	2.7	16.0	8.0	80
20	0	63.2	8.2	76	58	4 850
	0.1	10.0	0.9	11.8	2.8	5.9

b) 400 hours at 1000°C

Cr %	Ce %	$\bar{d}$ μm	s_d μm	d_{max} μm	t_{max} μm	$\bar{a}$ μm
12	0	112.5	30.5	168	168	∞
	0.1	59.6	25.1	122	117	25 400
14	0	107.0	21.4	149	139	35 800
	0.1	58.0	10.0	78	51	1 700

In this table $\bar{d}$ is the mean oxide thickness, s_d the standard deviation, d_{max} is the maximum oxide thickness and t_{max} is the maximum thickness of defective oxide as assessed using the rules outlined in Section 2.

Clearly the addition of 0.1wt% Ce reduces the rate of scale formation for all the alloys examined. This behaviour is consistent with the many papers reporting that active elements modify oxide growth processes by blocking short circuit diffusion paths[9-15] and/or providing vacancy sinks[21,24-26]. Both of these mechanisms will reduce the rate of scale formation.

After 70 hour exposure, oxide scales formed on alloys containing 15% Cr or less can be considered to be cracked through. Many scales contained through thickness cracks while the remainder contained sufficient interacting defects as to be considered as through thickness cracks. After 400 hours protective oxides were found on the 12% Cr and 14% Cr alloys containing 0.1wt% Ce and also on the 14% Cr alloy without cerium. Scale thicknesses were always significantly reduced with the cerium additions, even after this extended exposure time.

Short duration (less than 10 minutes) exposure studies by Rhys-Jones et al.[57] have shown that at least 16wt% Cr is necessary for the early formation of Cr_2O_3 scales. Furthermore, when cerium is present this addition has been shown to favour the formation of Cr_2O_3 scales and to suppress the formation of Fe-oxides.[57]

Where protective oxides are formed, the equivalent through thickness crack length ($2\bar{a}$) is found to depend on both chromium content and the presence of cerium additions. This is best illustrated in Figure 4, which plots $\bar{a}$ as a function of the chromium content of the alloys. The beneficial effect of a 0.1wt% Ce addition in reducing the effective defect size is clearly evident. For example, a Fe-15wt% Cr-0.1wt% Ce alloy would be expected to be more resistant under cyclic oxidation than an Fe-22% Cr alloy. It forms a more perfect oxide scale. For alloys forming a protective oxide, a defect-free region was observed at the metal/oxide interface. This thin interfacial layer was between 5-15 μm thick and depended on the chromium and cerium content. Based on the observations of Rhys-Jones et al.[57] this layer is thought to be a continuous Cr_2O_3 subscale. No continuous subscale was observed for the alloy containing 10wt% Cr. The cross-hatched areas in Figure 4, at low chromium contents, indicate regions where the formation of a defect-free region within the scale is time dependent.

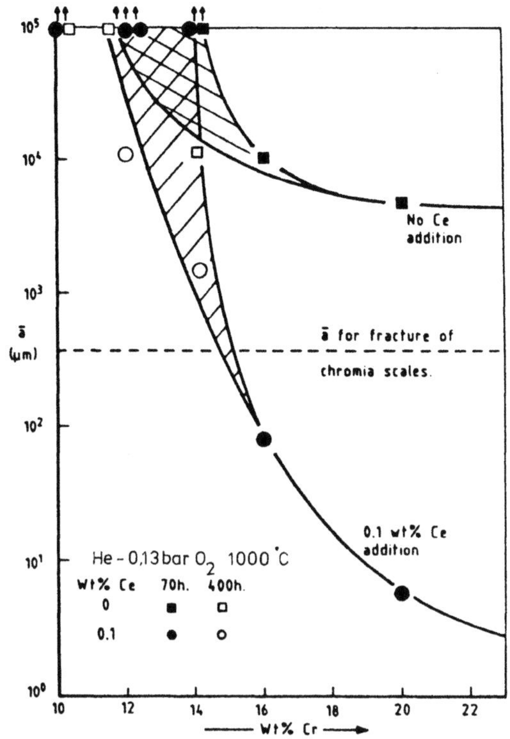

Figure 4: Values of effective defect length $\bar{a}$ present in iron-chromium alloys with and without additions.

From this example, it is clearly evident that small active element additions promote the formation of a more perfect oxide scale. These observation give strong support to the vacancy condensation mechanisms[21,24-26] that have been proposed to account for the beneficial effects of active elements. The observed synergism between the chromium content of the alloy and the presence of the active element is thought to be due to the active element addition promoting the formation of a protective Cr_2O_3 scale by enhancing scale nucleation processes. Such behaviour is consistent with observations in the literature[5-8] and with the direct observation by Rhys-Jones et al.[57] of a protective Cr_2O_3 scale on the high Cr, Ce-containing alloys after only 2 minutes oxidation.

As a second example, micrographs of scales on an Fe-20Cr-25Ni-Nb stainless steel oxidised for 250 hours at temperatures of 950, 1000 and 1050°C in carbon dioxide[58] were analysed using the method outlined in Section 2. Accurate measurement of the defect distribution in scales formed at 900°C proved impossible because of the limited oxide thickness produced at this temperature, particularly for the Y-implanted alloy.

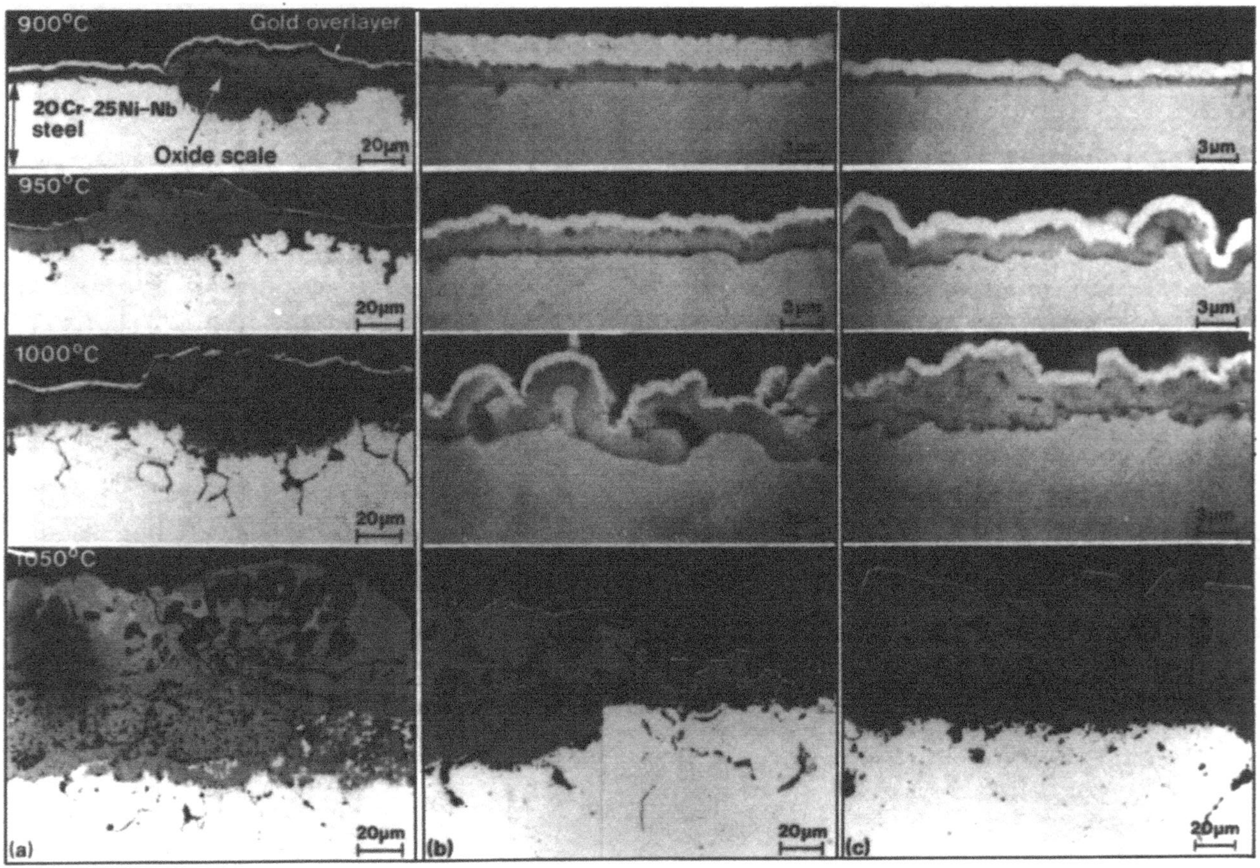

Figure 5: Influence of Y and Ce implantation on the oxidation behaviour of Fe-20Cr-25Ni-Nb alloys (a) without addition (b) with Ce implantation (c) with Y implantation (after Bennett et al.[58]).

Figure 5 reproduces the micrographs of Bennett et al.[58] and clearly shows the benefit of ion implantation with active elements in modifying the scale morphology.

In this study either 10^{17} ions cm ^{2} of cerium (Fig. 5b) or yttrium (Fig. 5c) were implanted. From comparison with the unimplanted reference alloy (Fig. 5a) it is evident that both implanted species improve the oxidation resistance at the lower temperatures (900-1000°C). The effect of implantation is much less marked at the highest temperature (1050°C). Visually it is difficult to assess which is the more effective addition, cerium or yttrium, although measurements of the mass of spalled oxide plus the mass gain of each sample[58] would suggest that at 900 and 950°C the yttrium-implanted alloy is better than the cerium-implanted alloy; while at 1000°C the roles would appear to be reversed.

Figure 6 represents the distribution of oxide thicknesses measured from one set of these micrographs, 250 hours exposure in CO_2 at 1000°C. Probability plotting for corrosion scales is indicative of the protective nature of the scale formed. For protective scales a single normal (or lognormal) distribution is observed,[59] while a bimodal or multimodal distribution is indicative of breakaway oxidation or contaminant corrosion, i.e. a non-protective scale. From Figure 6 it is evident that the total scale thickness of the unimplanted alloy is bimodal indicating that breakaway oxidation has occurred for this alloy.

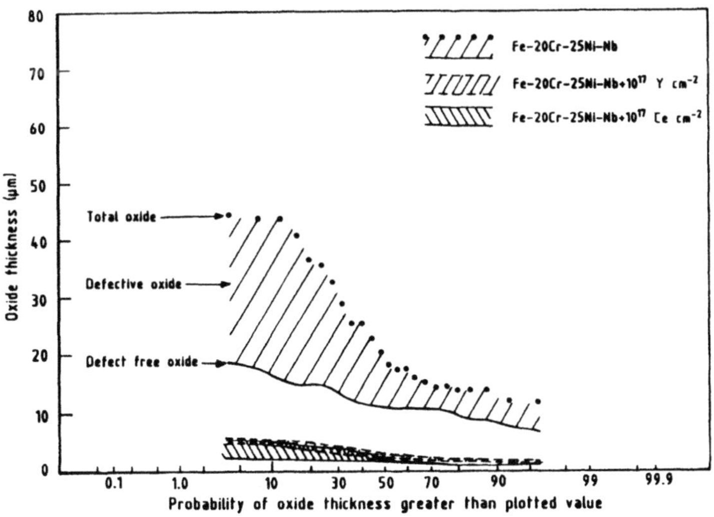

Figure 6: Distribution of oxide thickness on samples oxidising for 250 hours in CO_2 at 1000°C.

The fact that the region of defect-free oxide of the unimplanted alloy is approximately normal would further suggest that at the time of measurement this region of the scale was reasonably protective. Similarly, both Ce- and Y-implanted scales can be considered protective, with scaling rates an order of magnitude lower than for the unimplanted alloy, as evident from data presented in Table 2. The morphology of the scales formed on the Ce-implanted alloy suggests that although it oxidises at similar rates as the Y-containing alloy, a larger fraction of its oxide scale contains defects.

Table 2

Measured thickness (μm) on ion implanted Fe-20Cr-25Ni-Nb stainless steel oxidised for 250 hours in CO_2.

a) Total Oxide

	No implantation		10^{17}ion cm^{-2} Ce		10^{17}ion cm^{-2} Y	
	$\bar{d}$	s_d	$\bar{d}$	s_d	$\bar{d}$	s_d
950°C	16.4	8.8	1.62	0.36	1.84	0.74
1000°C	24.1	11.3	2.68	1.32	3.18	1.28
1050°C	87.4	9.9	42.6	26.4	63.6	6.2

b) Defect Free Oxide

	No implantation		10^{17}ion cm^{-2} Ce		10^{17}ion cm^{-2} Y	
	$\bar{d}$	s_d	$\bar{d}$	s_d	$\bar{d}$	s_d
950°C	11.75	6.13	1.62	0.36	1.84	0.74
1000°C	11.73	2.88	1.59	0.31	2.52	1.08
1050°C	8.64	2.79	17.14	9.46	22.33	4.86

$\bar{d}$ is the mean scale thickness and s_d the standard deviation in scale thickness readings. From these oxide thickness measurements, $\bar{a}$ can be calculated. In Table 3 $\bar{a}$ has been calculated for the worst case defect, i.e. t in equations (2) and (3) has been taken as a maximum

Table 3

Equivalent through thickness defect size $\bar{a}$ (μm) in oxides on ion implanted Fe-20Cr-25Ni-Nb stainless steel after 250 hours oxidation in CO_2.

	No implantation	10^{17}ion cm^{-2} Ce	10^{17}ion cm^{-2} Y
950°C	4 415	6.8	2.6
1000°C	5 770	508	397
1050°C	30 750	22 400	5 816

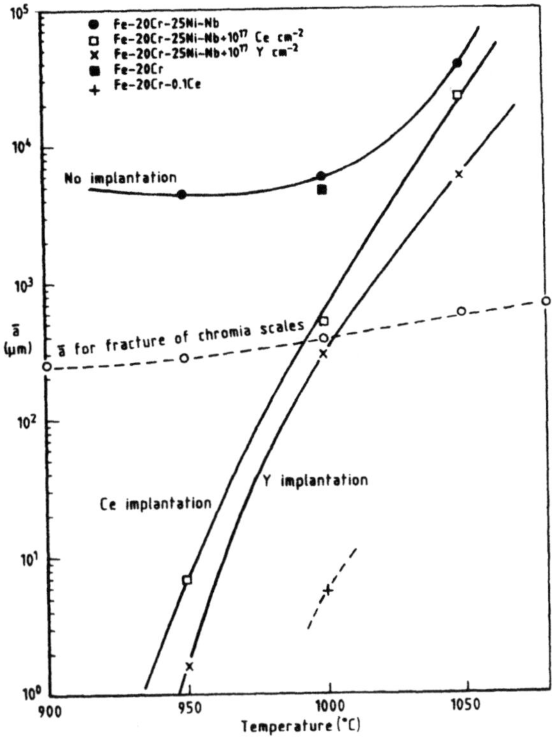

Figure 7: The effect of oxidation temperature on the effective defect length of surface oxides.

Figure 7 plots the value of $\bar{a}$ versus the oxidation temperature for the unimplanted and both Ce- and Y-implanted alloys. The benefit of active element additions in reducing the defective nature of the oxide scales is clearly evident, particularly at temperatures below 1000°C. At 950°C, the improvement amounts to a reduction in the effective defect size by three orders of magnitude, with the yttrium-implanted alloy being marginally better at all test temperatures.

Superimposed on Figure 7 is the data measured for an Fe-20wt% Cr alloy with and without a 0.1wt% Ce addition, taken from Table 1. The good agreement between the equivalent through thickness defect sizes calculated from these two studies[7,57,58] for a Fe-20wt% Cr alloy without active element additions implies that the defect nature of the scale does not depend on the environment, provided a stable Cr_2O_3 subscale can form. As both environments result in similar scale morphologies, it is possible to compare the effects of ion implantation with the direct addition of active elements to the alloy. From Figure 7 it can be seen that small Ce additions are more effective than ion implantation in reducing the equivalent through thickness defect size ($\bar{a}$). A reduction in defect size by some two orders of magnitude is observed when 0.1wt% Ce is added to the alloy, compared to that produced by ion implantation using a 10^{17} ion cm^{-2} dose rate.

Bennett et al.[58] concluded from SIMS studies through the oxides formed on these ion implanted alloys that the active elements were incorporated within the outer spinel layer, and at the interface with the inner chromia scale. This incorporation of active elements in the outer spinel results in active element segregation at grain boundaries,[60] reducing cation transport through these boundaries, which therefore inhibits scale growth.

In summary, these studies, have shown that active elements modify the equivalent through thickness defect size for chromia, and chromium-rich scales by a combination of three mechanisms:

a) The proportion of gross defects (voids, pores etc.) are reduced due to active element oxides within the scale acting as vacancy sinks. The provision of vacancy sinks also limits the diffusional growth of these scales.
b) Active element additions have been shown to encourage the early formation of a protective chromia subscale,[57] and
c) active elements limit oxide growth mechanisms by blocking outward cation short circuit diffusion paths in these scales.[50]

Reduced scale thicknesses have a direct effect on the equivalent through thickness defect size ($\bar{a}$).

Having demonstrated that active element additions modify the defect distribution within the scale, one must ask, 'What is the critical value of equivalent through thickness crack length at which scale failure will occur?' To determine this value one must examine how active elements modify both the fracture toughness of the scale, and the stress generated within the scale.

3.2. Scale Fracture Toughness

Fracture toughness data for oxide scales can be measured directly[39] using methods similar to those reviewed in Section 2 or can be estimated from measurements on bulk oxides.[48]

Little data is available in the literature as to the effect active element additions have on the fracture toughness of bulk oxides and no data is available for oxide scales. Where data is available it relates to the fracture toughness of ZrO_2 and Y_2O_3-

stabilised ZrO_2,[40] which indicates that Y_2O_3 marginally increases the fracture toughness of ZrO_2 from 7.5 MPa $m^{1/2}$ to 10 Mpa $m^{1/2}$ due to phase transformation toughening mechanisms within the ceramic. Protective scales contain predominantly chromia or alumina oxides. It is relevant to determine whether active element additions can be expected to modify the fracture toughness of these oxides.

Fracture toughness (K_{1c}) data for chromia scales has been determined by the current authors[39] from measured fracture stresses for Cr_2O_3 scales formed on Nimonic 75, following a detailed assessment of the equivalent through thickness defect size as outlined in Section 2. From linear elastic fracture mechanics, scale failure occurs when $K_1 > K_{1c}$ where $K_1 = \sigma_T(\pi \bar{a})^{1/2}$ and σ_T is the in-situ tensile stress on the scale. Values of K_{1c} for Cr_2O_3 scales determined using this approach are plotted as a function of temperature in Figure 8.[39] This figure shows that, as the temperature increases, the fracture toughness of the scale increases. However this effect is small until such temperatures where scale plasticity is observed (i.e. > 700°C).

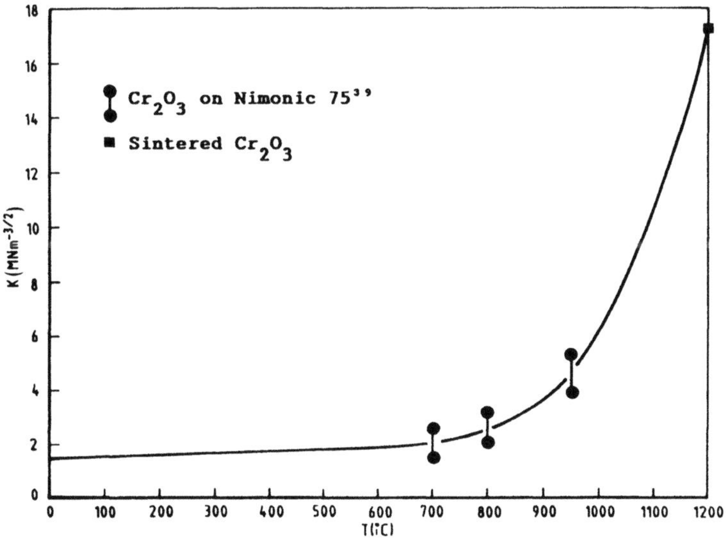

Figure 8: The influence of temperature on the stress intesity factor of chromia scales.

Following the original ideas of Griffiths,[61] the fracture toughness K_{1c} is a function of the strain energy necessary to form two new surfaces, hence,

$$K^2 = 2E\gamma \qquad (5)$$

where γ is the surface energy required to form a newly extended crack. In other words, ΔK under elastic fracture conditions is only related to interatomic bonding within the bulk of the oxide (or across the oxide/substrate interface, should scale fracture occur along this interface). As small active element additions have only a minor influence on elastic properties,[62] the fracture toughness of the scale would not be expected to change with active element additions at temperatures below 700°C.

At temperatures above 700°C, significant scale plasticity can occur and this is reflected in the increases in K_{1c} for Cr_2O_3 scales (Fig. 8). Under these conditions equation (5) must be modified to take into account the plastic work of fracture (p), hence,

$$K^2 = (2\gamma + p) \cdot E = G \cdot E, \qquad (6)$$

where G is the strain energy release rate for the material, which is dependent on the plasticity of the oxide scale. Hence, at temperatures where significant plasticity can occur (> 700°C) the active elements could modify the scale fracture toughness if their addition increased scale plasticity.

There is no conclusive evidence that scale plasticity is increased through the addition of active elements.[9] Measurements on sintered polycrystalline chromia containing Y_2O_3 additions have also demonstrated that little or no increase in scale plasticity is observed for a 0.09wt% Y_2O_3 addition when tested at 1200°C with a strain rate of 5×10^{-6} sec^{-1} (Fig. 9).[63] As a result of this study, King et al.[63] concluded that chromia scales offer limited plasticity at high temperatures and that the presence of yttrium in the oxide does not enhance plasticity compared with the addition-free oxide.

Based on these limited studies, one must conclude that active elements do not alter the fracture toughness (K_{1c}) of chromia scales significantly. Hence, references in the literature to active elements producing an increase in fracture toughness[22,58] should more correctly refer to active elements reducing the inherent defect size present in the oxide scales.

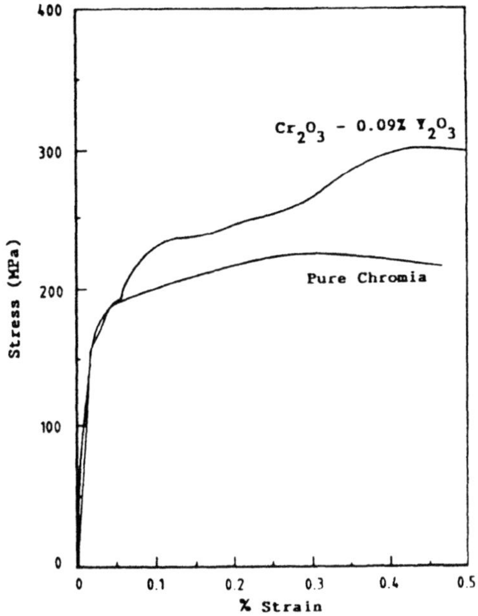

Figure 9: The tensile behaviour of chromia scales at 1200°C (after King et al.[63]).

3.3. <u>The Stress to cause Scale Fracture</u>

Stress generation within oxide scales has been the subject of a number of recent review papers[3,15,39,44] and therefore will not be discussed in detail here, unless directly relevant to the role of active elements. Suffice to say that these stresses may be generated as a result of the mode of oxide growth, as a result of geometric effects, or may be externally applied either by direct loading or thermal cycling.

Scale fracture results when the summation of stresses generated within the scale exceeds some critical value. This critical value may be directly measured or can be predicted by as discussed in Section 2. This section examines the direct measurement of oxide fracture stresses, as this permits a critical defect size for oxide fracture to be evaluated. In addition, the role of active elements in modifying growth stresses is also reviewed as this will influence the in-situ stress necessary to cause oxide failure.

Measurements of the tensile stress to cause oxide fracture have been examined by the authors for Cr_2O_3 scales on Nimonic 75.[39] Results from this study are plotted in Figure 10, for temperatures between 700-950°C. Superimposed on this figure are the fracture stresses measured by King et al.[63] for sintered chromia and for yttrium-modified chromia.

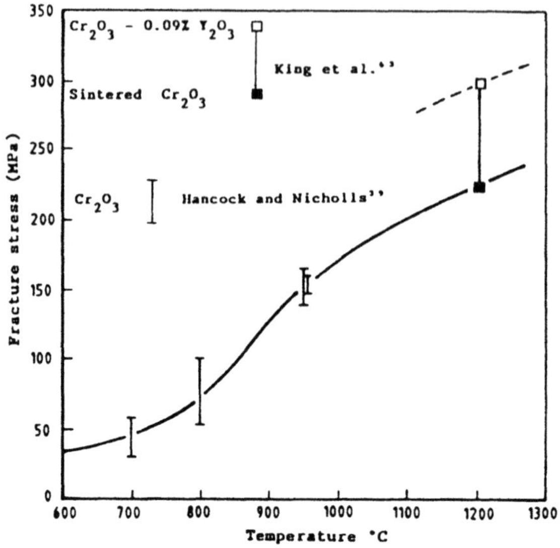

Figure 10: Influence of temperature on the tensile stress to fracture of chromia scales.

Both the proof stress and fracture stress of the sintered Cr_2O_3 are increased with the 0.09% Y_2O_3 addition by 11 and 33%, respectively. Good agreement exists between the extrapolated data of Hancock and Nicholls[39] and the data determined by King et al.[63] Hence by assuming that the fracture stress for Cr_2O_3 scales at 1200°C will be similar to values determined by King et al. on sintered bulk Cr_2O_3, the measured fracture toughness data for Cr_2O_3 scales[39] can be extended to 1200°C, see Figure 8.

Using this fracture toughness data and measured fracture stresses for chromia scales (Fig. 10), values for the critical defect size to cause scale failure can be calculated for chromia scales formed at temperatures other than those reported in reference 39. These calculated equivalent through thickness crack lengths ($\bar{a}$) are plotted in Figures 4 and 7 as broken lines. Clearly, scales with a value greater than that necessary to cause scale fracture will be prone to spallation and through thickness cracking when moderately stressed, while scales with $\bar{a}$-values well below this critical level will exhibit good adhesion and high strain tolerance. As $\bar{a}$ depends on the defective nature of the scale and on scale thickness, scales may initially be protective (when thin) but as they thicken with time, $\bar{a}$ will increase until some critical level is reached where scale failure occurs due to the level of internal stress generated within the scale. This condition corresponds to the onset of breakaway oxidation.

For the text conditions examined in Figures 4 and 7, the alloys without active element additions would be highly susceptible to breakaway oxidation, and this is clearly evident from the micrographs in Figs. 3 and 5. Even when modified by active element additions, low chromium (below 15wt%) alloys, are still susceptible to breakaway oxidation. Long term exposure of active element implanted alloys above 1000°C would equally be expected to be problematic, and this was observed with yttrium-implanted Fe-20Cr-25Ni-Nb showing early signs of breakaway oxidation after 250 hours exposure in CO_2 at 1000°C.[58] At 1050°C, none of the alloys examined by Bennet et al.[59] remained protective, which lead the authors to conclude that implantation had no significant beneficial influence on the oxidation behaviour of stainless steel above 1050°C.

As stated earlier, the stresses generated in oxide scales may result from direct loading, thermal loading, geometric effects or from the mode of oxide growth. Active element additions are not expected to modify the stresses generated by either direct loading or thermal cycling as any changes in the elastic modulus or expansion coefficient of the oxide by the active element addition will be very small. Geometric effects are in the main related to the mode of oxide growth[39-44], and it is in this area of modifying the mode of oxide growth that active element have been reported to play a major role.[9,10,16,17]

Direct measurement of oxide growth stress has been undertaken by Huntz and co-workers[2,3,13,15,37] using a bending strip technique in which only one side of the specimen is permitted to oxidise, by coating the second one. Figure 11 illustrates two examples, the first for an Fe-45Ni-25Cr alloy and the second for a Ni-34Cr alloy with and without yttrium implantation. Both have been oxidised at 1000°C. Scales were duplex with an inner Cr_2O_3 layer and an outer scale that was either $Ni(Fe,Cr)_2O_4$ or NiO + $NiCr_2O_4$, depending on the alloy tested.[2]

Scales formed on the yttrium-implanted alloys were generally thinner, as can be seen in Figure 11. The compressive stresses generated in the oxide reach a maximum during early scale formation on both alloys, with values between 60-70 MPa for the unimplanted alloys. These internal stresses reduce to more stable values of 15-20 MPa for the FeNiCr alloy and 44-48 MPa for the NiCr alloy.

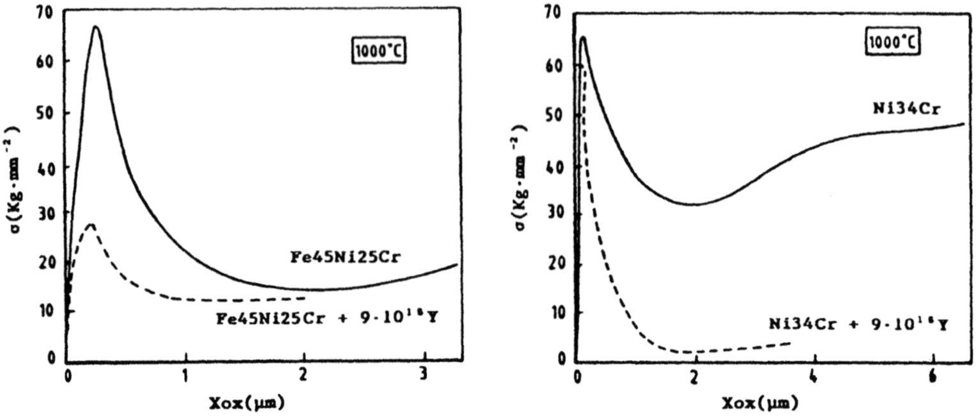

Figure 11: Oxide growth stress as a function of oxide thickness and Y implantation (after Huntz et al.[2])

The implantation of yttrium (9×10^{15} ion cm^{-2}) reduces this initial peak stress for the FeNiCr alloy (Fig. 11a) although after prolonged oxidation a stable residual compressive stress of 12-14 MPa was again observed. For the NiCr alloy (fig. 11b) the initial peak stress level changes little, although with growing oxide thickness the stable residual stress level is significantly reduced (2.5-4 MPa).

The implications of these studies to scale fracture are:

i) Alloys with yttrium implantation would be more resistant to spalling than alloys without this addition. This results from the fact that both the growth stress and the thermally induced stresses are compressive, increasing the driving force for spalling.

ii) The most resistant alloy to spalling is expected to be the yttrium-implanted Ni-34Cr alloy as this generates the lowest compressive stress.

iii) The high levels of compressive stress measured during early oxide formation are unlikely to cause oxide failure as the inherent defect size in the oxide scales is expected to be small under these conditions, as it is directly related to oxide thickness.

4. CONCLUSIONS

In a recent review of scale growth and adhesion, Stott and Wood[35] concluded that the controversy regarding the role played by active elements would continue until a theory was produced that could explain all the observations relating to the beneficial effects of active elements on scale growth, and more particularly scale adhesion.

This paper offers an alternative view of the role active elements play in modifying scale properties. Through the use of a fracture mechanics approach the effect of active element additions on scale fracture is quantified and hence a link between scale growth, scale morphology and adhesion is established. It has been shown that:

1. The beneficial effect of active elements in modifying scale mechanical properties is directly related to the role these elements play in reducing gross defects such as voids, crack, pores, etc. present within the scale.

2. The defective nature of oxide scales may be quantified by calculating the "Equivalent through thickness crack length". This parameter is related to both the thickness of the scale and the proportion of defective oxide within a scale. Therefore, it incorporates the beneficial effects of active elements in modifying both the growth rate and morphology of the oxide scales.

3. Active elements do not alter the scale fracture toughness or plasticity. Hence, references in the literature to active elements producing an increase in fracture toughness or plasticity should more correctly refer to active elements reducing the inherent defect size present in the scale.

4. Active element additions reduce the compressive growth stresses within the scale.

[N.B. A peak compressive stress is observed when scales are thin, whether active elements are present or not. This peak compressive stress is however not expected to cause spallation for scales containing active elements as the equivalent through thickness crack length is small when scales are thin.]

5. Scale failure results when the applied stress exceeds a critical value. This critical stress level can be calculated by applying Linear Elastic Fracture Mechanics concepts to the scale.

Finally, this paper has concentrated on the role active elements play in modifying the mechanical properties of chromia scales. Similar behaviour is expected for alumina scales, but difficulties in measuring defect distributions in these thin scales makes this more difficult to quantify.

ACKNOWLEDGEMENTS

The authors wish to express their thanks to T. Rhys-Jones and M. Bennett for providing micrographs for this analysis and for permission to reproduce them in this paper.

REFERENCES

1. F.A. GOLIGHTLY, F.H. STOTT and G.C. WOOD, Werkst. und Korros., 1979, 30, 847-491.
2. D. DELAUNAY, A.M. HUNTZ and P. LACOMBE, Corros. Sci., 1980, 20, 1109-1117.
3. A.M. HUNTZ, Mater. Sci. and Tech., 1988, 4, 1079-1088.
4. A.M. HUNTZ, G. MOULIN and G.B. ABDERRAZIK, cited in reference 15.
5. J. STRINGER, B.A. WILCOX and R.I. JAFFEE, Oxid. Met., 1972, 5, 11-47.
6. J. STRINGER, A.Z. HED, G. WALLWORK and B.A. WILCOX, Corros. Sci., 1972, 12, 625-636.
7. T. RHYS-JONES, H.J. GRABKE and H. KUDIELKA, Corros. Sci. 1987, 27, 49-73.
8. P.Y. HOU and J. STRINGER, Oxid. Met., 1988, 29, 45-73.
9. D.P. WHITTLE and J. STRINGER, Phil. Trans. Royal Soc. (London), 1980, A295, 309-329.
10. T.A. RAMANARAYNAN, R. AYER, R. PETKOVIC-LUTON and D.P. LETA, Oxid. Met., 1988, 29, 445-472.
11. F.A. GOLIGHTLY, F.H. STOTT and G.C. WOOD, J. Electrochem. Soc., 1979, 126, 1035.
12. J. JEDLINSKI and S. MROWEC, Mater. Sci. Eng., 1987, 87, 281-287.
13. G.B. ABDERRAZIK, F. MILLOT, G. MOULIN and A.M. HUNTZ, J. Am. Ceramc. Soc., 1985, 68, 302-307.
14. E.W.A. YOUNG and J.H.W. DE WIT, Oxid. Met., 1986, 26, 351.
15. A.M. HUNTZ, Mat. Sci. Eng., 1987, 87, 251-260.
16. F. GOLIGHTLY, F.H. STOTT and G.C. WOOD, Oxid. Met., 1976, 10, 163-187.
17. T.A. RAMANARAYNAN, M. RAGHAVAN and R. PETKOVIC-LUTON, Oxid. Met., 1984, 22, 83-100.
18. C.S. GIGGINS, B.H. KEAR, F.S. PETTIT, and J.K. TIEN, Met. Trans., 1974, 5, 1685-1688.
19. I.M. ALLAM, D.P. WHITTLE and J. STRINGER, Oxid. Met., 1978, 12, 35-66.
20. H. HINDAM and D.P. WHITTLE, J. Electrochem. Soc., 1982, 129, 1147-1149.
21. Y. SAITO and T. MARUYAMI, Mat. Sci. Eng., 1987, 87, 275-80.
22. D.P. Moon, UKAEA Report no. AERE R 12930, 1988.
23. H. PFEIFFER, Werkst. und Korros., 1957, 8, 574-579.
24. J.K. TIEN and F.S. PETTIT, Met. Trans., 1972, 3, 1587-1599.
25. C.H. YANG, G.E. WELSCH and T.E. MITCHELL, Mater. Sci. Eng., 1985, 69, 351-357.
26. J.D. KUNEZLY and D.L. DOUGLASS, Oxid. Met., 1974, 8, 139-179.
27. G.C. WOOD, Werkst. und Korros., 1971, 22, 491-503.
28. A.B. ANDERSON, S.P. MEHANDRU, J.L. SMIALEK, J. Electrochem. Soc., 1985, 132, 1695-1701.

29. A.W. FUNKENBUSCH, J.G. SMEGGIL and N.S. BORNSTEIN, Met. Trans., 1985, 16a, 1164-1166.
30. J.G. SMEGGIL, A.W. FUNKENBUSCH and N.S. BORNSTEIN, Met. Trans., 1986 17a, 923-938.
31. J.G. SMEGGIL, Mater. Sci. Eng., 1987, 87, 261-265.
32. J.G. SMEGGIL, at this meeting (1989).
33. M.J. BENNETT, in: Proc. Conf. 'High Temperature Corrosion' (ed. R.A. RAPP) NACE 6, 1983, 145-154.
34. G.C. WOOD and F.H. STOTT, ibid 1983, 227-250.
35. G.H. STOTT and G.C. WOOD, Mater. Sci. Eng., 1987, 87, 267-274.
36. G. BERANGER, F. ARMANET and M. LAMBERTIN, at this meeting (1989).
37. A.M. HUNTZ, at this meeting (1989).
38. M.J. BENNETT and D.P. MOON, at this meeting (1989).
39. P.J. HANCOCK and J.R. NICHOLLS, Mat. Sci. and Tech., 1988, 4, 398-406.
40. P. HANCOCK and R.C. HURST, in: 'Advances in Corrosion Science and Technology' (eds. R.W. STAEHLE and M.G. FONTANA) 1974, pp. 1-84, Plenum Press, N.Y.
41. M. SCHÜTZE, Mater. Sci. and Tech., 1988, 4 (5), 407-414.
42. M.I. MANNING in: 'Corrosion and Mechanical Stress at High Temperatures' (ed. V. GUTTMANN and M. MERZ), 1980, pp. 323-338, Applied Science Publ., London,
43. 'Stress Effects and the Oxidation of Metals', Ed. J.V. CATHCART, AIME, 1974.
44. M.I. MANNING, in: 'Aufbau von Oxidschichten auf Hochtemperatur-Werkstoffen und ihre technische Bedeutung' (ed. A. RAHMEL) 1983, pp. 283-297, Oberursel, Deutsche Gesellschaft für Metallkunde.
45. A.G. EVANS, G.B. GRUMLEY and R.E. DEMARAY, Oxid. Met., 1983, 20, 193-216.
46. H.E. EVANS and R.C. LOBB, Corros. Sci., 1984, 24, 209-222.
47. W. CHRISTH, A. RAHMEL and M. SCHÜTZE, Mater. Sci. Eng., 1987, 87, 289-293.
48. P. HANCOCK, 1985, MRS-Europe, 163-179.
49. M. SCHÜTZE, Oxid. Met., 1985, 24, 199-132.
50. M. SCHÜTZE, Oxid. Met., 1986, 25, 409-422.
51. J.R. NICHOLLS and K. MAHMOOD, to be published in Mater. Sci. and Tech., 1989.
52. H.E. EVANS and R.C. LOBB, in: Proc. 9th International Congress on Metallic Corrosion, 1984, Vol. 2, P-46-53, Toronto, Canada.
53. J. ARMITT, D.R. HOLMES, M.I. MANNING, D.B. MEADOWCROFT and E. METCALFE, EPRI Report FP-686, 1978.
54. D.J. STEPHENSON, J.R. NICHOLLS and P. HANCOCK, Corros. Sci., 1985, 25, 1181-1192.
55. D.J. STEPHENSON, J.R. NICHOLLS and P. HANCOCK, Wear, 1986, 111, 31-39.

56. PD6493 - British Standards Institution, London.
57. T. RHYS-JONES, H.J. GRABKE and H. KUDIELKA, Werkst. und Korros., 1987, 38, 62-72.
58. M.J. BENNETT, H.E. BISHOP, P.R. CHALKER and A.T. TUSON, Mater. Sci. Eng., 1987, 90, 117-190.
59. J.R. NICHOLLS and P. HANCOCK, in: Proc. Conf. 'High Temp. Corrosion' (ed. R.A. RAPP) NACE-6, 1983, pp. 198-210.
60. M.J. BENNETT, B.A. BELLAMY, C.F. KNIGHTS, N. MEADOWS and N.J. EYRE, Mater. Sci. Eng., 1985, 69, 359-373.
61. A.A. GRIFFITHS, Phil. Trans. Roy. Soc., 1921, A221, 163.
62. J. NOROWIK, Oxid. Met., 1982, 18, 1-17.
63. W.E. KING, J.H. PARK, J.L. ROUTBORT and K.C. GORETTA, Oxid. Met., 1988, 29, 217-223.

EFFECT OF RARE EARTH METALS ON OXIDATION RESISTANCE AND CREEP STRENGTH

Sven Darnfors and Rune Johansson

Research and Development

Avesta AB, 774 01 Avesta, Sweden

ABSTRACT

Addition of Rare Earth Elements (REM) to Fe-base 20Cr 10Ni-steel has increased the oxidation resistance and the creep strength. The influence of manganese content on the oxidation resistance of this type of steel is also investigated.

1. INTRODUCTION

The use of active elements in steel processing is very promising. AVESTA uses rare earth metals (REM) in high temperature steels which have been commercially available since 1973. The development of these steels started within AVESTA in the late sixties. Information in the literature suggested that alloying nickel-base alloys with REM improved the oxidation properties of resistance wire. When Wenderott[1] published the evidence of the improvement achieved by REM on the oxidation resistance of a powder material, the interest of using REM in heat-resistant austenitic stainless steels further increased. The argument put forward was that REM contained in the matrix as fine dispersed oxides might strengthen the matrix and thereby increase the creep strength.

2. EXPERIMENTS

The first oxidation tests were performed on 30x60 mm specimens of the various alloys which were cycled 9 times from the exposure temperature (1050-1200°C) to room temperature during the 72 hours holdtime in a furnace. The weight loss was calculated from weighing the specimens before and after pickling in Hooker-salt.

Later intermittent oxidation tests were performed using a thermal balance where the weight increase was recorded during 45 hours. During the test the specimen was withdrawn 5 times from the furnace and cooled to room temperature for a duration of 10 minutes.

The creep tests have been carried out in standard creep testing machines at temperatures between 600 and 1000°C.

The tested steels had chemical compositions according to Table 1. REM addition to the heats was done in the form of Mischmetal, containing cerium, lanthanum and minor amounts of other lanthanides; only the content of cerium has been analysed.

Table 1: Chemical composition (in w%) of alloys used in this investigation; Fe balance.

	C	Si	Mn	Cr	Ni	N	Ce
Alloy 1	0.10	1.7	1.5	20	10	0.12	--
Alloy 2	0.10	1.7	1.5	20	10	0.12	0.04
Alloy 3	0.10	1.7	0.5	20	10	0.12	0.04
V 133	0.06	1.8	0.6	21.4	10.1	0.15	--
V 134	0.08	1.7	0.6	21.2	10.2	0.13	0.07
153 MA	0.05	1.3	0.6	18.5	9.5	0.15	0.05
253 MA	0.09	1.7	0.6	21	11.0	0.17	0.05

3. RESULTS

Figure 1 shows the amount of weight loss for three Fe20Cr10 Ni-steels after oxidation tests and pickling. These first results can be summarized as:

- the heat without REM addition had a high scaling rate already at 1050°C,
- addition of REM to the base composition lowered the scaling rate drastically,
- the manganese content had a marked influence on the oxidation resistance.

Figure 2 gives some preliminary results from creep testing of two heats (V 133 & V 134). The testing of these first experimental heats showed that the heat with the REM addition had a higher creep resistance than the melt without REM addition.

Today AVESTA has two austenitic heat-resistant steels with REM additions on the market. These steels, 153 MA and 253 MA, combine high oxidation resistance and good creep strength, as follows from Figs. 3 and 4 comparing their properties to those of other commercial steels.

REFERENCE

1. WENDEROTT, Z Metallkunde, 1965, 56, 2.

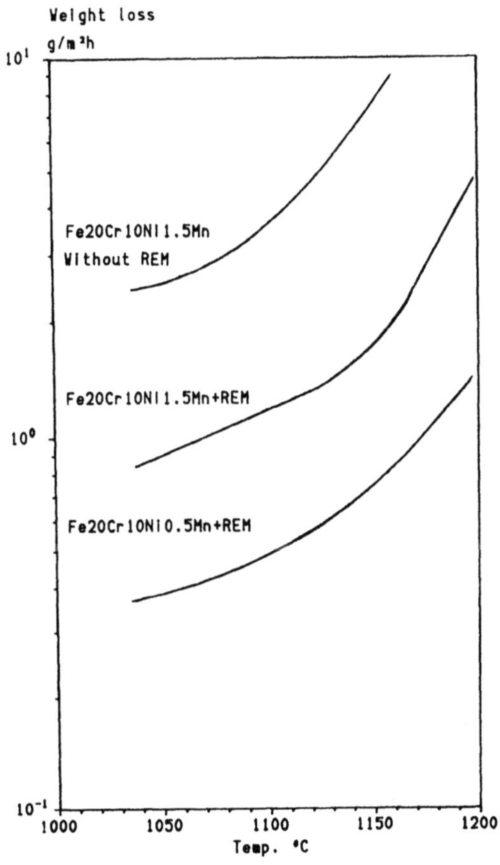

Figure 1: Weight loss after intermittent oxidation during 72 hours and pickling in Hooker-salt. Fe20Cr10Ni-steels (Alloy 1-3).

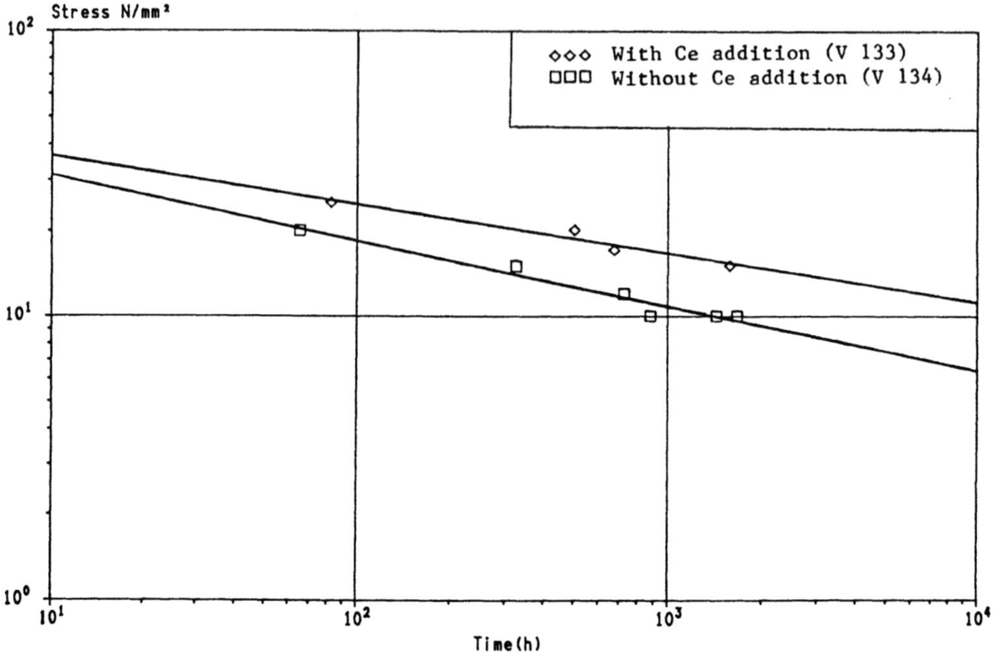

Figure 2: Creep strength for 1 percent creep strain at 900°C. Fe20CR10Ni-steels (Alloys V 133, V 134), heats with and without addition of REM.

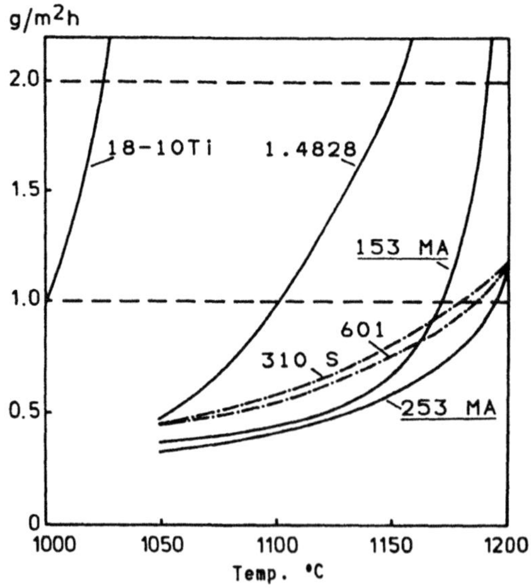

Figure 3: Weight increase for AVESTA 153 MA and 253 MA as compared to other standard steels. Cyclic oxidation tests in a thermal balance for 45 hours.

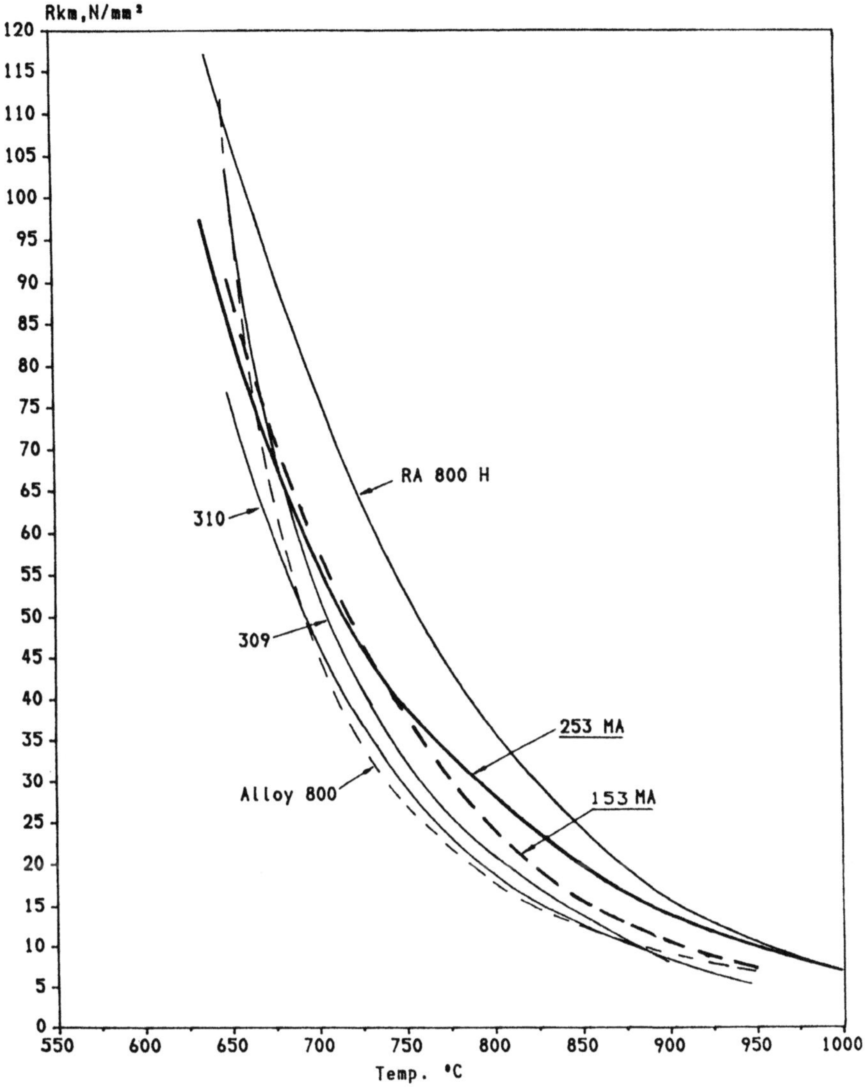

Figure 4: Creep rupture strength for 10.000 hours. AVESTA 153 MA and 253 MA are compared to other standard steels.

THE EFFECTS OF PRECIOUS METALS ON THE OXIDATION AND HOT CORROSION OF COATINGS

by

J. Schaeffer, G. M. Kim*, G. H. Meier and F. S. Pettit
Materials Science and Engineering Department
University of Pittsburgh
Pittsburgh, PA 15261

* Current Address: Materials Engineering Department
Chungnam National University
Daejon, Korea

ABSTRACT

Conventional and platinum modified diffusion aluminide coatings on nickel base superalloys have been compared in tests designed to establish conditions representative of those existing in gas turbines operating over a range of applications. The resistances of these coatings to oxidation, high temperature hot corrosion, and low temperature hot corrosion have been compared. Platinum has been found to significantly improve the resistance of diffusion aluminides to all these forms of degradation. Substrate composition is also a factor affecting coating lives in oxidation attack and high temperature hot corrosion. Platinum improves diffusion aluminide coating performances by causing such coatings to more effectively utilize aluminum in the selective oxidation process.

1. INTRODUCTION

Coatings that provide protection for high temperature applications can be formed by a variety of techniques with chemical vapor deposition (CVD)[1], physical vapor deposition (PVD)[2,3] and plasma spraying being the most commonly used. Pack cementation is a common CVD process used to produce diffusion aluminide coatings.[4,5] In this process a part is immersed in a pack mix and heated in a retort purged with an inert gas. The pack mix includes an aluminum source, an inert filler, and an activator to accelerate Al transfer from the source to the substrate. The final microstructure depends on the temperature of the pack, the pack mix, and any pretreatment of the part to be coated. Several descriptions of the pack process and resulting microstructures are available.[6-10]

Aluminide coatings are degraded via reaction with the environment (corrosion) and by interdiffusion with the substrate. Both processes result in depletion of aluminum from the coating. A significant advance in the technology of diffusion aluminides was made when other elements, such as Cr, Ta, Si, and most notably Pt, were incorporated into aluminide coatings.[11,12] The platinum modification to straight or normal diffusion aluminides resulted in a much more degradation resistant coating.[12-19] Platinum modified aluminide coatings are used commercially in the hot section of turbine engines.[18]

Various processing schemes have been developed to add platinum to diffusion aluminide coatings which produce coatings with different microstructures. In most of these processes, such as electroplating, platinum is formed initially as a thin layer ($\approx$ 3-10μm). The platinum aluminides formed using high activity packs involve aluminum diffusion through the platinum layer. If the Pt content of the layer is high enough at the surface a continuous $PtAl_2$ phase will form. Lower Pt contents may cause NiAl(β) precipitates to form in $PtAl_2$ or vice versa. When the Al reaches areas in the Pt layer where the Pt content is lower than that needed to form $PtAl_2$, a hyperstoichiometric NiAl (β) phase forms. Al is mobile in this phase until the Al content decreases to where Ni atoms become the mobile species. The resultant structure is similar to that for a straight aluminide except for of zones containing the $PtAl_2$. Some inclusions of alumina may also be present in this zone as a result of the grit blasting of the substrate before Pt electroplating and are excellent markers of the original surface. The platinum aluminide structures formed with a low activity pack evolve in a different manner. The Al still moves through the Pt layer eventually forming a hypostoichiometric NiAl phase in regions where the Pt content is lower. Nickel atoms are the only mobile species in this phase. The continued growth of the coating is mainly by Ni movement outward which results in a two zone structure. The first zone at the surface of the coating contains $PtAl_2$ precipitates in hyperstoichiometric NiAl. The second zone is a Ni-rich NiAl that forms via interdiffusion.

Platinum improves the lives of aluminide coatings. However, coatings can be degraded by different processes such as oxidation, high temperature hot corrosion and low temperature hot corrosion. This paper is concerned with defining the types of degradation for which platinum is an effective means to extend coatings lives. In addition, the mechanisms by which platinum produces such beneficial effects will be described.

2. EXPERIMENTAL

The alloy substrates investigated were Mar M200, IN738, Rene 80, PWA 1480 and CMSX-3. The nominal compositions of these alloys are presented in Table I. PWA 1480 and CMSX-3 are single crystal superalloys and the others are polycrystalline alloys frequently used in the hot section of gas turbines. Two state-of-the-art pack cementation coating processes were used for comparison, namely PWA 73 and CoDep. The platinum modified coatings were prepared by Turbine Components Inc.. Coupons of the alloys (12 x 10 x 1.5mm) were grit blasted with alumina particles prior to the deposition of a 3-7μm electroplated Pt layer. To deposit the platinum layer a thin platinum wire was welded to each coupon to provide electrical contact with the bus line. After electroplating, the coupons were exposed to an out-of-the-pack aluminizing process. This process produced an aluminide coating of the discontinuous $PtAl_2$ type. A limited number of Pt aluminide coatings prepared by Howmet were also studied.

Table I

Compositions of Alloy Substrates

Sample	Ni	Cr	Al	Ti	Co	Ta	Mo	W	Hf	B	Zr	C	Nb
IN738	Bal.	16	3.4	3.4	1.7	1.7	1.7	2.6	-	.01	-	.17	.9
MAR M200	Bal.	9	5	2	10	-	-	12	-	.015	.05	.15	1
RENE 80	Bal.	14	3	5	9.5	-	4	4	*	.015	.32	.17	-
CMSX-3	Bal.	7.7	5.5	.9	4.9	5.8	.6	7.8	.1	-	-	-	-
PWA 1480	Bal.	9.8	5.1	1.5	4.7	11.9	-	4	-	-	-	-	-

*EDS indicated the presence of Hafnium.

Cast specimens of Ni-32.3Al-2Pt(wt %) and Ni-32.2Al were also prepared and exposed in some tests to compare with the coated specimens. These alloys were single phase β alloys. The coated specimens and the cast specimens were exposed to different tests to observe and compare the effects of platinum.

Isothermal and cyclic oxidation tests were conducted at various temperatures (1100 - 1200°C) in air. The isothermal oxidation tests showed the platinum effect on oxide growth. The cyclic oxidation tests imposed thermal cycles on the growing scales. The thermal cycles caused cracking and spalling of oxide scales resulting in premature failures. Cyclic oxidation tests with careful microstructural observations can yield information on oxide scale formation and adherence. These tests were designed to show how Pt affects these processes.

Acoustic emission tests were performed on various samples at 1100°C in air. The test measures elastic waves that are emitted when an oxide cracks or spalls. This measurement requires the use of a wave guide that transmits the elastic energy from the hot zone of a furnace to a transducer that processes the signal. The acoustic emission test assesses the amount of scale damage that occurs when an oxide grows isothermally or when thermal stresses are imposed on the scale.

Diffusion anneals were carried out at 1200°C in low oxygen environments to observe diffusional interactions of a coating-substrate system without rapid oxidation. The samples were sectioned periodically to obtain metallographic specimens. These tests demonstrate how Pt affects the diffusional stability of coatings.

High temperature hot corrosion tests were performed using the cyclic oxidation procedure but with 1 mg/cm^2 of Na_2SO_4 applied once every 20 hours. The specimens were cycled from 1000°C to room temperature once every hour. Every 20 hours the specimens were water washed, visually examined and weighed. A fresh Na_2SO_4 deposit was then applied and the test continued. These tests were performed in air and produced degradation microstructures typical of those for high temperature hot corrosion.

Low temperature hot corrosion must have SO_3 present in the gas phase and to apply deposits which are liquid at the test temperature of 700°C. Low temperature hot corrosion tests were performed by applying deposits of Na_2SO_4-5C mole percent $NiSO_4$ and using gases with an initial room temperature composition of O_2 with 100 ppm SO_2. These gas mixtures were passed over a platinum catalyst at 700°C which established a SO_3 pressure of about 10^{-4} atm. These conditions produce degradation microstructures typical of low temperature hot corrosion. It is important to emphasize that the amount of attack of these coatings at 700°C without hot corrosion attack was extremely small and usually not detectable by the current state-of-the-art microbalances. Hence, the observation of weight changes of 0.2 mg/cm^2 or more indicated that hot corrosion attack was occurring. As in other tests, it is imperative that results from these tests be corroborated by metallographic examination of polished cross-sections of the exposed specimens.

Oxidation morphologies were evaluated by optical and scanning electron microscopy(SEM) and phase identification was performed using x-ray diffraction(XRD) using Cu-Kα radiation. Cross-sections were mounted using standard procedures, polished to 1μm, and etched.

3. RESULTS AND DISCUSSION

3.1 As-Coated Microstructures

Micrographs showing the microstructures of typical as-processed coatings are presented in Figures 1-4. The light discontinuous phase evident at the surface of the coating in Figures 3 and 4(TR,BR) is $PtAl_2$ which lies in a matrix of (Pt,Ni) Al. Carbides exist in the inner zone. Alumina particles from the grit blasting are evident in the platinum coating seen in Figure 4. These particles are in the inner zone and are from 1-4 um in diameter. Alpha-refractory metal precipitates can be observed beneath the $PtAl_2$ layer. The PWA 73 coating had a three zone structure (Figure 2) and the CoDep exhibited only a two zone structure.

3.2 Isothermal Oxidation

Isothermal oxidation tests were performed on several of the coating systems at 1100°C in air. The platinum modified coatings usually had smaller weight increases than the conventional aluminide coatings, however, some of the conventional aluminides also exhibited very small weight increases. This test was further complicated by some weight losses that may have occurred on the platinum modified samples due to PtO_2 vaporization from the platinum tabs used to support the coupons for electroplating.

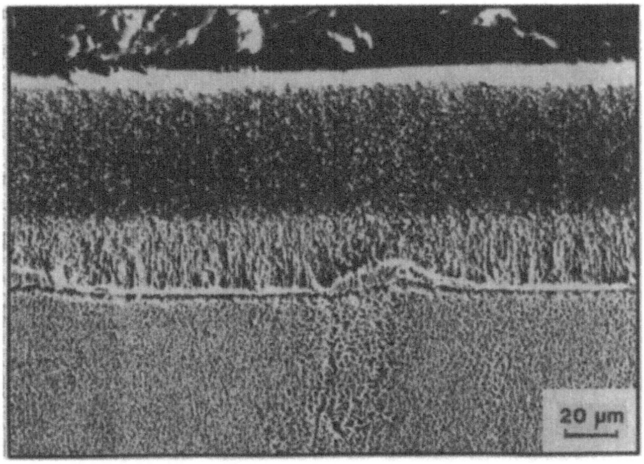

Figure 1. SEM micrographs showing the diffusion aluminide coating produced by the CoDep coating process on PWA 1480.

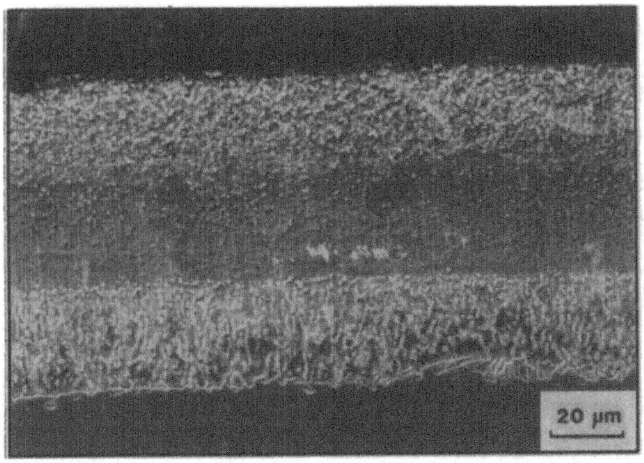

Figure 2. SEM micrograph of the diffusion aluminide coating produced by the PWA 73 process on PWA 1480.

The surface of the platinum aluminide on MAR M200 is shown in Figure 5. Visually the sample exhibited only slight discoloration. The microstructure in Figure 5 shows that the scale has a web structure typical of pure alumina scales. The EDS analysis of the ridge indicated that it was essentially pure alumina. The EDS analysis of the area between ridges showed Al along with Ni, Cr, Co, Ti, and Pt. The elements other than Al are thought to result from interaction of the electron probe with the underlying substrate. This results from the thinness of the scale in this region. The scale showed no signs of spallation.

Figure 3. SEM micrograph of Pt modified aluminide coating process on PWA 1480.

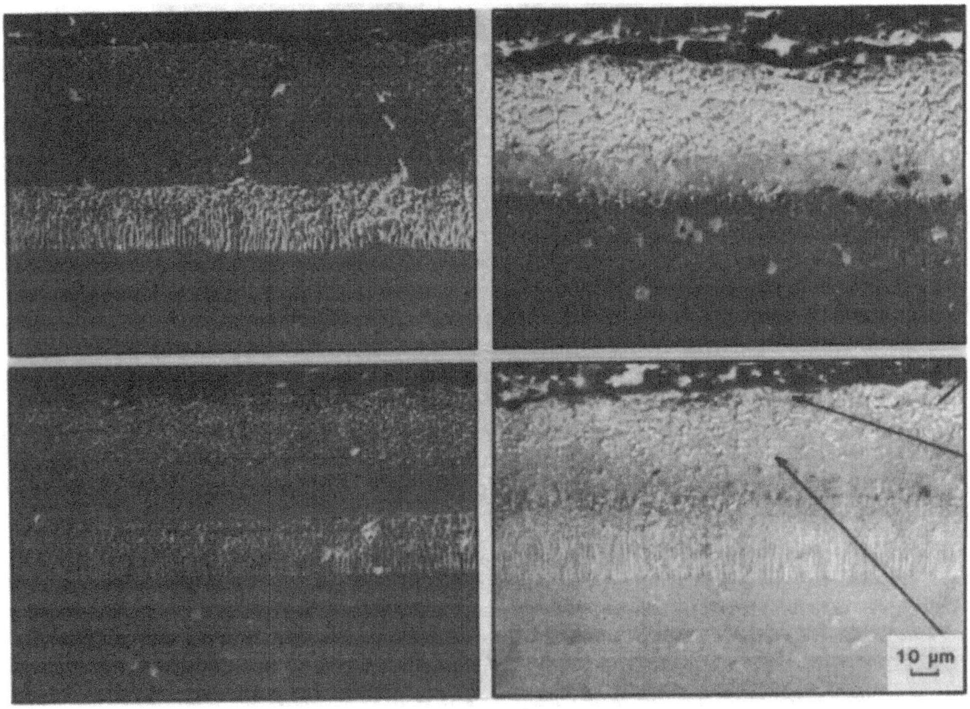

Figure 4. SEM micrographs of initial diffusion aluminide coating microstructures on various systems. Top left (TL) is PWA 73 on IN738. Top right (TR) is Pt modified IN738. Bottom left (BL) is PWA 73 on MAR M200. Bottom right (BR) is Pt modified coating on MAR M200.

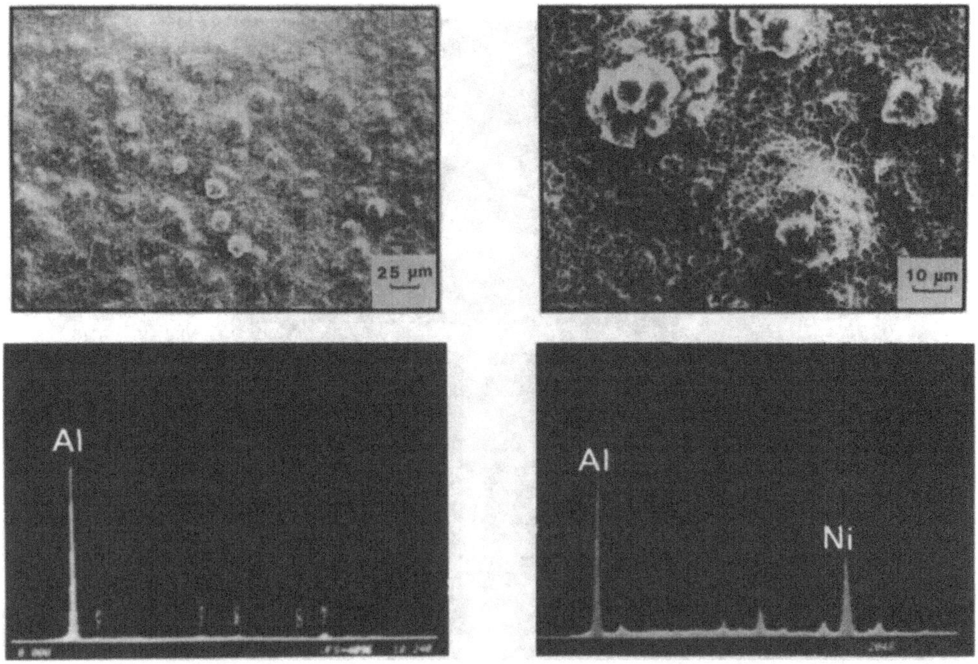

Figure 5. SEM micrograph of the surface of the Pt mod. MAR M200 after 1 week of isothermal oxidation in air at 1100°C. The alumina has a web or lacy morphology. The EDS spectra on the left analyzes the composition of the thicker ridges. The EDS on the right shows the composition of the thinner valleys.

The surface of the straight aluminide on MAR M200 is shown in Figure 6. Visually the surface was multicolored with some green indicating the presence of transient oxides. Spallation was evident as indicated by the appearance of shiny facets on the surface. The unspalled scale had large amounts of impurities such as Ni, Cr, Co, Ti, and Nb. A spalled region in Figure 6 shows bare coating with oxide grain imprints. The dark scale in this micrograph is alumina. An EDS of the bare coating is also shown in the figure.

Cross sections of the scales of both coating types on MAR M200 are presented in Figure 7. The scale upon the straight aluminide coating is about 4 times thicker than that on the Pt modified coating. These thicknesses were used to calculate parabolic rate constants. Values of 2.8×10^{-12} and 2×10^{-13} g^2/cm^4-sec were obtained for the oxides formed upon the conventional and platinum aluminides, respectively. The value of 2×10^{-13} is in good agreement with alumina growth on platinum-aluminum alloys.[20]

The microstructures of the two coatings after isothermal oxidation are compared in Figure 8. The difference in the amount of degradation between the coatings is evident. The straight aluminide had formed an appreciable amount of

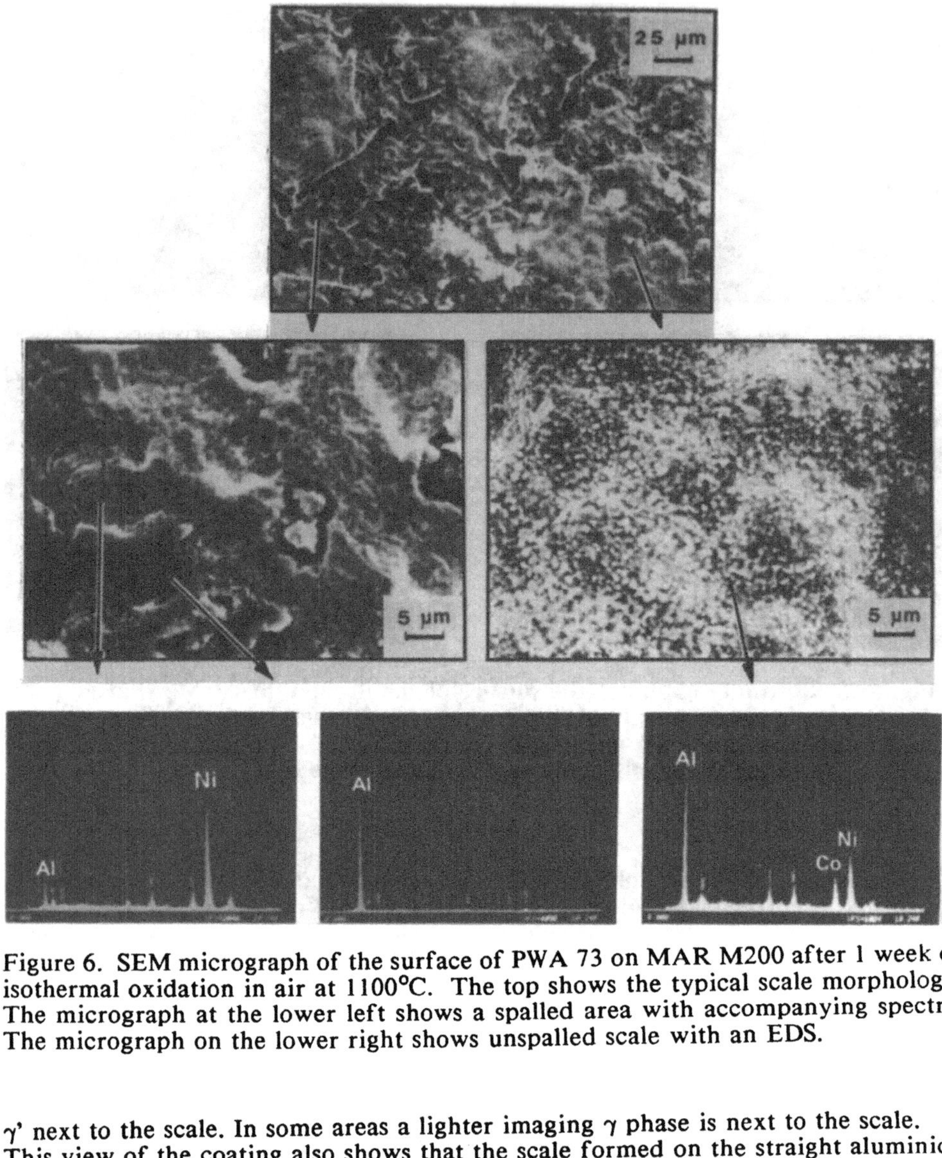

Figure 6. SEM micrograph of the surface of PWA 73 on MAR M200 after 1 week of isothermal oxidation in air at 1100°C. The top shows the typical scale morphology. The micrograph at the lower left shows a spalled area with accompanying spectra. The micrograph on the lower right shows unspalled scale with an EDS.

γ' next to the scale. In some areas a lighter imaging γ phase is next to the scale. This view of the coating also shows that the scale formed on the straight aluminide is much thicker than that shown in Figure 7 in some places. The platinum aluminide on the other hand displays a continuous layer of β phase adjacent to the scale with another phase, γ', appearing at the grain boundaries.

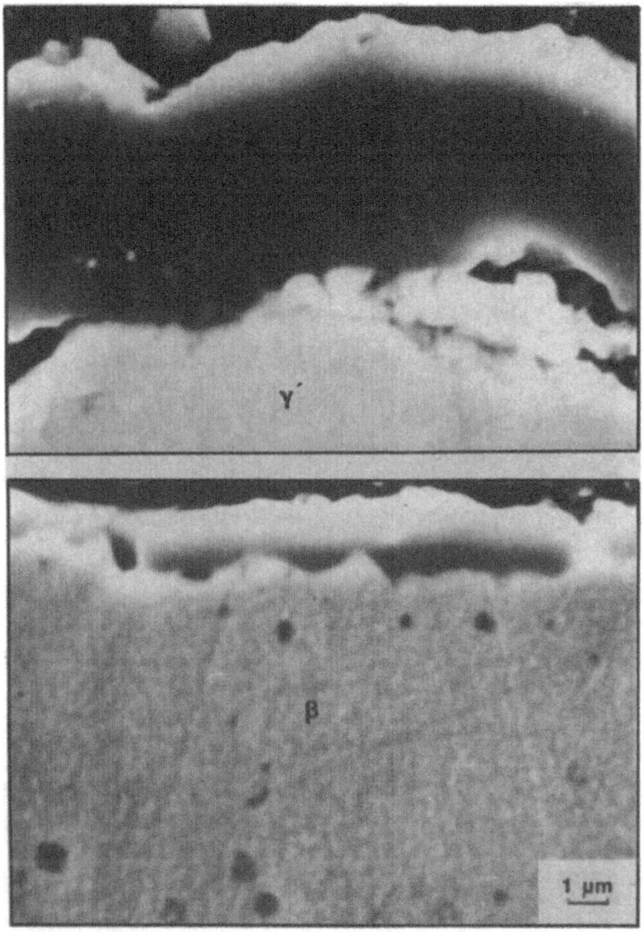

Figure 7. SEM micrographs of scale cross-sections after isothermal oxidation in air for 1 week at 1100°C. The top shows PWA 73 on MAR M200 while the bottom shows the Pt modification on MAR M200.

3.3 Cyclic Oxidation

Cyclic oxidation tests were performed at 1100, 1135 and 1200°C and weight change versus time measurements from these tests are presented in Figures 9, 10 and 11 for different coating systems. In all of these tests the platinum modified coatings performed better than the conventional aluminides. In particular, protective alumina scales were maintained for longer times on the platinum modified coatings compared to the conventional coatings. Some platinum modified coatings exhibited longer lives than others because substrate elements influenced the coatings performance.

Figure 8. Micrographs of the coating cross-sections of PWA 73 MAR M200 (top) and the Pt modified MAR M200 (bottom) after isothermal oxidation at 1100°C. The Pt modified section shows only β phase.

Metallographic examination of specimens after test showed less degradation of the platinum modified coatings than the straight aluminides after the same exposure. Figure 12 shows cross sections of the IN738 samples after the cyclic test. The Pt modified substrate exhibited two β fronts (arrows). Near the surface of the coating large voids are present. The spheroidized particles near the lower front are carbides, probably MC in nature. The straight aluminide in Figure 12 shows no sign of a β phase. All that is present are extremely thick scales, carbides, and some intergranular oxidation. In Figure 13 low magnification micrographs of the surfaces of coatings on Mar M200 and IN738 after cyclic oxidation at 1100°C are presented. The coating degradation involves localized attack and pit

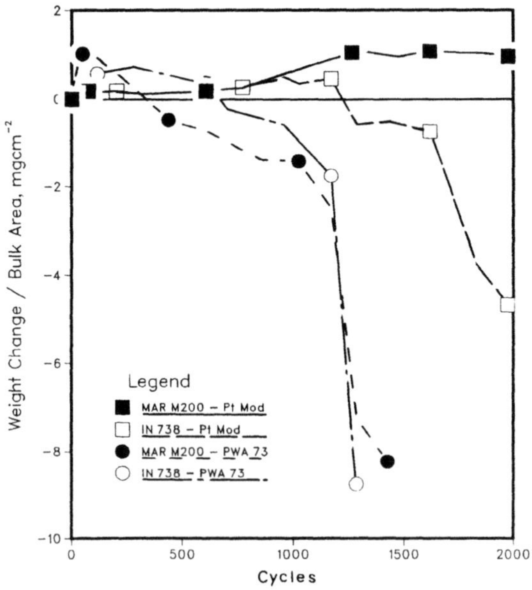

Figure 9. Cyclic oxidation kinetics of various systems at 1100°C in air. One thermal cycle is 45 minutes hot and 15 minutes cool.

development. The scales on all the samples appeared greenish but a blackish overtone was noted on the straight aluminides. Only one pit is shown by the Pt modified coating on MAR M200 after more than 2000 hours of testing while the straight aluminide has numerous pits after 1500 hours. The IN738 samples both show pits of the same size and number but this is deceiving. The pits on the Pt modified coating developed at a later stage in the testing and may have been the result of some inhomogeneity noticed on the surface of the sample after the first 100 cycles of testing. The other major morphological difference is that edges of the straight aluminides were more heavily attacked. A further important visual observation was that the platinum aluminides seemed to have a purer alumina resulting from the absence of other nonprotective oxides when compared to the straight aluminides. This observation is based on the uniformity of the scales as well as the whiteness of the oxide flakes that spalled from the platinum modified coatings. Spalling occurred from both coating types but appeared to be less frequent and in smaller areas on the Pt aluminides.

Figures 14 and 15 show micrographs of the surface of the samples after 322 cycles of testing. The Pt aluminide coatings, Figure 14, display alumina whiskers and outgrowths upon a web or lacy structure. Little variation in the scale morphology is observed. The EDS of the general scale on a Pt aluminide indicates a scale rich in Al and Cr. Figure 15 shows the micrographs of the surfaces on the unmodified coatings. There is a considerable amount of relief and no one oxide morphology dominates the appearance although IN738 appears to be more uniform. EDS analyses of the scale on PWA 73 on MAR M200 indicate the presence of W, Nb, Ti, Co, and Ni. The white particles on the PWA 73 on the MAR M200 scale are rich in Ti. The EDS analyses of the scales on IN738 are similar to that of the Pt modified coating except for a higher amount of Ti.

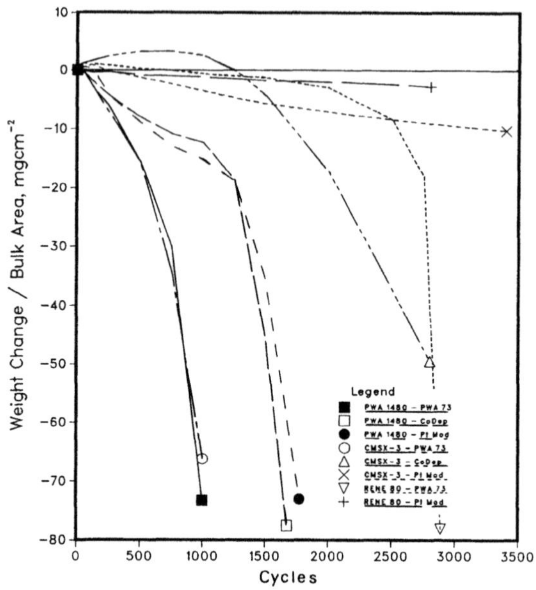

Figure 10. Cyclic oxidation kinetics of various systems at 1135°C in air. One thermal cycle is 45 minutes hot and 15 minutes cool.

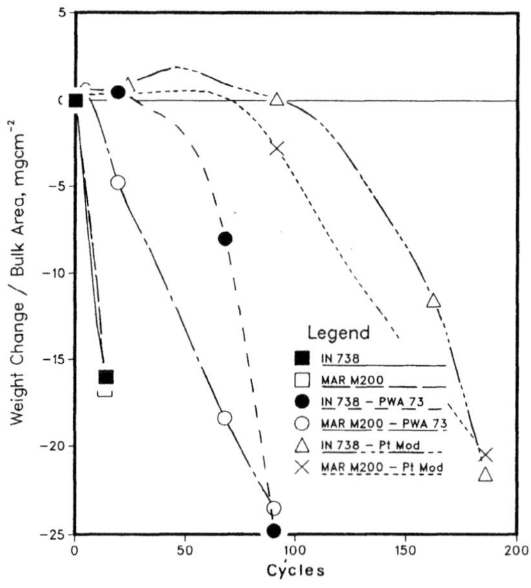

Figure 11. Cyclic oxidation kinetics of various systems at 1200°C in air. One thermal cycle is 45 minutes hot and 15 minutes cool.

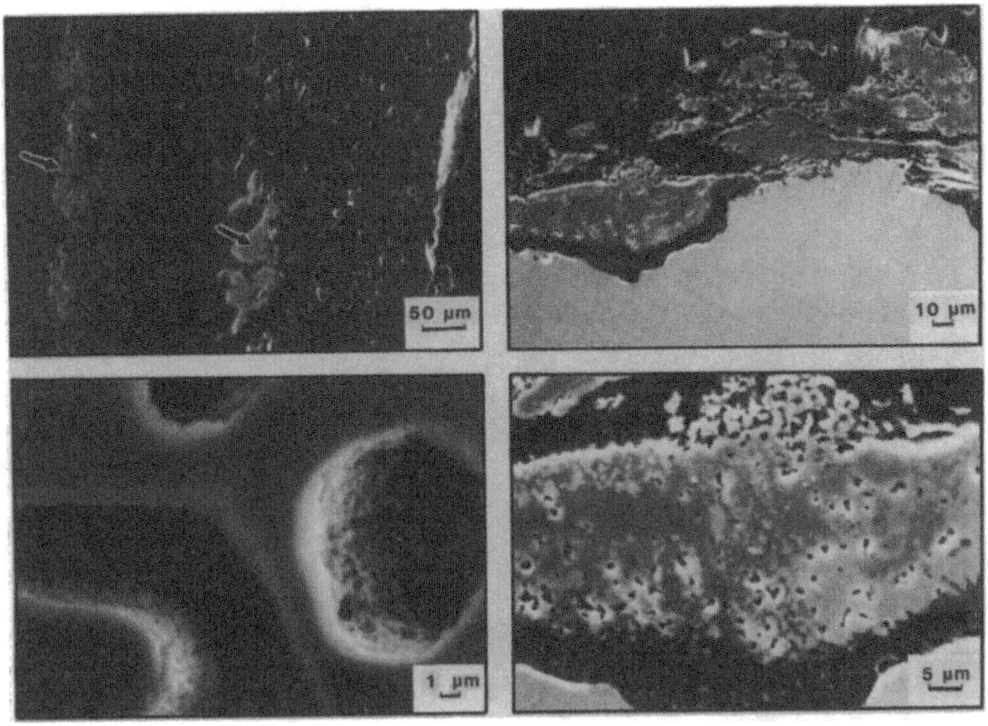

Figure 12. Cross-sections of Pt mod. IN738 (left) and PWA 73 IN738 (right) after cyclic oxidation at 1200°C in air. Arrows indicate β fronts.

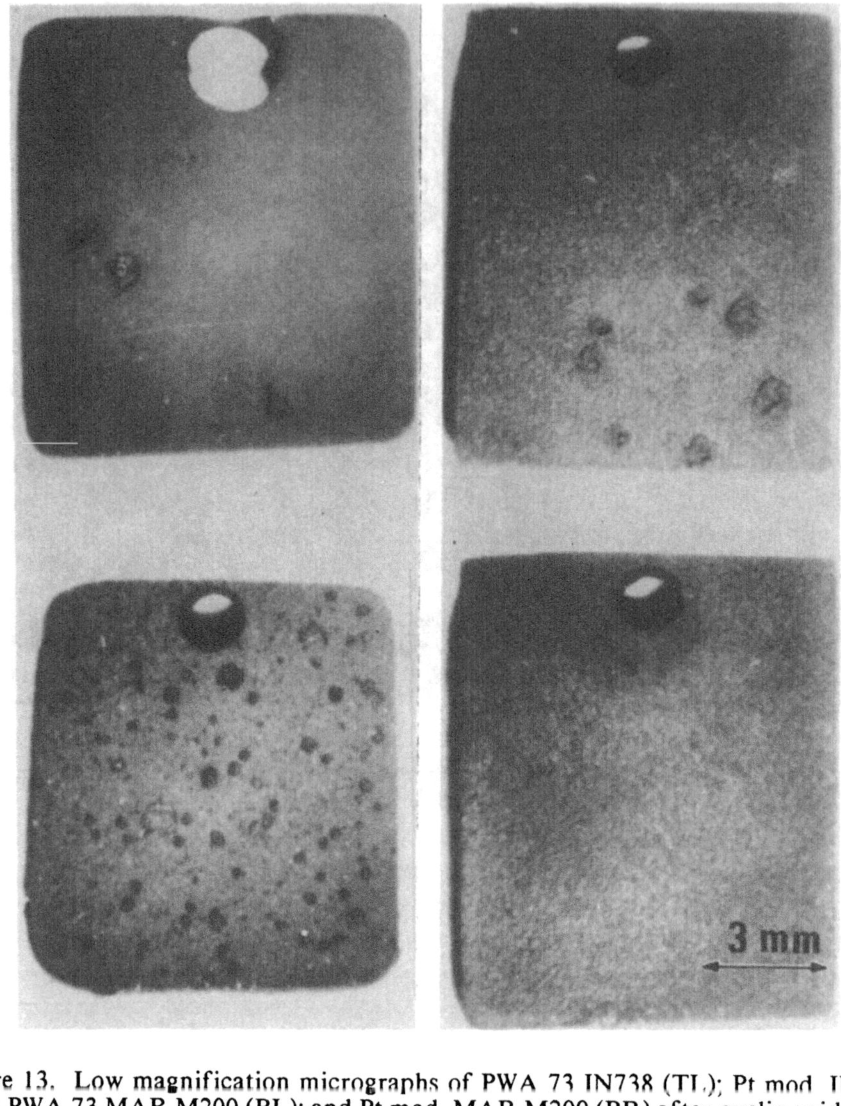

Figure 13. Low magnification micrographs of PWA 73 IN738 (TL); Pt mod. IN738 (TR); PWA 73 MAR M200 (BL); and Pt mod. MAR M200 (BR) after cyclic oxidation at 1100°C. All samples are shown at the same magnification.

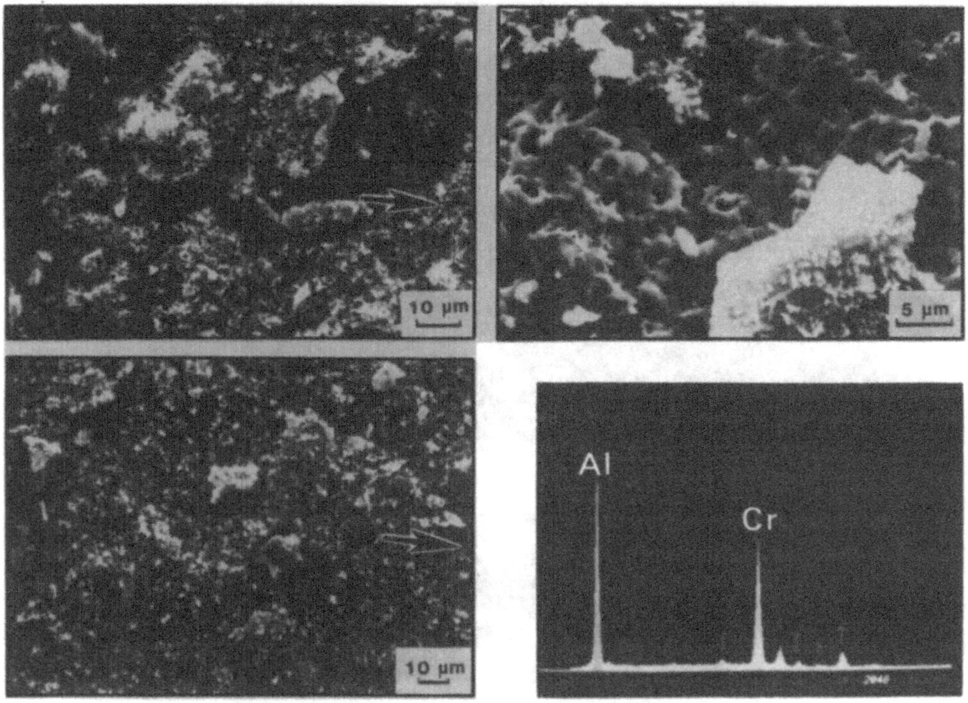

Figure 14. Surface micrographs of the Pt coatings on IN738 (top) and MAR M200 (bottom) after 322 cycles of cyclic oxidation at 1100°C. Spallation is present but the scales are richer in Al than their unmodified counterparts.

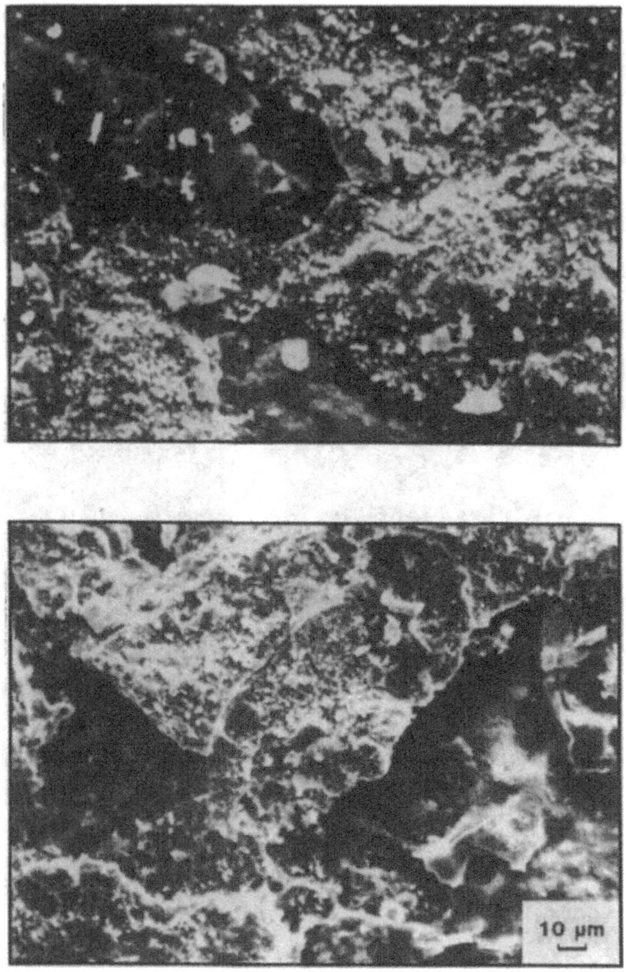

Figure 15. Surface micrographs of PWA 73 process on IN738 (top) and MAR M200 (bottom) after 322 cycles of cyclic oxidation at 1100°C. Both samples show spalling and subscale formation.

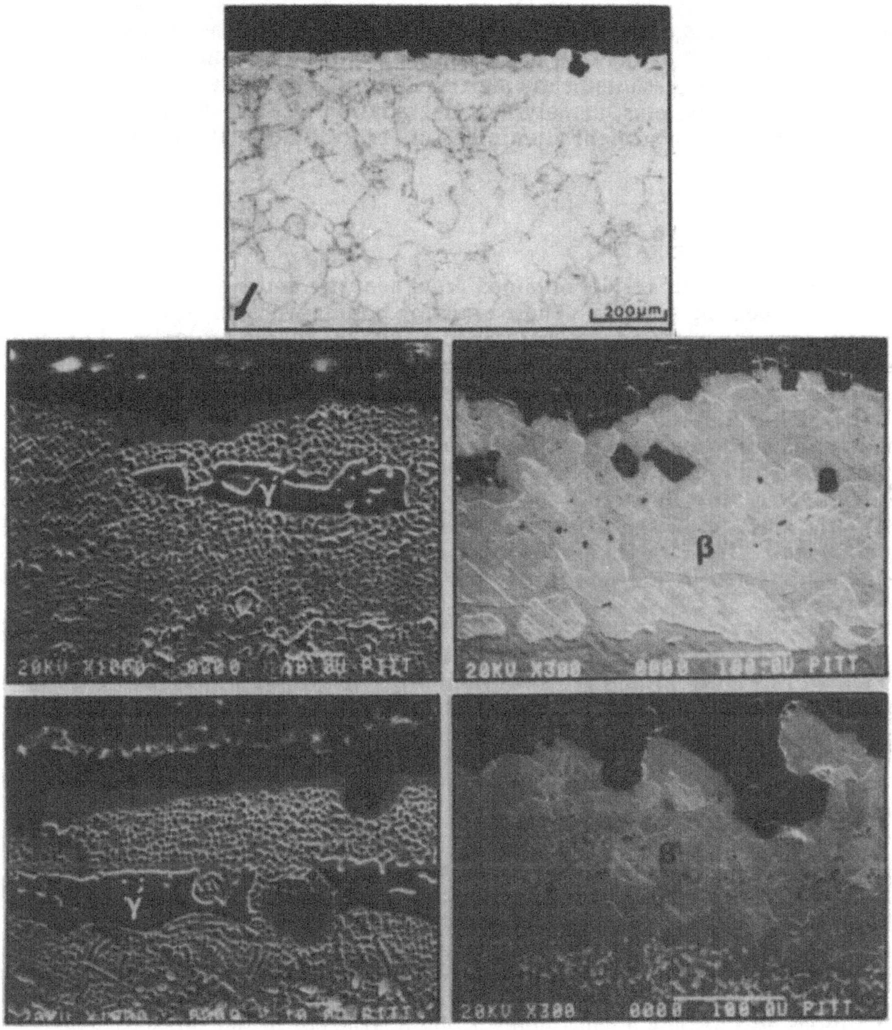

Figure 16. Coating cross-sections of samples after cyclic oxidation in air at 1100°C. The Pt modified coatings (IN738 (TR) and MAR M200 (BR)) show β phase while the unmodified coatings show γ' (IN738 (TL) and MAR M200 (BL)). Optical micrograph at the top shows the coating thickness of PWA 73 on IN738.

In Figure 16 cross sections of IN738 and MAR M200 after cyclic oxidation at 1100°C are presented. The straight aluminides show no β phase remaining while some β phase is still evident on the platinum modified coatings. In Figure 17 and 18 micrographs are presented showing results similar to those described above but for other coating systems, namely, less degradation of the platinum modified coatings compared to straight aluminides under cyclic oxidation conditions.

3.4 Acoustic Emission

Figure 19 presents results obtained from acoustic emission measurements with some of the coating systems. These emission counts were taken during the 15 minutes cool down period. A difference of 30.000 counts represents a gradation in behavior between the coating-substrate systems. In general a lower number of counts represents less scale damage. The results show that a binary Ni-32.2Al and a ternary Ni-32.2Al-2Pt alloy undergoes less damage than any combination of coating and substrate. The results also indicate that the IN738 substrate-coating systems undergo less damage from the thermal cycle than the MAR M200 samples. The type of coating apparently has little effect on this general result. Figure 20 shows the surfaces of the tested coating systems. The Pt aluminides display a thin alumina web. The straight aluminides had a varying scale morphology with other elements detected in addition to Al. Spalling of the oxide scales was evident on all the samples.

Results obtained from acoustic emission tests with other coating systems are presented in Figure 21 and are consistent with the proposal that the platinum modified coatings do not exhibit stronger oxide scale adherence than the straight aluminide coatings.

3.5 Diffusion Annealing

Results from some of the diffusion annealing tests are presented in Figures 4 and 22-25. The as-processed microstructures are presented in Figure 4 for comparison to structures developed after time at 1200°C.

Figure 22 shows the specimens after 2 hours at 1200°C in Ar. The Pt aluminides have lost their $PtAl_2$ phase and are composed of a β phase. The straight aluminides also display a continuous β phase adjacent to the surface. The α precipitates have disappeared on both of the PWA 73 samples.

Figure 23 depicts the samples after 6 hours at 1200°C. The platinum aluminides still display a continuous layer of β at the surface, and a new phase is beginning to appear at the grain boundaries in the β. The diffusion of Ni out to the surface has created new β under the inner zone of $PtAl_2$ on IN738. The IN738 at this point does not display the fingers seen in the as-received state. Diffusion of Ni seems to have lengthened the inner zone of the Pt modified coating on MAR M200.

The straight aluminides shown in Figure 23 exhibit signs of degradation from oxidation and interdiffusion. For the PWA 73 on MAR M200, a 7 μm thick scale

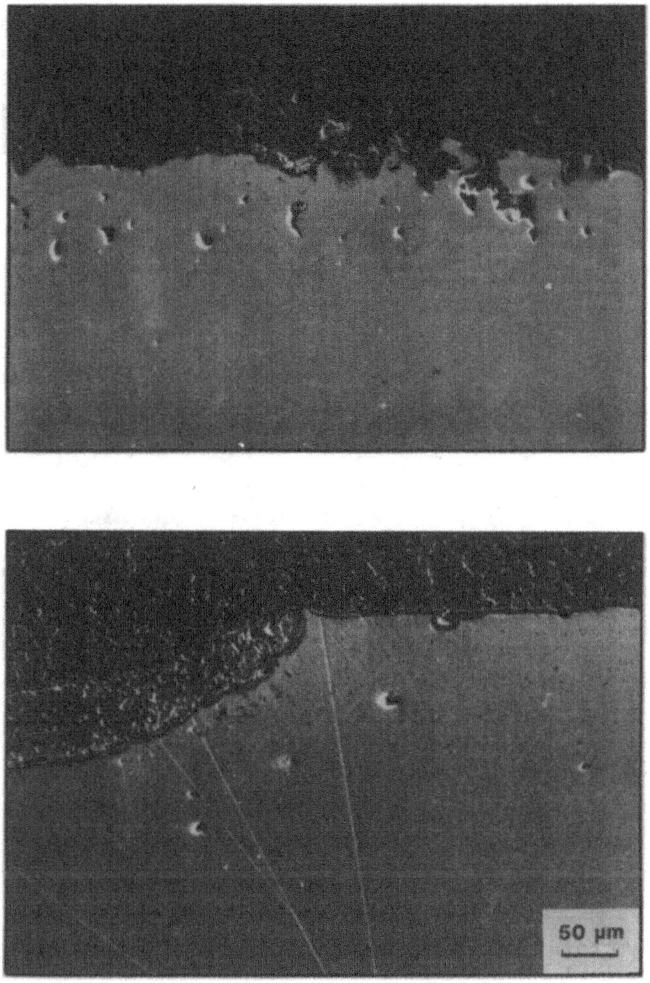

Figure 17. Coating cross-sections after cyclic oxidation at 1135°C in air. Top shows Pt mod. CMSX-3 after 3414 cycles while the bottom shows PWA 73 CMSX-3 after 1007 cycles.

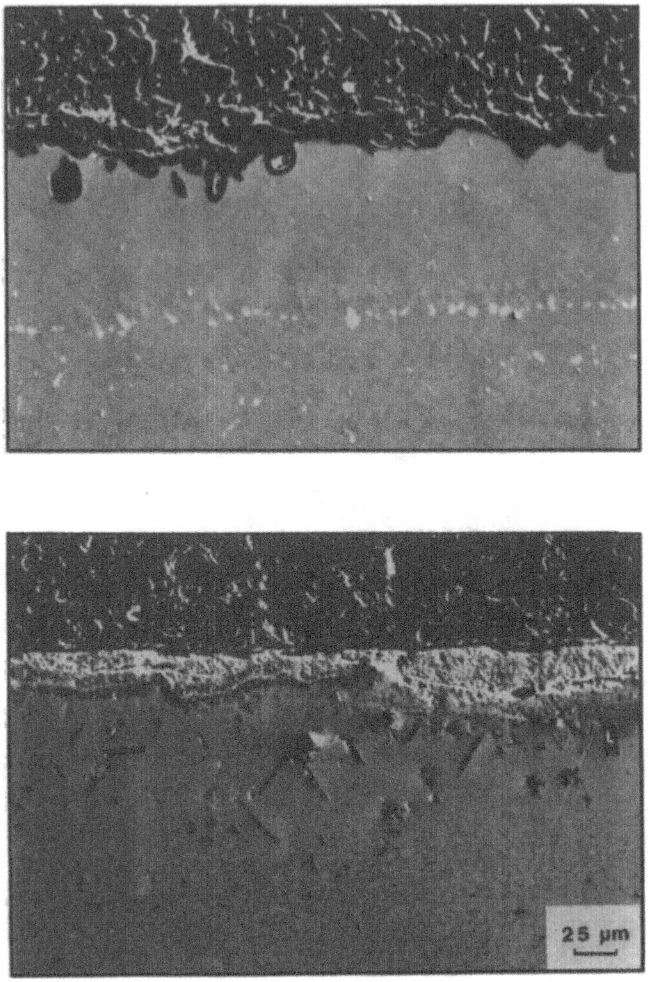

Figure 18. CoDep. coating cross-sections after cyclic oxidation at 1135°C in air. Top shows CMSX-3 substrate after 2806 cycles while the bottom shows PWA 1480 substrate after 1706 cycles.

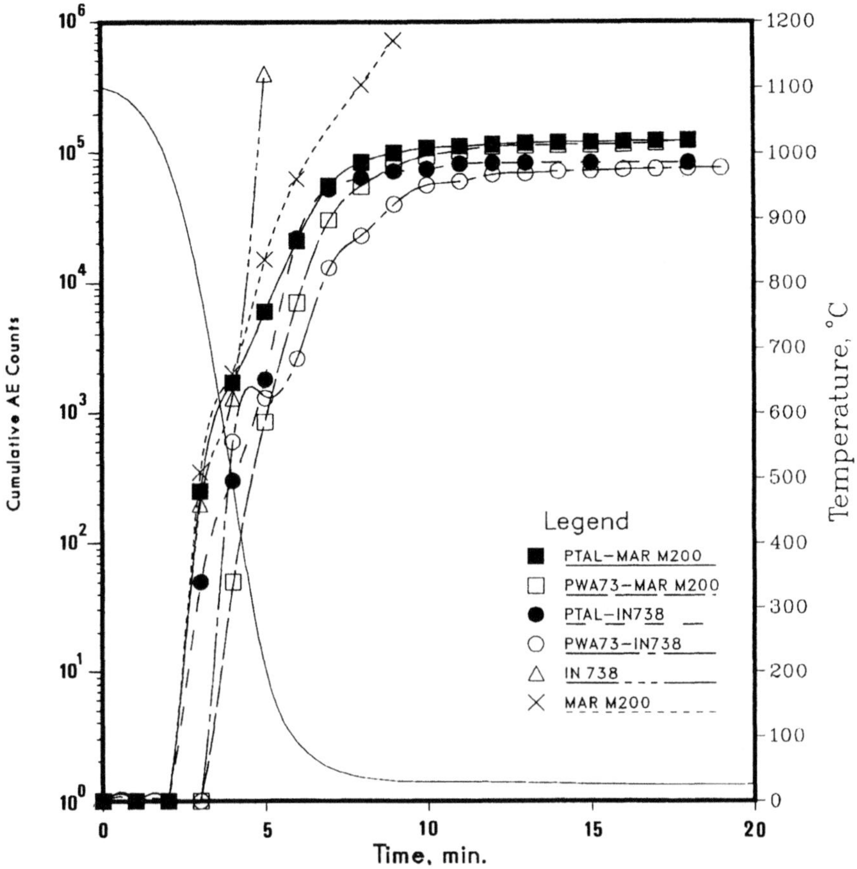

Figure 19. Cumulative acoustic emission counts taken from a 15 minute cool down from 1100°C in air. (Temperature profile is also plotted.)

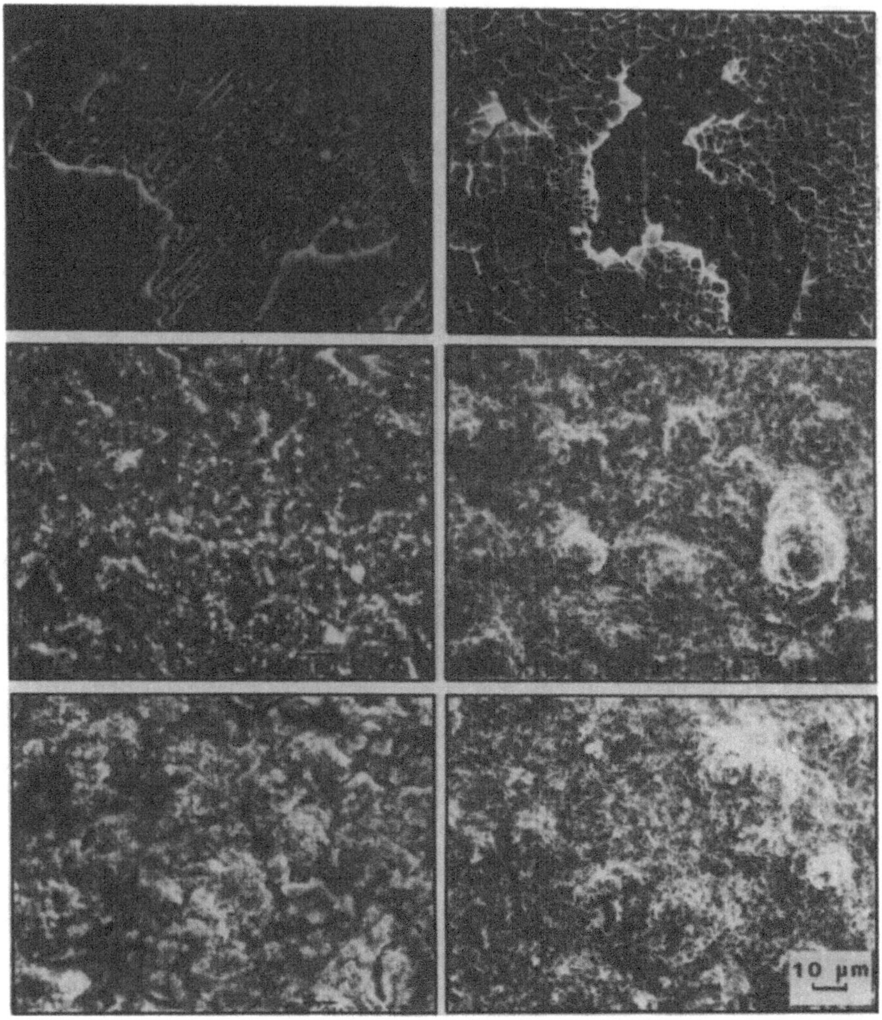

Figure 20. The surfaces of the samples after the acoustic emission tests at 1100°C. The cast NiAl (TR) and the cast PtNiAl (TL) alloys and both Pt modified coatings (MR & BR) exhibit thin alumina scales. The unmodified coatings (ML & BL) had multicomponent scales.

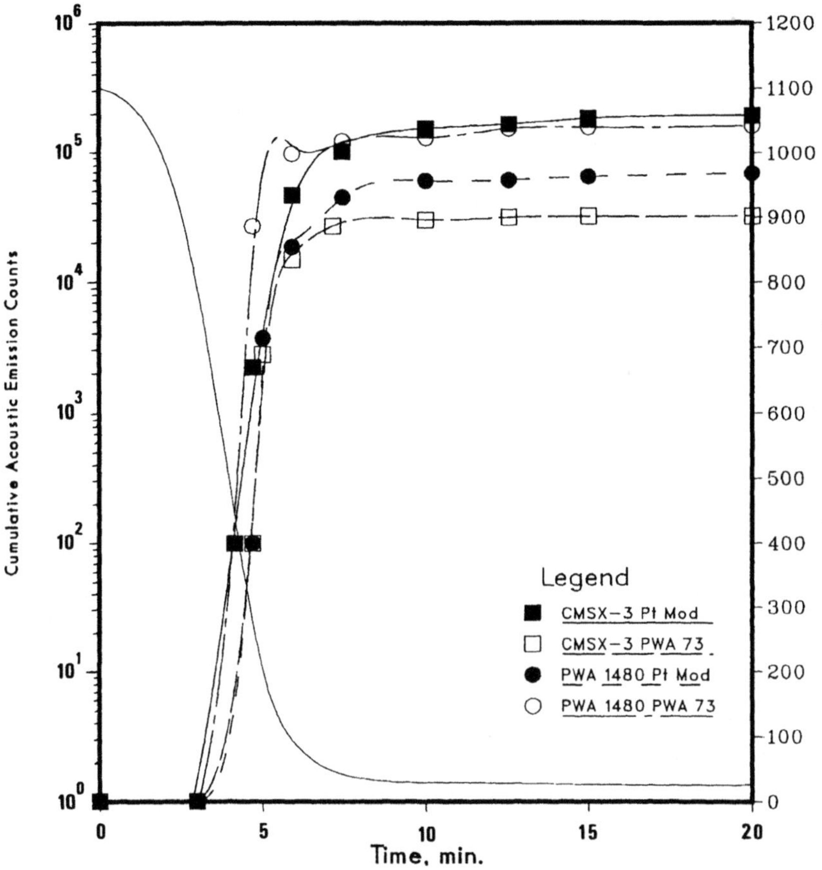

Figure 21. Acoustic emission results for various other coatings-single crystal substrate systems.

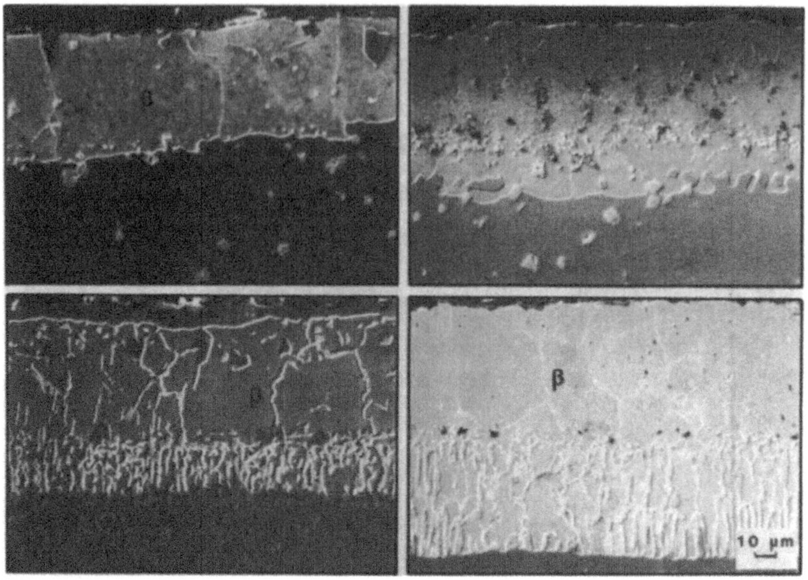

Figure 22. SEM micrographs of PWA 73 IN738 (TL), PWA 73 MAR M200 (BL), Pt mod. IN738 (TR), and Pt mod. MAR M200 (BR) after diffusion annealing in Ar at 1200°C for 2 hours.

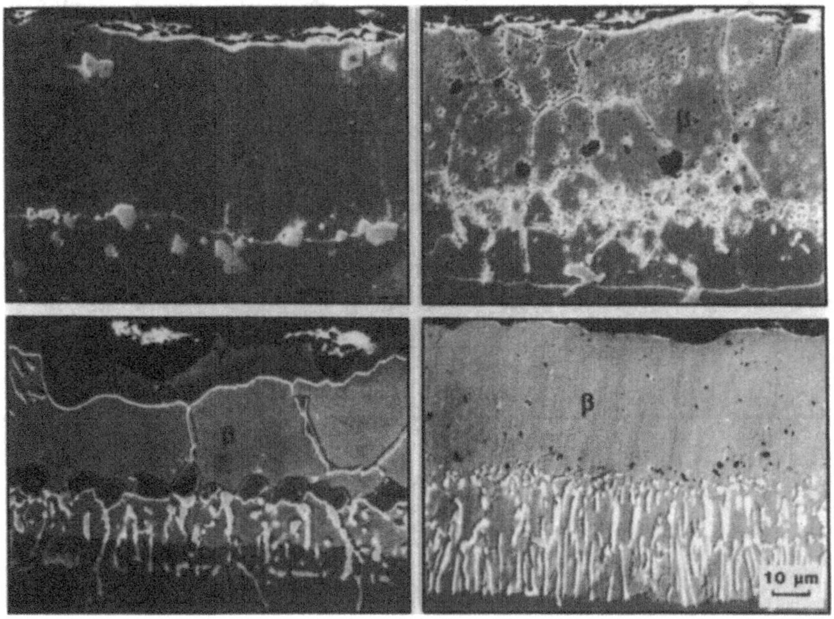

Figure 23. SEM micrographs of PWA 73 IN738 (TL), PWA 73 MAR M200 (BL), Pt mod. IN738 (TR), and Pt mod. MAR M200 (BR) after diffusion annealing in Ar at 1200°C for 6 hours.

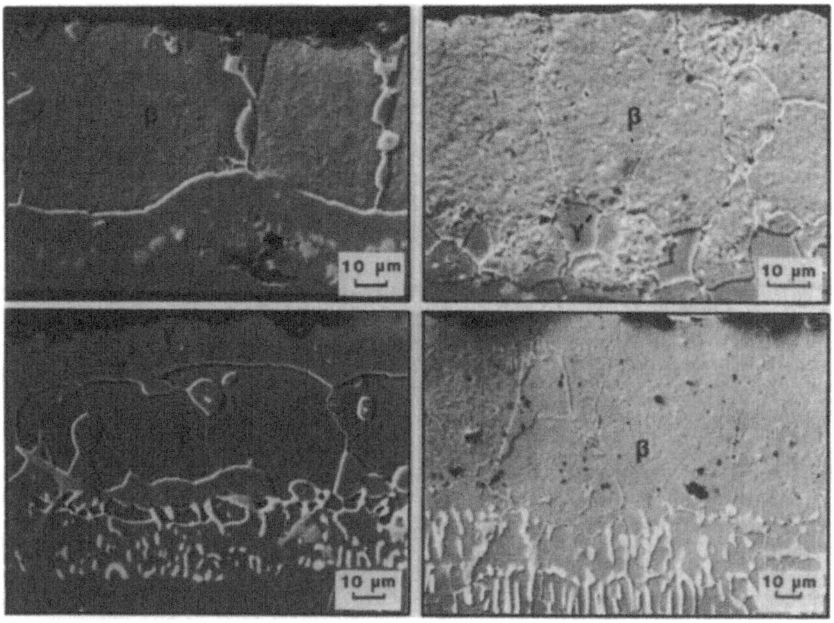

Figure 24. SEM micrographs of PWA 73 IN738 (TL), PWA 73 MAR M200 (BL), Pt mod. IN738 (TR), and Pt mod. MAR M200 (BR) after diffusion annealing in Ar at 1200°C for 19 hours.

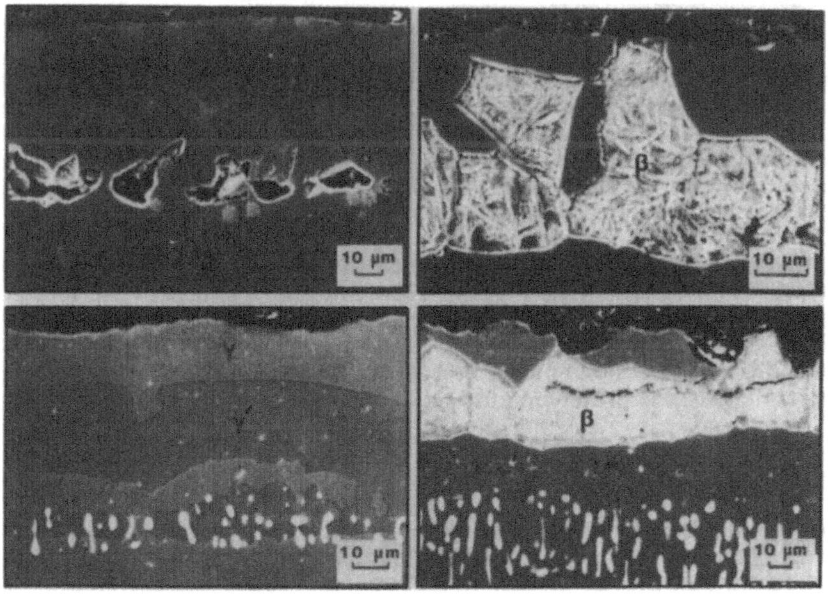

Figure 25. SEM micrographs of PWA 73 IN738 (TL), PWA 73 MAR M200 (BL), Pt mod. IN738 (TR), and Pt mod. MAR M200 (BR) after diffusion annealing in Ar at 1200°C for 161 hours.

was observed to be produced in the Ar environment. Beneath the scale lies the γ phase. γ' lies directly beneath the γ phase. The γ' extends through the β phase along grain boundaries to a layer of γ' islands. Under the γ' islands new β has formed by Al enrichment. This will be referred to as enrichment β phase. The loss of Ni or gain of Al has caused the envelopment of the α fingers by γ'. The PWA 73 on IN738 displays a different morphology. In this case the γ' manifests itself as grains growing from the surface of the coating towards the substrate. One γ' grain is adjacent to a carbide. The rest of the layer is β. Growth of substrate phases seems to have engulfed the carbides present in the inner zone after 2 hours.

Figure 24 shows the Pt modified systems after 19 hours. The coatings still display the β phase adjacent to the surface. The coating on IN738 has a grain boundary phase extending from the surface to γ' islands present in the layer 10 μm from the substrate interface. Under the γ' islands, enrichment β phase, created by the loss of Ni or a gain in Al content, is being converted to γ'. The MAR M200 structure is similar to that in Figure 23.

The straight aluminides in Figure 24 display microstructures similar to those at 6 hours. In the IN738 substrate γ' grains have linked up with the grain boundary γ' to isolate the β phase into islands. All signs of an inner zone have been swept away by the growth of substrate phases. In the MAR M200 a large γ and γ' front exists at the surface of the coating. The β is isolated by γ'. The enrichment β is being consumed by γ'.

The straight aluminides, after 73, hours display degraded morphologies. No β phase is present in the MAR M200 sample. All that remains of the once Al-rich coating is γ and γ'. On the IN 738 substrate the coating has suffered a mechanical failure. β phase is still observed but γ' seems to surround it with appreciable thicknesses.

Figure 25 shows the samples after 161 hours at 1200°C. The β phase is present in the platinum aluminide coatings on both substrates and resides close to the surface. The IN738 shows complete envelopment of the β by γ'. The MAR M200 has a continuous layer in some places but not in others. A crack is seen through the phase. The straight aluminides are nothing more than Al enriched substrates. Voids appear in the diffusion-affected zone of the IN738.

One set of samples were annealed tested for 48 hours in Ar at 1200°C without sectioning. These samples had only one thermal cycle. The cross sections are consistent with those previously discussed meaning that the thermal cycling did not significantly accelerate the degradation. The surfaces of the samples showed profuse spalling. Diffusion anneals performed with other alloy substrates showed similar results as those presented above.

3.6 High Temperature Hot Corrosion

Results obtained from the high temperature hot corrosion tests are presented in Figure 26. These data show conclusively that the platinum modified coatings are better than coatings that do not contain platinum. Typical microstructures of specimens exposed to cyclic hot corrosion conditions are presented in Figures 27-29. Even though the platinum modified coatings have been exposed for longer times, the amount of degradation is much more extensive in the coatings that do not contain any platinum. Examination of the data presented in Figure 26 shows

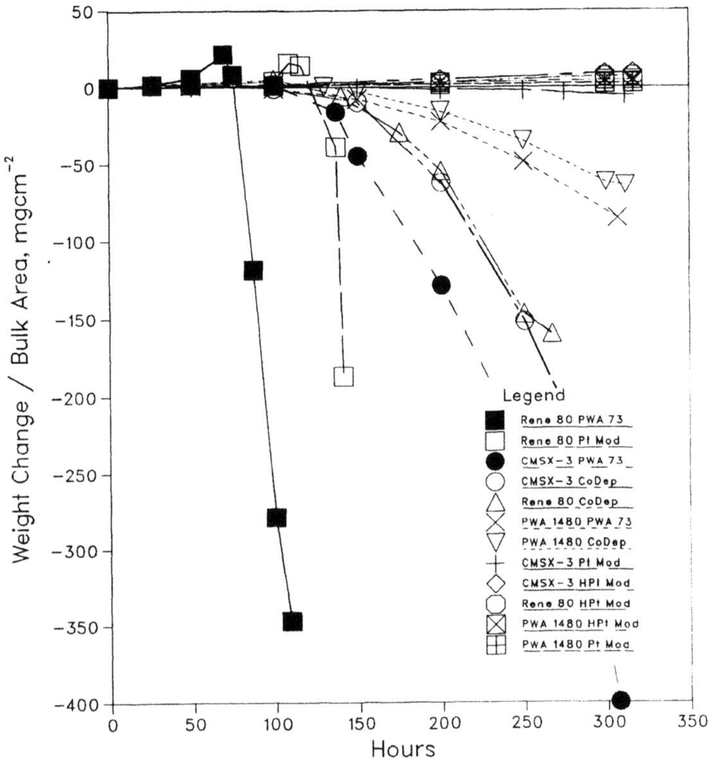

Figure 26. Cyclic high temperature hot corrosion kinetics for various systems at 1000°C with 1 mg/cm^2 Na_2SO_4 deposits.

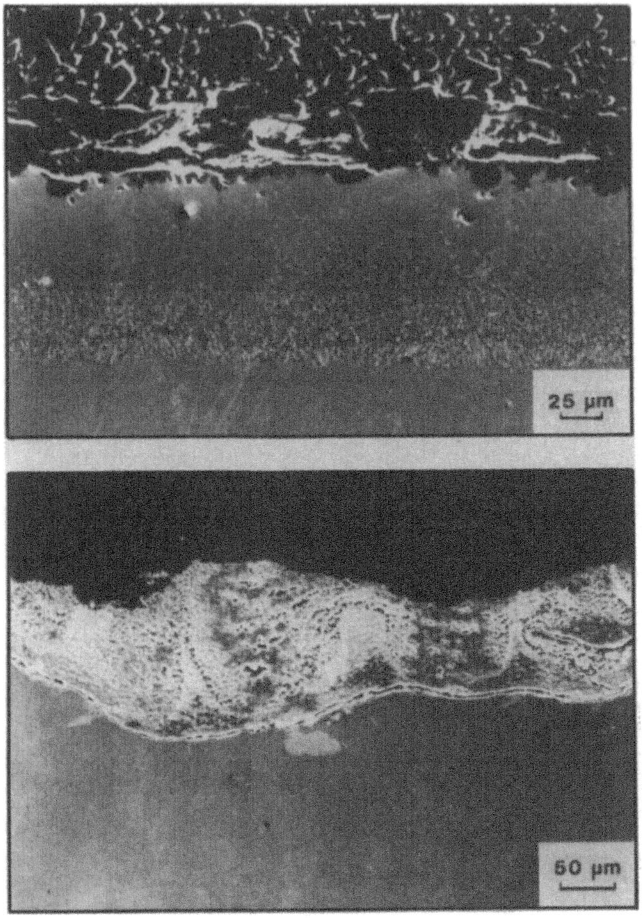

Figure 27. Cross-sections of coated PWA 1480 after 300 hours of cyclic hot corrosion in air at 1000°C. Top is Pt modified while bottom is PWA 73.

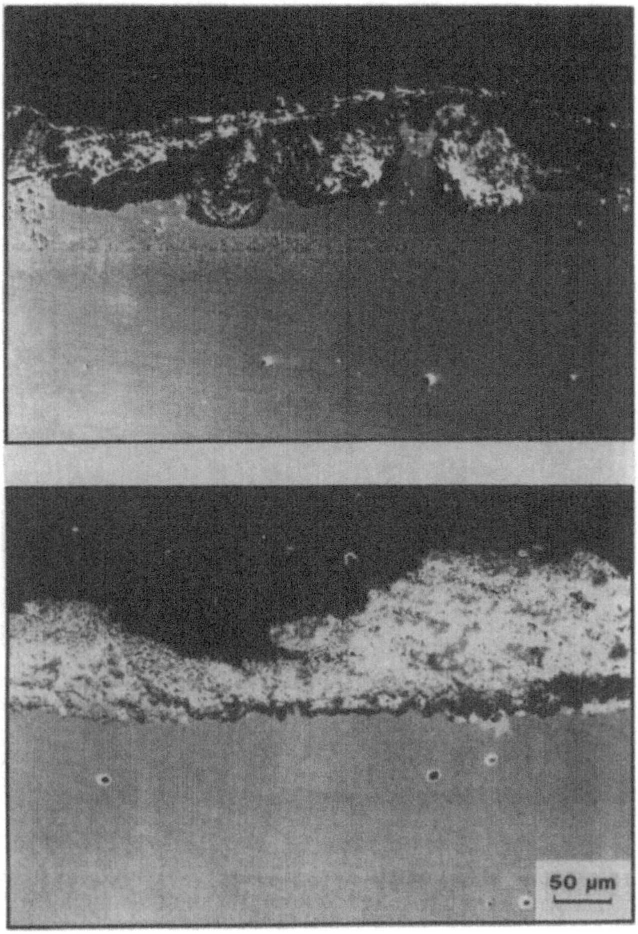

Figure 28. Cross-sections of coated CMSX-3 after 300 hours of cyclic hot corrosion in air at 1000°C. Top is Pt modified while bottom is PWA 73.

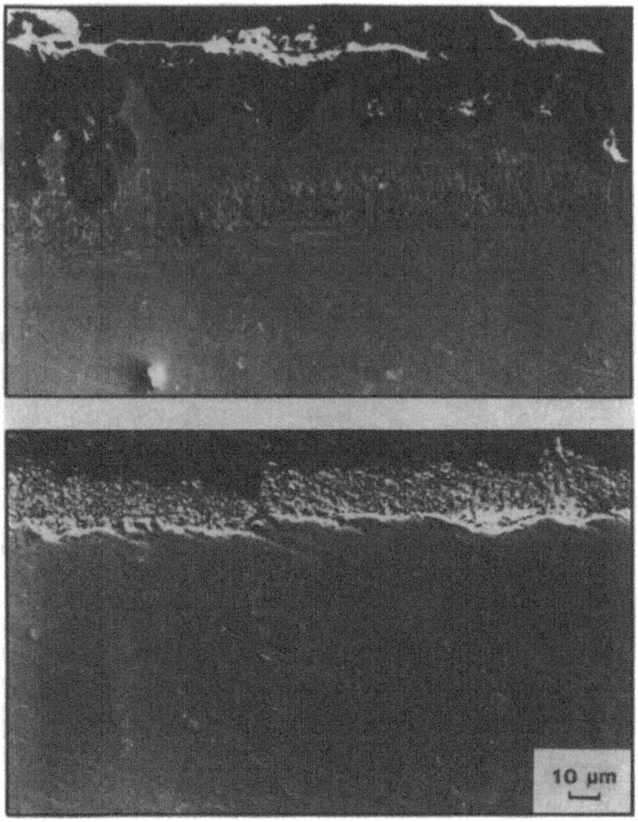

Figure 29. Cross-sections of coated Rene 80 after 300 hours of cyclic hot corrosion in air at 1000°C. Top is Pt modified (150 hours) while bottom is PWA 73 (100 hours).

that there also is an effect of substrate composition. Coatings have shorter lives on Rene 80 compared to CMSX-3, which in turn are not as good as coatings on PWA 1480. The significant difference between these substrates is the amount and type of refractory elements. Rene 80 contains both Mo and W. CMSX-3 contains mainly W and PWA 1480 contains predominantly Ta. In the hot corrosion literature[21], it is well documented that refractory elements, such as Mo and W, can have a profound effect upon hot corrosion behavior. The results obtained for the high temperature hot corrosion tests show that platinum extends the lives of diffusion aluminide coatings. Optimum lives, however, depend on the composition of the superalloy substrate.

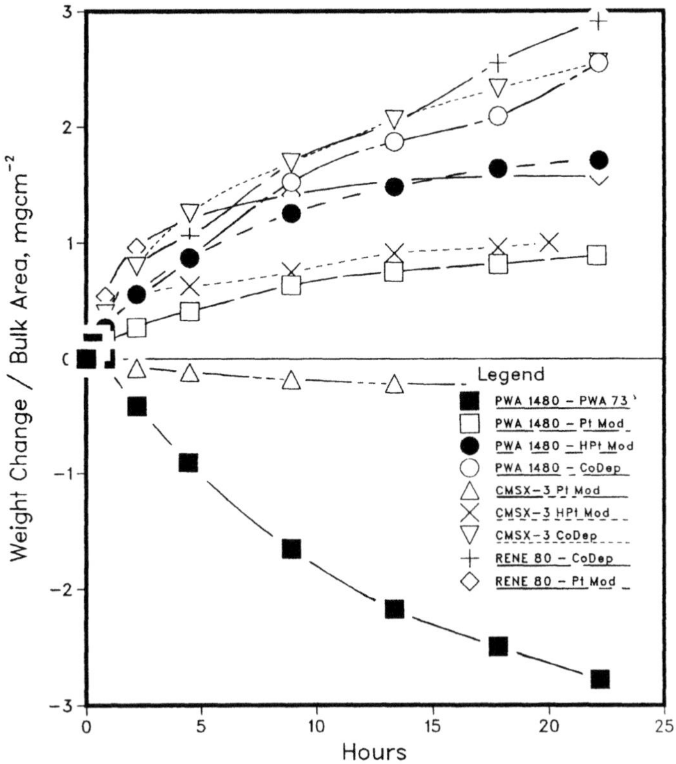

Figure 30. Low temperature hot corrosion kinetics at 700°C and 10^{-4} atm. SO_3 for various systems.

3.7 Low Temperature Hot Corrosion

The weight change data for coating systems exposed to low temperature hot corrosion conditions are presented in Figure 30. The weight changes are not as large as those of the high temperature hot corrosion test but still very much larger than would have been observed without a deposit and without any SO_3 in the gas phase. Examination of these results show that the weight changes for the coatings which contain platinum are less than those which do not contain any platinum. This observation is also supported by metallographic examination of the exposed specimens. In Figure 31, less attack of the coating on CMSX-3 with platinum has occurred than of the coating with no platinum. No obvious influence of substrate composition was evident in the low temperature hot corrosion tests.

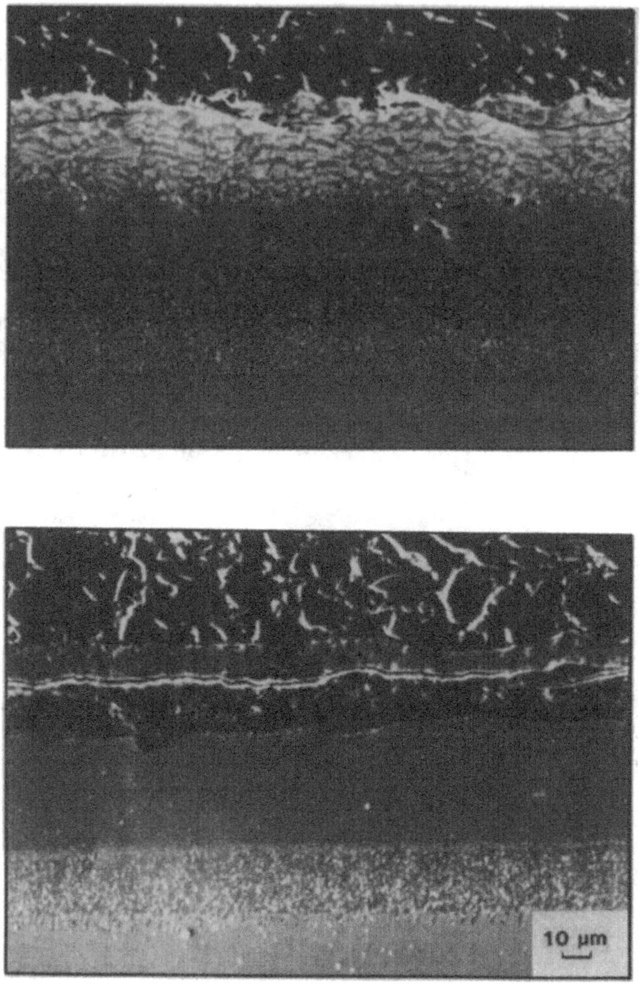

Figure 31. Cross-sections of CMSX-3 after 24 hour exposure to low temperature hot corrosion conditions at 700°C. Top is Pt mod. and bottom is PWA 73.

4. MECHANISMS BY WHICH PLATINUM AFFECTS THE LIVES OF DIFFUSION ALUMINIDE COATINGS

The cyclic oxidation results imply that the platinum modified aluminide coatings can reform alumina scales for longer periods than substrates with straight aluminide coatings. The superiority of platinum aluminides in these tests cannot be totally attributed to a Pt-related adherence effect. The justification for this statement is that spalling of pure alumina scales was observed on the platinum aluminide samples at early stages in the cyclic oxidation tests. Profuse spalling also occurred on the Pt modified coatings during cool down from the 1200°C diffusion anneal.

When platinum is present on the surface of a superalloy, higher atomic percentages of aluminum are incorporated into the coating for a fixed aluminum activity of the aluminizing process compared to the same nickel base superalloy with no platinum layer. To illustrate this point MAR M200 was aluminized under the conditions used to prepare platinum aluminides but no platinum layer was put on its surface. A hypostoichiometric NiAl phase was formed with X_{Al} no larger than 0.49 compared to X_{Al} equal to 0.67 on the platinum modified coating. The high affinity of platinum for aluminum was also evident when a platinum coated MAR M200 specimen was annealed in vacuum at 1000°C. A two phase structure was observed to form in the platinum layer. One phase was richer in aluminum than the other. XRD was not conclusive but peaks for β phase [(Ni,Pt)Al] were observed at low intensity. A continuous layer of alumina was formed upon the specimen after 24 hours. These results indicate that aluminum diffuses from the γ-γ'substrate of MAR M200 to form PtAl in the platinum-rich layer. Jackson and Rairden [22] also noted that the platinum increases the capacity for aluminum assimilation in a pack aluminizing process.

The growth direction of the coatings formed with and without a Pt layer at the same Al activity is different. Reference to Figure 4 shows that the original Pt layer still resides at the gas/coating interface. This means that the coating forms inwardly by Al atoms diffusing through the Pt layer. When no Pt layer is present, the coating forms outwardly by Ni atoms diffusing to the gas/coating interface. This observation necessarily requires that the Al atoms are mobile in $PtAl_2$ and in other high atomic fraction Al-compounds that form when Pt is present in the coating in large enough concentrations. If the Al atoms were not mobile in these phases, then the Ni would be able to diffuse through the Pt layer to react with Al at the gas/coating interface. This sequence would not result in Pt-rich phases being present at the gas/coating interface. The result of aluminizing alloys with thin platinum layers is that coatings are formed whose surfaces contain predominantly platinum and aluminum.

The Pt layer in the pack process may also be a physical barrier to impurities diffusing out to the gas/coating interface. Consider the case when no Pt layer is present before a pack process. In this case the bare substrate is exposed to the pack. This means that if a high activity process is used, the impurities are already present at the gas/coating interface. Later oxidation of the coating may involve the formation of transient oxides from these impurities, i.e., elements other than Al. Consider the case when a Pt layer is electroplated on a substrate that is exposed to a high activity pack. The Al, as already established, diffuses rapidly through the Pt layer. In the meantime, elements other than Al in the substrate competitively diffuse through the Pt layer toward the gas/coating interface. Since the diffusion of impurities is slow relative to Al diffusion in the Pt layer, the impurities will be physically separated from the gas/coating interface by the platinum aluminides that form during the coating process. Jackson and Rairden[22] refer to this effect and conclude that the exclusion of refractory elements from the surface of the coating results in the improved hot corrosion resistance of Pt modified coatings.

It was observed that the platinum modified coatings remained alumina formers for longer times than the conventional aluminides, and the alumina scales contained less substrate elements and grew slower on platinum modified coatings. The purity of the scales formed on platinum aluminide coatings implies that selective oxidation of Al is more efficient in these coatings. This can be accounted for in two ways. First, the Pt does not form a solid oxide at the temperatures of interest, therefore, a solid Pt oxide is not involved in the alumina formation. The purity of these scales indicates that other elements like Ni, W, Co, Ti, etc. are not involved in the formation of a continuous alumina scale. The second reason is that Al is very mobile in $PtAl_2$ and in high atomic fraction Al-compounds in general. Thus a platinum aluminide can establish a continuous alumina layer rapidly since Al diffuses rapidly to the gas/coating interface before other elements. An adherence effect, if present, is an indirect result of the purer, slower growing alumina scales that form on a platinum aluminide coating.

The results from the diffusion anneals showed that β phase is maintained by the platinum aluminide coatings for longer periods than by straight aluminide coatings. Such observations indicate that platinum could lower the aluminum activity as interdiffusion occurs and consequently decreases the driving force for diffusion. Such a condition would also result in coatings remaining alumina-formers for longer periods of exposure. The results from the diffusion anneals must be used with some reservation, however, because the conventional aluminides were formed under different aluminizing conditions than the platinum modified coatings.

The improved cyclic oxidation resistance of the platinum modified coatings is due primarily to their ability to form comparatively pure alumina scales for long periods of exposure. This results from the zone enriched in platinum and aluminum at their surfaces. Other secondary effects may also contribute to extending coating lives such as diffusional stability and less cracking and spalling of the alumina. Finally, it is important to emphasize that some substrate elements do evidently improve platinum aluminide performance. For example platinum modified coatings were observed to have longest lives on substrates that contained hafnium. Hafnium is an element that can improve the adherence of alumina on coatings.[23]

The improved high temperature hot corrosion resistance of the platinum modified coatings compared to the conventional aluminides can be attributed to the improved cyclic oxidation resistance since cracking and spalling of alumina is necessary before sufficient sulfur from the molten sulfate can enter the coating and cause degradation. The temperature of this test is sufficient, however, to permit substrate elements to affect coating performance. Those substrates with large concentrations of elements such as molybdenum and tungsten have shorter coating lives. Even though substrate elements such as hafnium improved cyclic oxidation behavior, in the cyclic hot corrosion tests, the refractory elements evidently exerted a more dominant effect.

The improved low temperature hot corrosion behavior of the platinum modified coatings is not as great of an effect as the improvements in cyclic oxidation resistance and high temperature hot corrosion resistance. Nevertheless the platinum modified coatings are better. It has been proposed that the $PtAl_2$[24] phase is more resistant than phases such as NiAl which is a major phase in conventional diffusion aluminides. During low temperature hot corrosion,[25] the acidic fluxing of oxides such as CoO and NiO is believed to play an important rôle. In the platinum modified coatings, replacement of nickel by platinum may cause the nickel fluxing reaction to be decreased. Moreover since this type of degradation occurs at low temperatures, substrate elements will not diffuse into the platinum modified coatings. On the other hand, it is worth noting that exposures to high temperatures (e.g. 1000°C) followed by exposure to low temperature hot corrosion conditions could result in loss of the observed improved low temperature hot corrosion resistance of the platinum modified coatings.

5. CONCLUSIONS

Platinum is a very useful element to incorporate into diffusion aluminide coatings on superalloys since it improves the resistance of such coatings to the principal forms of degradation encountered in high temperature service. The important characteristics and properties of platinum modified diffusion aluminide coatings are:

1. Platinum must be incorporated into diffusion aluminide coatings such that the coating surfaces contain predominantly platinum and aluminum. This can be readily achieved by applying the platinum as a thick layer followed by an appropriate aluminizing treatment.

2. Platinum modified coatings possess improved cyclic oxidation resistance because the platinum enriched zone of the coating promotes selective oxidation of aluminum with minimized effects of other elements in the coatings. The purer alumina scales have slower growth rates, which lead to longer exposure times before thicknesses are reached where spalling of the oxide occurs. Long time exposures under cyclic conditions result in substrate elements affecting oxidation behavior, and substrates containing Hf have longer coating lives.

3. The high affinity of platinum for aluminum is one of the necessary characteristics which results in the observed improved coating performances. This results in platinum aluminum phases that favor selective oxidation, exclude other elements from the coating surfaces, and may also impart some diffusional stability to the aluminide coatings.

4. Platinum improved the resistance of diffusion aluminide coatings to attack via high temperature hot corrosion. This improvement results from the improved selective oxidation characteristics of these coatings. After long exposures however substrate elements do affect coating performances and coating lives are shorter on the substrates with higher concentrations of molybdenum and tungsten.

5. Platinum improves the low temperature hot corrosion resistance of diffusion aluminide coatings. It is proposed that this results from exclusion of substrate elements from the surfaces of platinum modified coatings.

The platinum effect that has been observed on diffusion aluminide coatings should be produced also by other precious metals. Such elements must have a high affinity for aluminum, form intermetallic compounds with aluminum for which aluminum is the most mobile species, have a low standard free energy of formation for their oxides, and not a lower melting point than the incipient melting points of the coatings.

ACKNOWLEDGEMENTS

The Materials Technology Laboratory(Department of the Army), Watertown, MA and NASA Lewis Research Center, Cleveland, OH are gratefully acknowledged for financial support of parts of the research reported in this paper. Turbine components Inc. (Branford, Ct), Howmet Inc. (Whitehall, MI), Pratt and Whitney (East Hartford, CT), General Electric Co. (Evandale, OH), and Garrett Engine Co. (Phoenix, AZ) supplied coatings and alloys for this program.

REFERENCES

1. C. Duret and R. Pichoir: in 'Coatings for High Temperature Applications', (ed E. Lang), pp.33-78, 1983, New York, Applied Science Publishers.

2. D.G. Teer: in 'Coatings for High Temperature Applications', (ed E. Lang), pp.79-120, 1983, New York, Applied Science Publishers.

3. H.-D. Steffens: in 'Coatings for High Temperature Applications', (ed E. Lang), pp.121-138, 1983, New York, Applied Science Publishers.

4. N.R. Lindblad, Oxid. Metals, 1969, 1, 143-170.

5. G.F. Slattery, Metals Tech., 1982, 10, 41-51.

6. S.R. Levine and R.M. Caves, J. Electrochem. Soc., 1974, 121, 1051-1064.

7. R. Sivakumar and L.L. Seigle, Met. Trans., 1976, 7A, 1073-1079.

8. G.W. Goward and D.H. Boone, Oxid. Metals, 1971, 3, 475-495.

9. G.W. Goward and D.H. Boone, Trans. ASM, 1967, 60, 228-241.

10. M.M.P. Jannsen and G.D. Rieck, Met. Trans., 1967, 239, 1372-1385.

11. D.A. Joseph, U.S. Patent No. 3,102,044, 1963.

12. H.H. Todd, U.S. Patent No. 33,494,748, 1970.

13. M.J. Fleetwood, Metals Sci., 1970, 98, 503-509.

14. P. Deb and D.H. Boone,"Microstructural Formation and Effects on the Performance of Platinum Modified Aluminide Coatings", Tech. Report, Naval Postgraduate School, 1985.

15. K. Bungardt et al., U.S. Patent No. 33,677,789, 1972.

16. G.J. Tatlock, T.J. Hurd, and J.S. Punni, Platinum Metals Rev., 1987, 31, 26-31.

17. G.R. Johnston and P.G. Richards: in 'Corrosion in Fossil Fuel Systems', (ed I.G. Wright), pp.456-461, 1983, Pennington N. J., The Electrochemistry Society.

18. R. Bauer, K. Schneider and H.W. Grünling, High Temp. Tech., 1985, 3, 59-64.

19. R. Streiff and D.H. Boone, J. Mater. Eng., 1988, 10, 15-26.

20. E.J. Felten and F.S. Pettit, Oxid. Metals, 1976, 10, 189-223.

21. F.S. Pettit and C.S. Giggins: in 'Superalloys II', (ed C.T. Sims et al.), pp.327-358, 1987, New York, Wiley.

22. M.R. Jackson and J.R. Rairden, Met. Trans., 1977, 8A, 1697-1707.

23. D.P. Whittle and J. Stringer, Phil. Trans. Roy. Soc. London, 1980, 309, 309-329.

24. W.T. Wu, A. Rahmel, and M. Schorr, Oxid. Metals, 1984, 22, 59- 81.

25. L.K. Luthra, Met. Trans., 1982, 13A, 1843-1864.

Session 4

INTERACTION OF SULPHUR AND CARBON IN OXIDATION

THE ROLE OF SULFUR IN THE GROWTH AND ADHESION OF OXIDE SCALES

J. G. Smeggil, A. J. Shuskus, N. S. Bornstein, M. A. DeCrescente

United Technologies Research Center

East Hartford CT 06108

ABSTRACT

Experiments are reported in which indigenous sulfur, present at low levels within a substrate alloy, is sufficient to produce oxide scale exfoliation effects. The amount of sulfur commonly found at levels less than approximately 50 ppm by weight in normally prepared alloys is sufficient to produce exfoliation. The reduction of such sulfur either through the use of high purity, i.e. low sulfur starting materials or the use of active element additions, is sufficient to produce adherent scales. By reacting with indigenous sulfur to form refractory sulfides, active element, e.g. yttrium, additions effectively reduce the amount of sulfur available to segregate to and weaken the crucial scale–metal interfacial bond.

Pores are frequently observed on substrate surfaces when oxide scales exfoliate. These pores have commonly been attributed to Kirkendall effects. By reacting with and annihilating vacancies resulting from Kirkendall effects at the scale–metal interface, active element oxides at the scale–metal interface have been suggested to promote scale adherence. Studies conducted here have shown that without active element additions small pores few in number are associated with exfoliation effects. Despite adherent scales, similar pores occur at the scale–metal interface. Hence the relationship between porosity at the scale–metal interface and protective scale adherence is unclear.

In a series of experiments involving implanted species that would be little soluble at elevated temperatures, exfoliating scales were produced despite the presence of yttrium in a Ni - 20 wt% Cr - 12 wt% Al alloy. Pores in exfoliated regions of such implanted specimens had all the characteristics normally associated with scale exfoliation effects in unimplanted materials. Hence it is suggested that the porosity commonly observed at surfaces of conventionally produced alloy substrates is rather due to the expulsion of contaminant levels, i.e. low ppm levels, of little soluble elements, e.g. argon, present at low levels.

Experimental results which have led to these conclusions are presented.

1. INTRODUCTION

Based on studies involving model alloy systems, studies have revealed important factors affecting the adhesion of oxide scales whose growth processes are strongly influenced by oxygen diffusion effects, i.e. alumina and chromia (e.g. [1]). Specifically sulfur indigenously present in most alloys at low levels, i.e. approximately 20 – 50 ppm, is sufficient to produce exfoliation by weakening the bond at the scale-metal interface. Alternatively either reducing the sulfur content of the initial alloy or controlling its sulfur activity by yttrium additions, in the case of alumina-formers, restored scale adherence. For the chromia-forming alloy Ni – 40 wt% Cr, the addition of yttrium did not substitute for sulfur reductions as regards affecting scale adherence in 1000°C thermal cyclic testing.[2]

Additionally ion implantation techniques have been quite useful in exploring processes occurring during oxidation. Not only have they been useful from the viewpoint of examining factors responsible for adherence, but they have also shed light onto factors related to the formation of pores commonly observed at the scale–metal interface. Here findings suggest that small amounts of low solubility gases, e.g. helium, argon, chlorine, etc., segregate to the scale–metal interface. They do not diffuse through the growing scale. Their presence at this interface during oxidation allows for a restructuring of the metal surface producing pores. Based on these findings, such pores, frequently termed Kirkendall pores, although frequently and extensively reported previously, do not grow by classical Kirkendall processes.

Results of experiments which have led to these conclusions are presented.

2. EXPERIMENTAL MATERIALS AND PROCEDURES

Alloys were prepared by standard techniques using nickel and aluminum of 99.9 wt% purity and chromium of 99.99 wt% purity. The proper elemental combinations were combined, vacuum melted in alumina crucibles and poured in 2.5 cm chilled copper molds. Base line alloy compositions included the following:

Ni – 20 wt% Cr – 12 wt% Al
Ni – 20 wt% Cr – 12 wt% Al – 0.1 wt% Y
Ni – 40 wt% Cr

Additionally experiments were conducted in which yttrium as yttrium sulfide was added to the NiCrAl composition. The additions were made at levels corresponding to those which would yield approximately 0.1 wt% yttrium if it had been added instead as the metal. Furthermore a composition was prepared which contained both yttrium and yttrium sesquisulfide. The thought behind this experiment was that, when Y_2S_3 was added to the metal alloy, the lowest reported yttrium sulfide, YS, would tend to form effectively liberating additional sulfur to an alloy already containing sulfur at indigenous levels of approximately 20 to 50 ppm. Extra yttrium then would be needed to form refractory YS with both the liberated as well as the indigenous sulfur to restore oxide scale adherence.

Small buttons of the base line Ni - 20 wt% Cr - 12 wt% Al and the Ni - 40 wt% Cr alloys were also prepared from high purity starting metals obtained from Materials Research Corporation. In this case alloy manufacture occurred by using an arc melter which was disassembled and cleaned prior to melting. Extra care was also taken in washing the cut metal fragments prior to melting to minimize sulfur contamination from normal machine shop fluids and oils. After casting or arc melting, all ingots and buttons were annealed in either vacuum or hydrogen for 24 hours at 1200°C to reduce solidification induced compositional gradients. The gross analyzed compositions for the above alloys have been reported elsewhere, e.g. [2,3].

To establish whether indeed low sulfur levels present within normally prepared metals and alloys were responsible for oxide scale exfoliation, a reliable analytical technique for measuring low (ppm) levels of elements of interest had to be identified.[4] Vendor- supplied analyses are hoped to be correct but are always suspect. In addition, analyses of metals as commonly supplied by the vendor do not normally take into account nonmetallic components. Furthermore, on the basis of prior experience, target alloy compositions, particularly with regard to low level additions, can be in serious error. Additionally traditional methods, techniques and procedures used to produce alloys can themselves introduce low levels of impurities such as sulfur at levels which historically have not been considered to be significant. It is specifically these low, previously considered inconsequential, ppm levels of contaminants (again primarily sulfur) which this work has shown to dramatically affect oxide scale adherence properties. Therefore techniques had to be identified and established that could reproducibly yield bulk sulfur analyses for concentrations of interest, i.e. less than approximately 50 ppm by wt, in the actually prepared and homogenized alloys.

Normally, fusion techniques have been used to analyze for bulk sulfur in metal alloys. Although widely successful in support of the ferrous metals industry, fusion techniques yielded totally unsatisfactory results in the sulfur analyses of metal matrices of interest here. The problem manifested itself in poor precision of reported sulfur contents for the same specimen either analyzed at the same time or at different times. Bulk sulfur values based on traditional fusion techniques had to be completely discounted. After many experimental studies, sulfur analyses based on Auger surface techniques were also discounted as being able to yield reliable or reproducible bulk sulfur contents of multiphase alloys such as the Ni - 20 wt% Cr - 12 wt% Al alloy discussed here. The problems involving Auger analysis techniques included, but were not restricted to, such observations as that sulfur preferentially segregated to the surface of the chromium-rich gamma phase in the two phase gamma - beta (NiAl) NiCrAl alloy.[5] Hence the reported sulfur content of the surface became highly dependent upon the microstructure of the alloy and on which part of the two phase microstructure the analysis was being conducted.

Glow discharge mass spectrometry analysis has been found to give adequate precision to differentiate the effects of differing low bulk alloy sulfur levels on oxide scale adherence. To independently check the precision of the sulfur analyses, specimens from both a well homogenized ingot of a high purity NiCrAl alloy and a powder metallurgy superalloy were repeatedly analyzed. The actual identity of these specimens was hidden from the analytical chemists involved. The analytical precision is attested to both by the repeated random analysis of the single MCrAl specimen and by the insertion of the powder metallurgy prepared materials supplied with each analysis. A large supply of these powder metallurgy specimens has been prepared from a single ingot. Because of their powder metallurgy genesis, ingot materials produced by powder metallurgy techniques should be sufficiently homogeneous to insure the precision of this chemical analysis.

Literally based upon hundreds of analyses, sulfur contents derived from glow discharge mass spectrometry procedures have reproducibly indicated sulfur levels down to 1 ppm and have shown that significant differences between sulfur levels, e.g. 1 - 2 ppm vs 10 - 20 ppm are readily and reproducibly distinguishable. Although both accuracy and precision are desirable in an analysis, the ability to reproducibly demonstrate differences in sulfur analyses between different well homogenized materials is paramount to these studies. The sulfur values reported here are derived from such glow discharge spectrometry measurements.

Sulfur contents reported for the standard and high purity Ni - 20 wt% Cr - 12 wt% Al alloy are 25 (+/-8) ppm for 7 separate analyses and 1.7 (+/-0.6) ppm for 4 separate analyses, respectively. On the other hand for the Ni - 40 wt%

Cr alloy, values of 40 and 57 ppm by wt sulfur and 1.45 and 1.81 ppm by wt sulfur were reported for standard and high purity alloys.

Concerning the ion implantation studies, Ni - 20 wt% Cr - 12 wt% Al and Ni - 20 wt% Cr - 12 wt% Al - 0.1 wt% Y alloys were implanted with yttrium, sulfur, helium, argon, and chlorine at levels ranging between 10^{16} and 10^{17} ions/cm^2 at implantation energies of 23 to 200 Kev, using a procedure previously described.[6] The implantation voltages were varied as a function of atomic weight of the implanting species in an attempt to keep the maximum of the implant dosage approximately 100 nanometers below the surface.[6] Approximately one quarter of each specimen was not ion implanted so as to have a reference surface to allow comparison of implanted and unimplanted areas of the same specimen.

Cyclic oxidation experiments were conducted in which specimens were heated to elevated temperatures for 55 minutes then exposed to room temperature for five minutes; each such exposure constituted a single cycle. Specimens would be initially measured after every approximately twenty cycles and less frequently as experiments progressed. Isothermal experiments were also conducted in which specimens were heated in laboratory furnaces and then removed for examination.

Specimens were examined by standard optical metallography, scanning electron microscopy and electron microprobe techniques.

3. EXPERIMENTAL RESULTS AND DISCUSSION

3.1 Alloy Purity (Sulfur Content) Effects

In deciding whether or not the form in which the yttrium is present is critical for oxide scale adhesion, experiments were conducted in which the standard purity Ni - 20 wt% Cr - 12 wt% Al alloy was modified in the usual technique by yttrium additions. As shown by the 1050°C cyclic oxidation data presented in Fig. 1 in agreement with commonly made observations, elemental yttrium additions to the standard NiCrAl alloy clearly improve oxide scale adherence as compared with the performance of the NiCrAl alloy without this addition.

However when yttrium is added instead as the sesquisulfide, clearly detrimental effects ensue despite the presence of the yttrium. Moreover not only scale was exfoliation observed, the scale exfoliation is much worse than was the case for the NiCrAl alloy to which no addition was made. When the yttrium sesquisulfide was added to the ingot, it may have partially decomposed liberating sulfur to the ingot according to the following reaction:

$$Y_2S_3(c) \longrightarrow YS(c) + \text{“S”}.$$

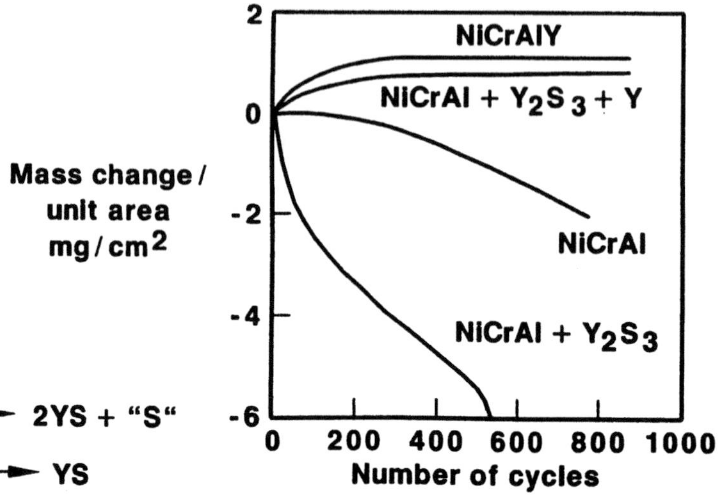

$Y_2S_3 \rightarrow 2YS + \text{"S"}$

$Y + \text{"S"} \rightarrow YS$

Fig. 1 Effect of yttrium added as elemental yttrium and yttrium sesquisulfide on the 1050°C cyclic oxidation behavior of Ni-20Cr-12Al.

If on the other hand, additional yttrium is added to the ingot, not only to react with the indigenous sulfur intrinsically present in the normally prepared alloy but also with the sulfur which has been liberated by the above mentioned reaction involving Y_2S_3, then beneficial oxide scale adherence properties are restored despite the fact that the alloy is replete with sulfides, Fig. 1.

The above experiments were conducted at 1050°C, a temperature deliberately selected so as not to be so high that if many different effects are occurring there is not chance at sorting them out and yet a temperature at which the alumina scale which forms is alpha alumina. Despite that concern, experiments at a temperature more commonly used to evaluate these kinds of materials, i.e. 1180°C, yielded exactly the same results, Fig. 2.

The above discussed experiments imply that sulfur is deleterious. However in each and every case where beneficial results ensue, elemental yttrium additions are to be noted in those alloys performing well, cf. Figs. 1 and 2. Experiments using normal and high purity Ni – 20 wt% Cr – 12 wt% Al alloys, where the purity here is specifically meant to be in regard to sulfur content, were conducted and are noted in Fig. 3. Here specific mention is to be made that beneficial adherence effects are present despite the fact that no yttrium addition has been made at all. Simply stated, yttrium additions are not necessary to achieve adherent alumina scales. Moreover although yttrium may be present,

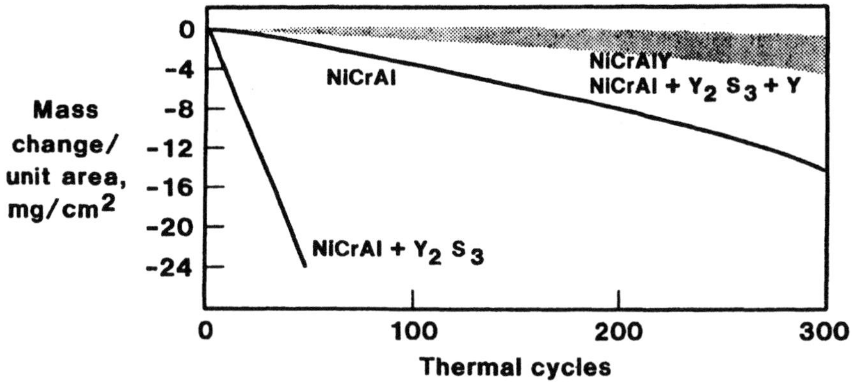

Fig. 2 Effect of yttrium added as elemental yttrium and yttrium sesquisulfide on the 1180°C cyclic oxidation behavior of Ni-20Cr-12Al.

unless it is there in a form such that it can react with indigenous sulfur, it is useless from the viewpoint of oxide scale adherence.

In cyclic experiments involving Ni – 40 wt% Cr, markedly different behavior was observed depending upon the temperature. In cyclic testing at 900°C, little difference was noted in performance between the normal and high purity specimens, Fig. 4. Analogously in 1100°C testing, both materials behaved

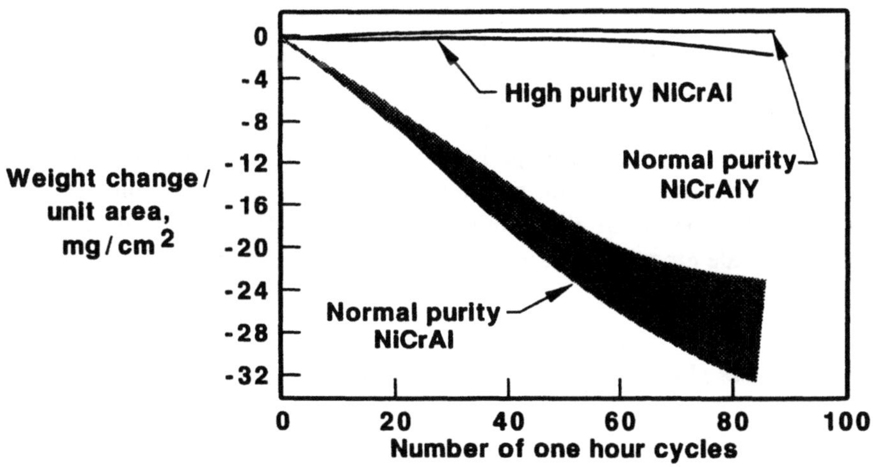

Fig. 3 Effect of indigenous sulfur content on the 1180°C cyclic oxidation behavior of Ni-20Cr-12Al. Sulfur content of normal purity material 20-40 ppm (by weight) and high purity material purity material approximately 1-2 ppm.

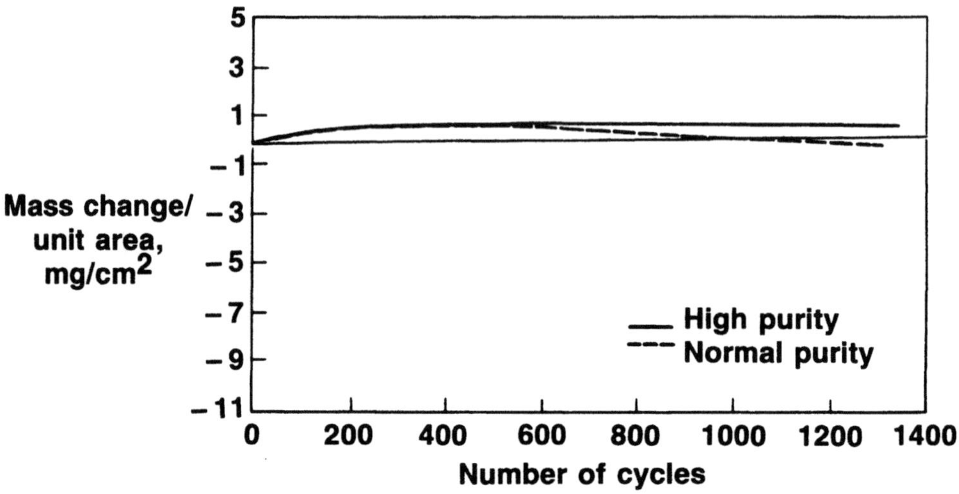

Fig. 4 A comparison of the 900°C cyclic oxidation behavior of normal and high purity, i.e. normal sulfur and low sulfur, Ni-40 wt % Cr alloys.

similarly, Fig. 5. However at 1000°C, the high purity Ni - 40Cr alloy exhibited superior performance over that exhibited by the standard purity alloy, Fig. 6. Moreover in isothermal testing conducted for the same total time as the cyclic experiments at 1000°C, weight change behavior indicated that the high purity alloy had gained less weight, Fig. 6. Optical metallographic examination of oxidized specimens confirmed this speculation, Figs. 7 – 8. The normal purity material had shown both typical grain boundary and internal oxidation effects normally expected for this alloy tested under these conditions, Fig. 7. On the other hand the high purity alloy had developed thinner (more protective) scales without any evidence of internal or grain boundary oxidation, Fig. 8. The thinner scales formed on the low sulfur alloy would suggest that indigenous sulfur also has an effect on scale growth processes.

3.2 Effects Producing Porosity at the Scale–Metal Interface

In the absence of active element additions to standard purity alloys such as Ni - 20Cr - 12Al, specimens are commonly observed, as stated above,to exfoliate protective scales either during cyclic testing or upon removal from the furnace after an isothermal test. Examination of isothermally tested specimens showing appreciable scale exfoliation frequently revealed porosity at the scale–metal

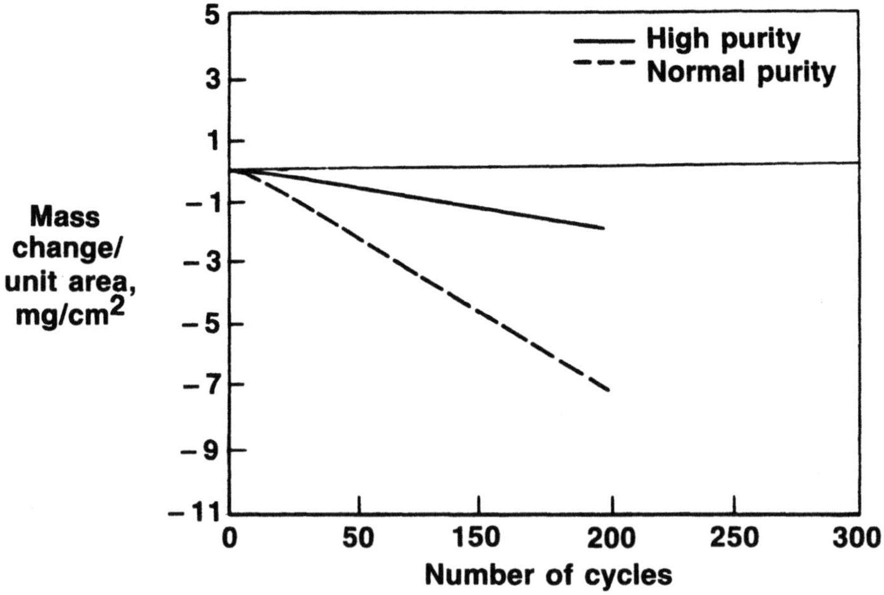

Fig. 5 A comparison of the 1100°C cyclic oxidation behavior of normal and high purity, i.e. normal sulfur and low sulfur, Ni-40 wt % Cr alloys.

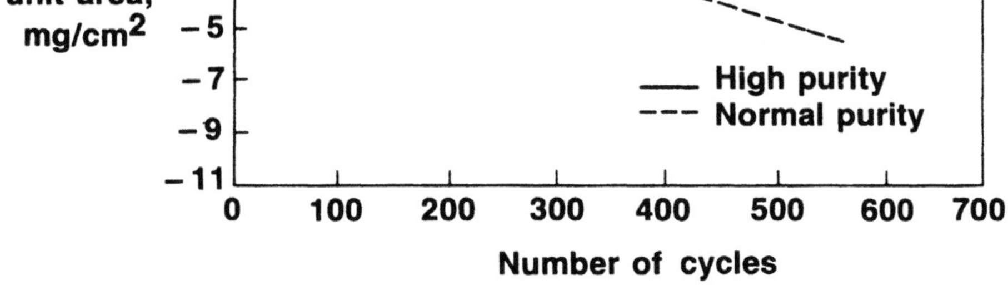

Fig. 6 A comparison of the 1000°C cyclic and isothermal oxidation behavior of normal and high purity, i.e. normal sulfur and low sulfur, Ni-40 wt % Cr alloys.

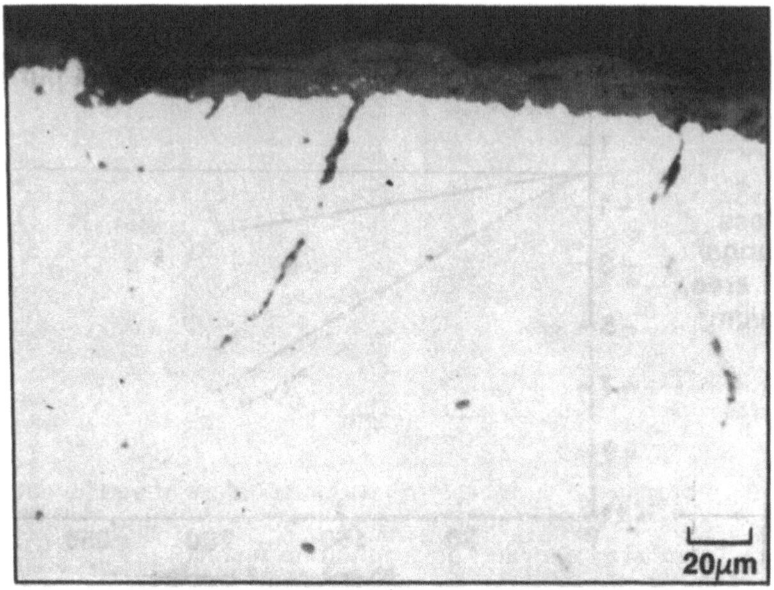

Fig. 7 Cross section of normal purity Ni-40 wt % Cr alloy after 576 hours of isothermal oxidation at 1000°C. Optical metallography.

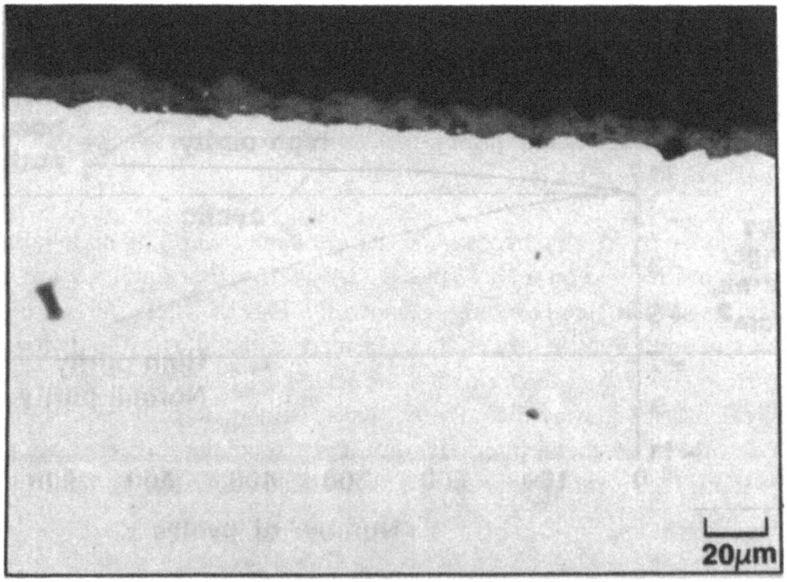

Fig. 8 Cross section of high purity Ni-40 wt % Cr alloy after 576 hours of isothermal oxidation at 1000°C. Optical metallography.

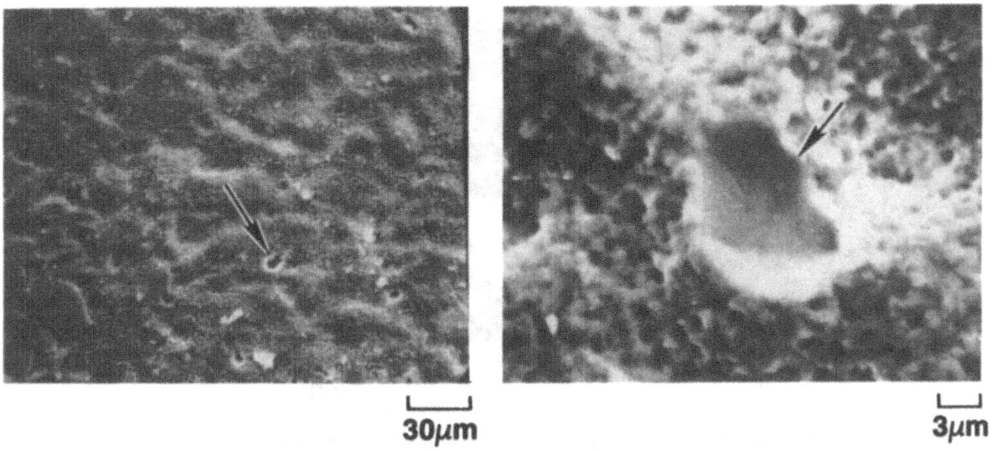

Fig. 9 Porosity on surface of Ni-20Cr-12Al alloy after 100 hours of isothermal testing in air at 1050°C. Alumina scale has exfoliated showing only substrate metal.

interface, e.g. Fig. 9. However it is to be specifically noted that these pores are relatively small in size (a few microns in diameter) and relatively few in number (at best perhaps twenty or so are to be found on the completely exfoliated metal surface shown in Fig. 9). However, because alloys with active element additions such as yttrium form adherent scales, it is commonly believed that such pores contribute to oxide scale exfoliation, e.g. [1]. Furthermore the active element addition results in the formation of oxides at the oxide-scale interface which serve as vacancy coalescence sites mitigating the formation of pores which would result in oxide scale exfoliation, e.g. [1].

Unfortunately for this hypothesis, the mechanical scribing of an adherent scale which has formed on a Ni - 20Cr - 12Al alloy with a small yttrium addition clearly shows similar small amounts of porosity, Fig. 10. These pores are just as common and equivalent in size as those formed on the alloy without yttrium additions. However despite this porosity, the scale here is quite adherent. [After mechanically scribing away the adherent scale, similar porosity has also been observed at the scale-metal interface of oxidized specimens of other MCrAlY alloys.] Hence the relationship between oxide-scale adherence and porosity at the scale-metal interface is unclear.

To investigate factors which can affect oxide scale adhesion, studies have been conducted involving ion implantation effects. [6] In these studies specimens have been implanted with various elements as indicated above in the Experimental Section and then oxidized at elevated temperatures.

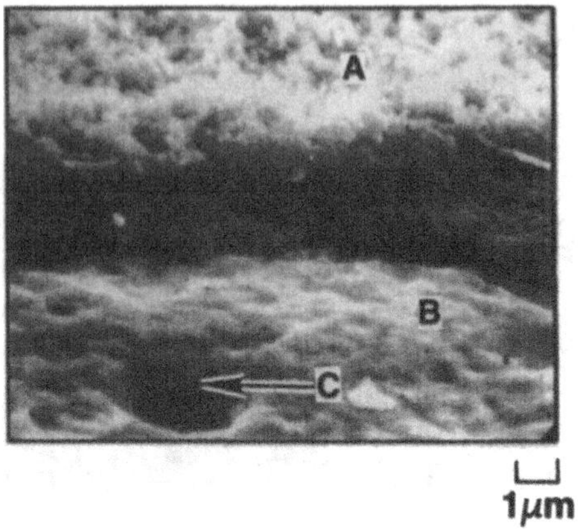

Fig. 10 Porosity on surface of a Ni-20Cr-12Al alloy after 100 hours of isothermal testing in air at 1050°C. Alumina scale (A) has been mechanically scribed to reveal the substrate metal (B). Porosity shown by arrow labeled C.

Blank specimens implanted with aluminum formed adherent scales upon subsequent oxidation at 1050°C. Hence the implantation process itself did not produce conditions which subsequently led to oxide scale exfoliation. However in instances where the implant consisted of helium, argon, neon or chlorine, non-adherent scales formed despite the presence of yttrium in the Ni - 20 wt% Cr - 12 wt% Al alloy. It is specifically to be noted that hydrogen implants did not produce exfoliation effects.

Energy dispersive analysis of the argon implanted Ni - 20Cr - 12Al - 0.1Y alloy clearly indicated the presence of argon, cf. Fig. 11. After 100 hours of oxidation at 1050°C in air, the unimplanted surfaces produced adherent scales in agreement with prior literature reports, e. g. [1].

However, despite the presence of yttrium the scales formed on the argon implanted surfaces exhibited considerable exfoliation, Fig. 12. Furthermore upon examination of the exfoliated surfaces at higher magnification, substantial porosity was observed at the scale-metal interface in regions which had previously been argon implanted. In addition, even higher magnification characterization indicated crystallographic features associated with this porosity, Fig. 13. Crystallographic features have not uncommonly been associated with porosity at the scale-metal interface. [Hydrogen may not have produced porosity effects for two possible reasons. In the first place, hydrogen could have diffused through the scale to the ambient atmosphere. Secondly, the implanted hydrogen could have reduced its free energy within the alloy by either of two mechanisms:

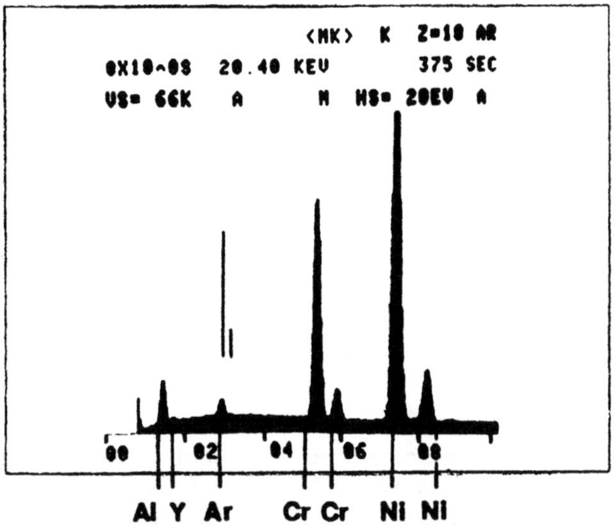

Fig. 11 Argon implanted into Ni-20 wt % Cr-12 wt % Al. Prior to oxidation. Implantation conditions: 10^{16} Ar^{+}/cm^{2} at 200 Kev.

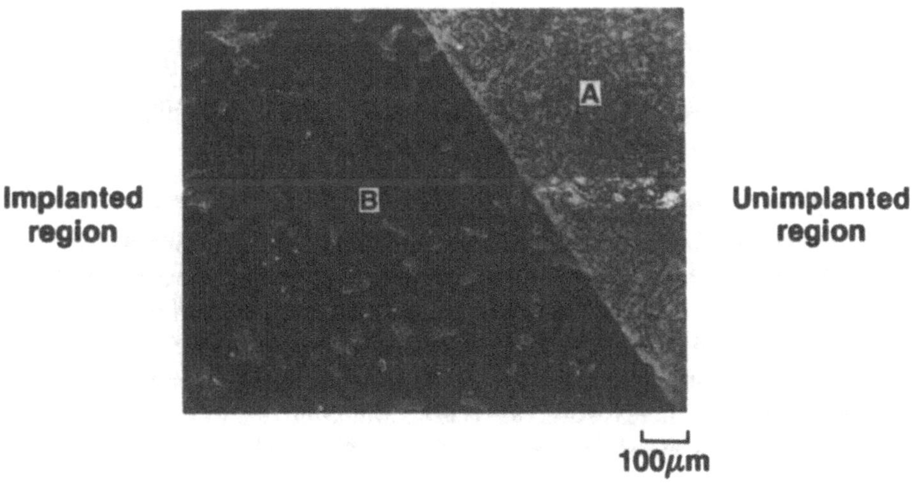

Fig. 12 Ni-20 wt % Cr-12 wt % Al-0.1 wt % Y oxidized for 100 hours in air at 1050°C.

A. Adherent scale formed on base alloy.

B. Non-adherent scale formed on argon implanted alloy, cf. Fig. 11 for implantation conditions.

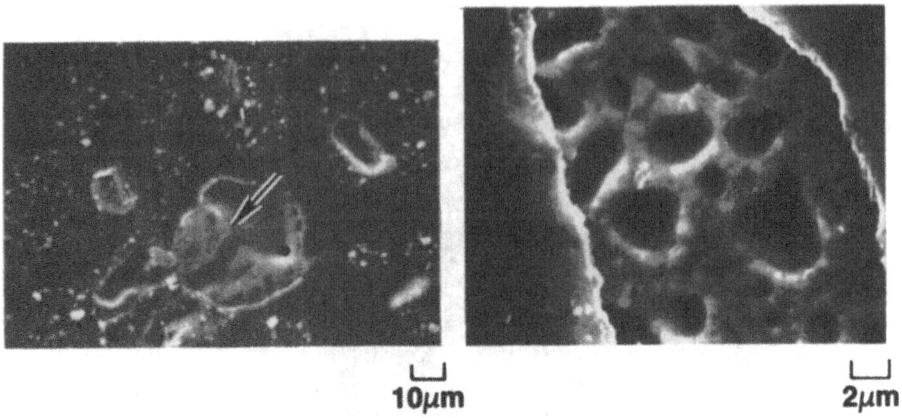

Fig. 13 Porosity at the scale-metal interface of argon implanted Ni-20 wt % Cr-12 wt % Al-0.1 wt % Y oxidized for 100 hours at 1050°C, cf. Fig. 11 for implantation conditions. Note geometric microstructural features associated with such porosity, cf. arrow

1) simple dissolution in the substrate alloy and 2) the formation of stable metal hydrides.]

Pores immediately adjacent to the scale have commonly been termed Kirkendall pores. However these pores are reported only to occur at the scale–metal interface, e.g. [1]. Reports of their presence away from the scale–metal interface deeper into the substrate metal are commonly lacking.

The findings presented here would suggest that chemical species present within normally prepared materials can segregate upon heating during oxidation to the scale–metal interface. At that interface these species can interact with the substrate alloy surface to lower its surface energy by the elimination of high energy planes and the subsequent growth of low energy planes. This surface re–organization process could upon cursory examination simply be interpreted as Kirkendall porosity whereas its actual genesis would be due to the presence of small amounts of surface active elements at the scale–metal interface.

4. CONCLUDING REMARKS

Experiments have been conducted which indicate that active element additions such as yttrium are not necessary for the formation of adherent chromia or alumina scales. Their addition may promulgate the formation of adherent films by the formation of refractory sulfides. Refractory sulfide formation would limit the amount of indigenous sulfur from the alloy which could segregate to and weaken the critical scale–metal interfacial bond.

In addition, pores are not infrequently observed on metal surfaces when oxide scales have exfoliated. These pores have been attributed to Kirkendall effects and have been suggested to produce oxide scale exfoliation. Active element oxides have been suggested to eliminate such porosity effects by acting as vacancy sinks. However these studies have shown that excessive scale exfoliation also occurs with tiny pores few in number. Moreover, similar amounts of porosity also occur when adherent scales form on MCrAlY alloys. Hence the relationship between porosity at the scale-metal interface and protective scale adherence effects is unclear.

As an alternative explanation of this porosity, ion implantation experiments have demonstrated that such pores can and do form when little soluble elements such as helium, argon, neon, chlorine are implanted into metal surfaces which are subsequently oxidized. Despite the presence of yttrium, exfoliation occurs and porosity is observed. The pores exhibit geometric microstructures similar to those observed for pores in unimplanted, yttrium-free specimens. Hence it is suggested that the pores at the scale-metal interface are not due to Kirkendall effects at all but rather result from interfacial effects resulting from relatively insoluble elements being expelled from the substrate alloy to the scale-metal interface during elevated temperature oxidation.

REFERENCES

1. D. P. Whittle and J. Stringer: *Philos. Trans. R. Soc. London*, 1980, *Ser.*, **A295**, 309–32.

2. J. G. Smeggil: "The Effect of Indigenous Sulfur on the Adherence of Chromia Scales" presented at the spring 1988 ASM-AIME meeting, Las Vegas, Nevada, submitted for publication in conference proceedings.

3. J. G. Smeggil: *Mater. Sci. and Eng.*, 1987, **87**, 261–266.

4. J. G. Smeggil, N. S. Bornstein and M. A. DeCrescente: *Oxid. of Metals*, 1988, **30**, 259–266.

5. J. G. Smeggil and G. G. Peterson: *Oxid. of Metals*, 1988, **29**, 103–119.

6. J. G. Smeggil, E. L. Paradis, A. J. Shuskus and N. S. Bornstein: *J. Vac. Sci. Technol., A.*, 1985, **3**, 2269–73.

THE INFLUENCE OF SULPHUR AND ITS INTERACTION WITH YTTRIUM ON THE COMPOSITION, GROWTH AND ADHERENCE OF OXIDE SCALES ON ALUMINA FORMING ALLOYS

A.S. Khanna[1], W.J. Quadakkers, C. Wasserfuhr and H. Nickel

Institut für Reaktorwerkstoffe, Kernforschungsanlage Jülich,

Postfach 1913, 5170 Jülich, F.R.G.

[1] Guest Scientist from IGCAR, Kalpakkam, India

ABSTRACT

The deleterious effect of small amounts of sulphur impurity on the oxidation behaviour of NiCrAl has been investigated. The oxidation of Ni-10Cr-8Al containing controlled amounts of sulphur was studied at 1000 and 1100°C. Conventional thermogravimetry, SEM, EDX and X-ray diffraction were used to study the scale growth and composition, whereas acoustic emission was utilised to obtain detailed information on the scale adherence. The results indicated that the dominant factor responsible for the deleterious effect of sulphur impurity on the oxidation of the NiCrAl alloy was a significant change in the growth rate and the composition of the scale. Addition of yttrium improved the oxidation behaviour, not only by increasing the scale adherence, but also by alleviating the enhanced scale growth initiated by the sulphur.

1. INTRODUCTION

It is well known that the presence of trace amounts of impurities such as sulphur, phosphorous, boron etc., can sometimes influence the corrosion behaviour of metals and alloys, e.g. their susceptibility to intergranular corrosion in aqueous media.[1,2] Recently, it has been found that trace amounts of sulphur also have a detrimental effect on the oxidation resistance of high temperature alloys. Smeggil and co-workers[3,4] and Smialek[5] have reported that trace levels of sulphur in an alloy decrease oxide scale adherence. They proposed that sulphur segregates to the metal/oxide interface resulting in a weakening of the substrate/scale bond. Further, it has been reported that active element additions to sulphur-containing alloys prevent scale spallation by gettering the sulphur . Luthra and Briant[6] initially agreed with this explanation, but their more recent results[7] showed that the scale/substrate bond on pure materials is intrinsically weak, irrespective of the presence of sulphur. Previous experiments from the present authors[8] confirmed that sulphur impurities have a detrimental effect on the oxidation behaviour of NiCrAl alloys. It was found that the presence of sulphur in these alloys altered the initial scale compositon, leading to an increase of the oxide growth rate rather than to a decrease in scale adherence. The observed tendency for spalling of scales on sulphur-containing alloys seemed not to be related to bond weakening by sulphur but to relief of growth stresses by thick oxide scales.

In the present work, the oxidation behaviour of Ni-10Cr-8Al alloys, containing various amounts of sulphur, was investigated at 1000°C in air. Furthermore, the mechanism by which the addition of yttrium influences the oxidation behaviour of sulphur-containing alloys was examined. Contrary to the previous experiments[8], all alloys were made from high purity metals. The scale growth and composition was studied by gravimetry and SEM/EDAX, the scale adherence by acoustic emission (AE) analysis.

2. EXPERIMENTAL

The alloys of base composition Ni-10Cr-8Al were prepared from high-purity nickel, chromium and aluminium. Sulphur was added in the form of chromium sulphide and yttrium in the form of pure metal. The alloy melting was carried out in an induction furnace. After casting the cylindrical ingots were chemically analysed. In the base alloy, without impurity addition, the sulphur content was below the detection limit of 10 ppm. The ingots were machined to a diameter of 15 mm from which specimens in the form of discs of 1 mm thickness were cut. These were then polished up to 600 grit finish, degreased and cleaned in acetone before exposure

Oxidation tests were carried out in air at 1000°C in a horizontal muffle furnace maintained at 1000 ± 2°C. Long-term exposures up to 1000 hour duration were performed to determine the rate of scale growth and the extent of spallation. Short duration tests (2 min. to 25 hours) were carried out for detailed analysis of scale morphology and composition before extensive spalling had occurred. Identification of the various oxides in the scale was carried out using X-ray diffraction, scanning electron microscopy (SEM) and energy dispersive X-ray analysis (EDAX).

More detailed information on the scale adherence was obtained by Acoustic Emission (AE) analysis using a Dunegan series 3000 equipment. All AE experiments were carried out during specimen cooling after 18 hours oxidation at 1100°C. Corresponding thermogravimetric experiments were carried out under similar conditions in order to correlate the results obtained by these two techniques.

3. OXIDATION KINETICS

Fig. 1 gives the linear plots of weight gain vs time for the oxidation of Ni-10Cr-8Al containing various levels of sulphur impurity. The alloy without sulphur addition shows spallation even after very short exposure times. Scale adherence on the samples with sulphur addition is significantly better, however, the scale growth rates are noticeably higher. The growth rate increases with increasing sulphur level and the alloy with the highest growth rate (i.e. highest sulphur-content) shows the earliest onset of scale spalling.

Fig. 2 shows the linear plots of weight gain vs time for the samples containing yttrium in addition to sulphur. A low level of yttrium (e.g. 0.01% Y in the 80ppm S sample) influenced the scale growth only to a limited extent. When a higher concentration of yttrium was present (e.g. 0.26% in the 200 ppm S sample) the scale growth rate decreased and the scale adherence was markedly improved. A further increase in the concentration of yttrium, as

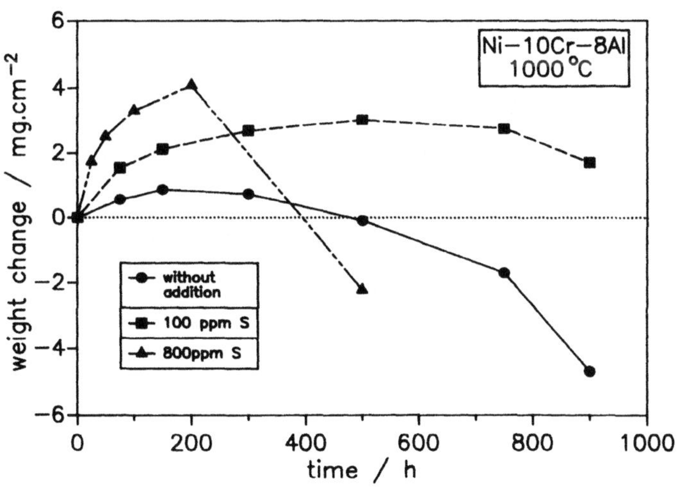

Fig. 1: Linear plots of weight change vs time during the oxidation of Ni-10Cr-8Al containing different levels of sulphur impurity at 1000°C in air. Weight changes were measured after cooling to ambient temperature.

in the case of the sample containing 0.53% Y and 70 ppm S, led again to an increase in the oxidation rate. Thus, an optimum concentration of yttrium is required to obtain the most beneficial effect on the oxidation behaviour. This optimum amount depends upon the sulphur level in the alloy.[7]

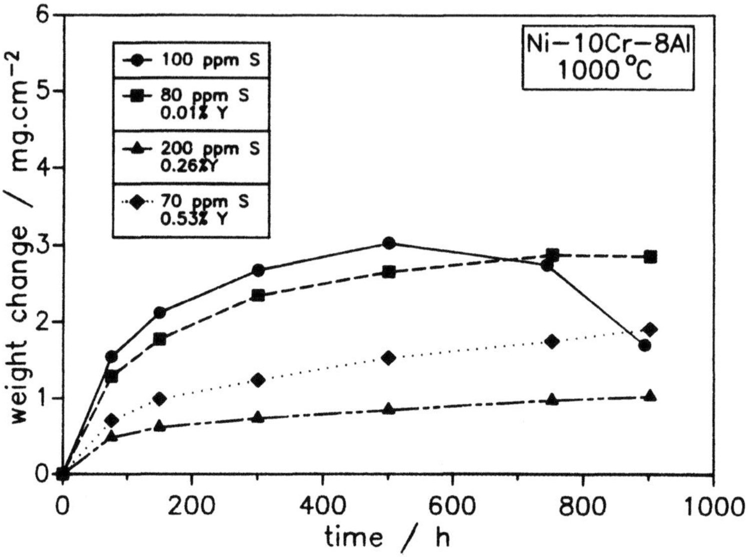

Fig. 2: Linear plots of weight change vs time, showing the influence of yttrium addition on the oxidation kinetics of sulphur-containing alloys at 1000°C in air. Weight changes were measured after cooling to ambient temperature.

4. OXIDE SCALE ANALYSIS

Fig. 3a depicts the surface morphology of the oxide formed on the sample without impurity addition. It is evident that the scale formed in the short oxidation period of 25 hours has spalled extensively. Corresponding EDAX analysis showed that the scale consisted mainly of alumina. Figs. 3b and c show SEM micrographs of the oxide scale formed on the samples containing 100 and 230 ppm sulphur, respectively. The oxide morphology appears to be similar for both alloys, except that more NiO crystals were seen on the sample with the higher amount of sulphur. EDAX analysis showed higher concentrations of Cr and Ni indicating more chromia and NiO being formed on the 230 ppm sulphur specimen than on the pure material and on that containing 100 ppm sulphur. Formation of small amounts of chromia on the sulphur containing alloys occurred along with the alumina, even in the very first stages of oxidation[7] as shown in Fig. 4. It is not quite clear whether the formation of chromia particles is related to chromium sulphides present in the alloy microstructure (Fig. 4).

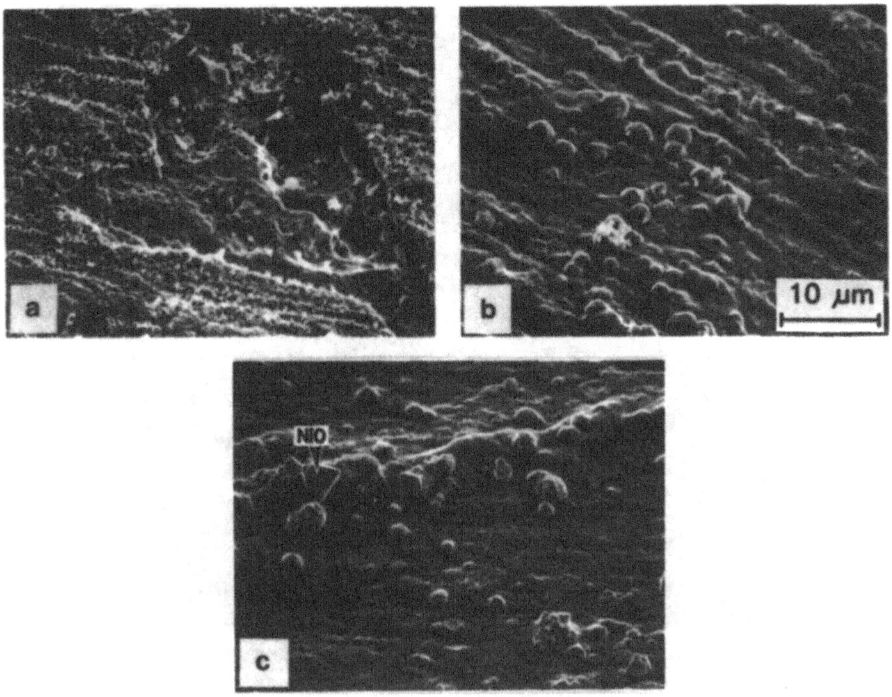

Fig. 3: SEM micrographs showing the influence of sulphur content on the scale morphology and composition of Ni-10Cr-8Al after oxidation in air at 1000°C for 25h:
a) without addition, b) 100 ppm S, c) 230 ppm S.

Fig. 5 shows the surface morphologies of the oxide scales formed on the samples containing yttrium in addition to sulphur. The oxide morphology on the sample containing 80ppm S and 0.01% Y (Fig. 5a) was very similar to that of the sulphur-containing sample without yttrium (Fig. 3b). This indicates that, in view of the high sulphur level of the alloy, the amount of yttrium was too small to cause any significant change in the oxidation behaviour. A larger amount of yttrium, as in the sample containing 70 ppm S and 0.53% Y, led to an entirely different type of oxide layer which consisted mainly of alumina and possessed good adherence. Apparently, yttrium addition not only improves scale adherence, as has been known for a long time, but in sulphur-containing alloys it also restores the selective oxidation of aluminium to the situation in sulphur-free alloys.

Fig. 6 shows the cross-sections of the oxide scales, formed after 25h oxidation on the sulphur-free sample, and on the samples

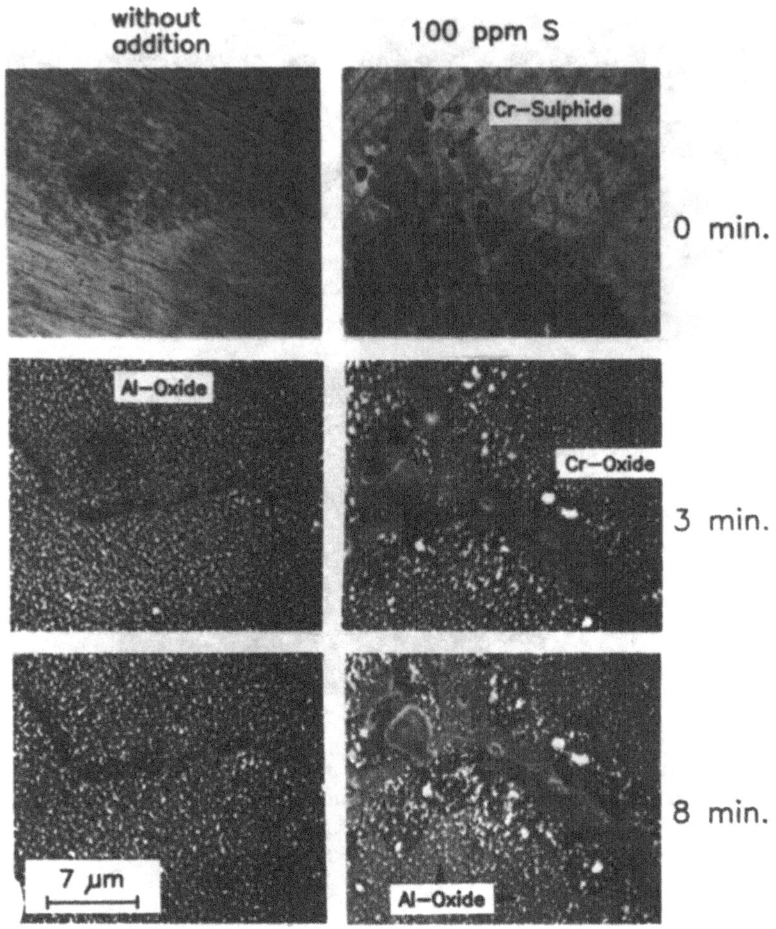

Fig. 4: SEM pictures showing the influence of sulphur content on the scale formation of Ni-10Cr-8Al in the initial stages of oxidation at 1000°C. The figures show exactly the same specimen area before oxididation and after two different oxidation periods.

containing 100 and 230ppm S. While there was a thin, non-adherent oxide layer on the sulphur-free sample, the sulphur-containing samples showed more adherent, but thicker oxide layers. Pores, which are seen within the matrix on the sulphur-free sample, were probably formed by the removal of internal aluminium oxides or nitrides during metallographic preparation. The oxide layer on the sulphur-containing samples consisted of a mixture of alumina, chromia and a spinel of Ni, Al and Cr.

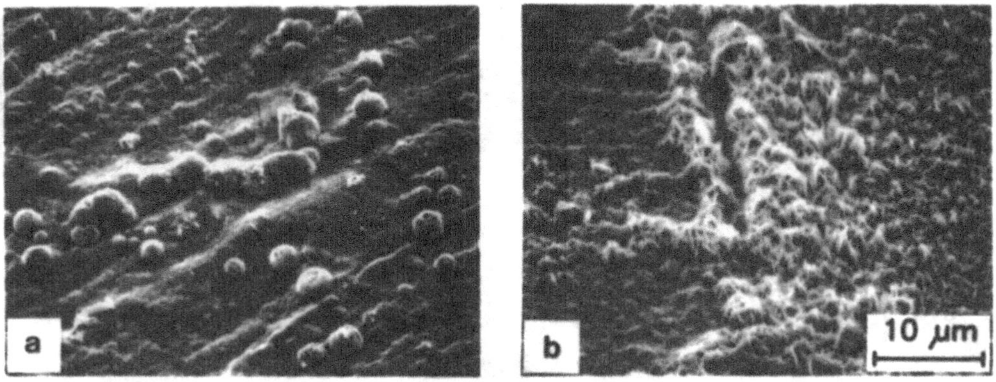

Fig. 5: SEM micrographs of the oxide scales formed on Ni-10Cr-8Al containing (a) 80ppm S + 0.01% Y (b) 70ppm S + 0.53% Y after oxidation in air at 1000°C for 25h.

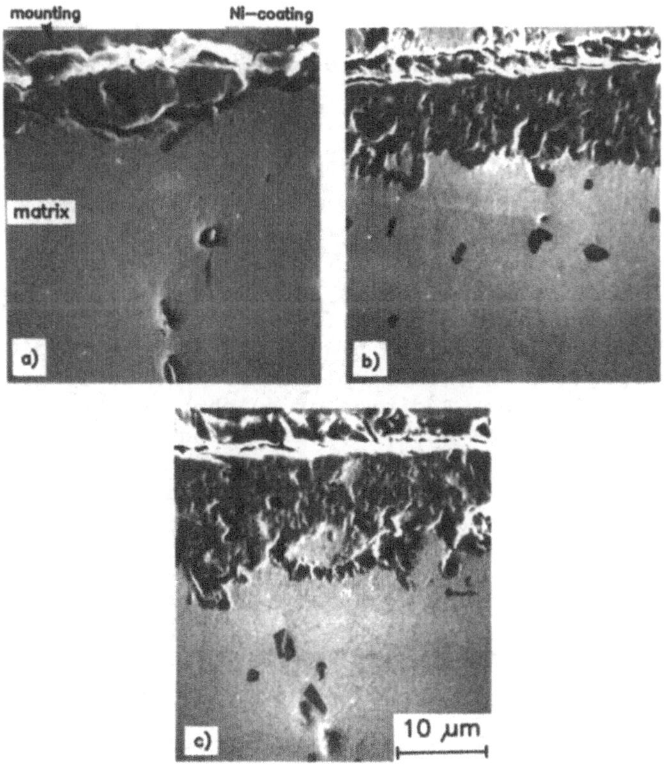

Fig.6: SEM micrographs showing cross-sections of the oxide scales formed on Ni-10Cr-8Al, oxidized in air for 25h: (a) without addition, (b) 100ppm S, (c) 230ppm S.

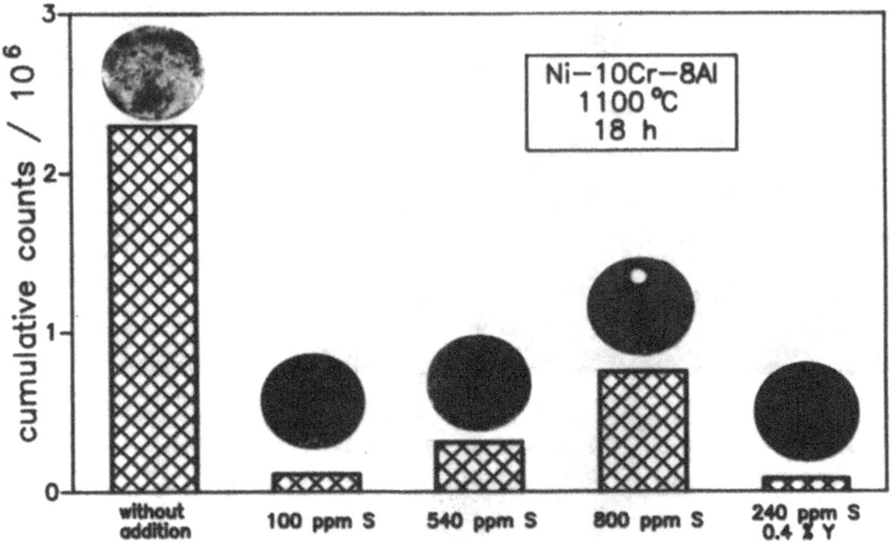

Fig. 7: Influence of sulphur and yttrium additions in Ni-10Cr-8Al on the cumulative acoustic emission counts, measured during specimen cooling after oxidation at 1100°C in air for 18h. The optical macrographs of the corresponding specimens are also shown.

5. ACOUSTIC EMISSION ANALYSIS

Fig. 7 shows the cumulative AE counts measured for various specimens during cooling after 18h oxidation at 1100°C. The sulphur-free sample gave the highest AE activity indicating more extensive spalling on this sample than on those containing sulphur and sulphur + yttrium. Addition of sulphur to the Ni-10Cr-8Al alloy decreased the AE activity indicating far less spallation than that on the sulphur-free sample as confirmed by the corresponding macrographs. Further increase in the sulphur level caused an increase in AE activity and also increased spalling of the scale. Addition of yttrium to the sulphur-containing material caused a drop in the AE level and also in the spalling of the scale. The weight change data during oxidation at 1100°C (Fig. 8a) showed results similar to those observed at 1000°C: an increase of the sulphur level of the alloy increased the growth rate of the scale rather than scale spalling, which is in agreement with the corresponding AE results. In the sulphur-containing alloys spalling occurred after longer times when a critical scale thickness was exceeded. Addition of 0.4% Y to the alloy containing 240ppm S decreased the scale growth and subsequent spallation only slightly (Fig. 8a), in contrast to sulphur-free alloys, in which even very small additions of yttrium are sufficient to obtain an excellent oxidation behaviour (Fig. 8b).

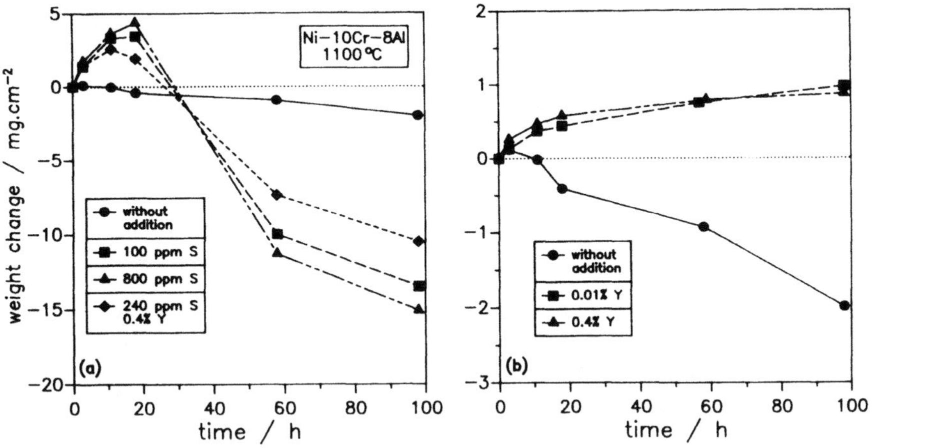

Fig. 8: Linear plots of weight change vs time, showing the effect of sulphur as well as of sulphur and yttrium additions on the oxidation of Ni-10Cr-8Al at 1100°C in air. Weight change data were measured after cooling to ambient temperature.

6. DISCUSSION

One of the main results of the present work is that the presence of sulphur impurity in a Ni-10Cr-8Al alloy caused a significant change in the growth rate and the composition of the oxide. This has been well demonstrated by the gravimetric results, the surface analysis of the oxide scales as well as the analysis of scale cross-sections. Both gravimetric and surface morphological results showed that the pure alumina scale formed on the sulphur-free sample not only spalled more extensively but also much earlier than the scales on specimens containing various levels of sulphur impurity (Figs. 1 & 3). This was confirmed independently using acoustic emission analysis which is a sensitive technique for detection of scale damage due to spalling and cracking.[9] Addition of small amounts of sulphur reduced the AE activity, indicating less spalling, in agreement with the macrographs and the gravimetric results (Figs. 7 & 8). This is probably caused by the observed change in the oxide composition, which is a mixture of alumina and chromia instead of pure alumina as on the sulphur-free alloy (Fig. 3). Increasing the sulphur content to very high levels significantly increased scale growth so that the scale reached a critical thickness faster, causing the scale to spall earlier. The present results indicate that the suggestion made by some authors[3-5] that the presence of sulphur in the alloy is the primary cause of poor scale adherence is not completely correct. According to their work, carried out on NiCrAl-based alloys containing less than 100 ppm of S, weakening of the oxide/metal bonding due to sulphur segregation at the interface was seen as the main reason for the scale spalling. If this explanation was correct, a sulphur content of more than 100 ppm, as used in the present investigations, should induce severe spalling due to heavy segregation at the metal/oxide interface. Both gravimetric and AE data show, however, that addition of sulphur did not necessarily lead to decreased scale adherence. In fact, the scale/metal interface of an alloy, made from high purity alloying elements without any deliberate impurity addition, appeared to be very weak, in agreement with the recent results from Luthra and Briant.[7]

The present investigations indicated that the main effect of sulphur impurity in the alloy was enhanced growth of the oxide scale. This caused the critical oxide thickness for the onset of spallation to be reached earlier. Since spallation of the thick oxide scale causes a higher weight loss than that of a thin scale, one might come to the conclusion that the scale on the sulphur-containing sample had a lower degree of adherence. However, the AE analysis, which gives more reliable information on oxide adherence than gravimetry, showed that this is not necessarily the case.

The results of the sulphur + yttrium containing alloys confirmed the proposal made by other investigators[3,5] that the detrimental effect of sulphur on the oxide scale properties can be prevented by gettering of sulphur by yttrium. This was verified by the fact that small amounts of yttrium in the alloy were not able to eliminate the deleterious effect of sulphur completely.[8] Excellent oxidation behaviour of the alloy was only obtained when the amount of yttrium was sufficient to bind all the sulphur (Figs. 2,5,8). Large amounts of yttrium addition were found to cause grain boundary precipitation of yttrium- and aluminium-rich phases, which oxidised internally, leading to grain boundary cracking. Therefore, an optimum amount of yttrium is required to completely prevent the deleterious effect of the presence of sulphur in the alloy without causing harmful internal grain boundary oxidation.

7. CONCLUSIONS

The main effect of sulphur in Ni-10Cr-8Al was found to be enhanced oxide scale growth. The associated thick oxide scales spalled after reaching a critical thickness resulting in a large weight loss, which might lead to the conclusion that sulphur addition decreases scale adherence. Neither from the thermogravimetric results, nor from the acoustic emission analysis was clear evidence found which would indicate that the mere presence of sulphur in the alloy results in weakening of the alloy/oxide bond.

Addition of yttrium not only improved adherence but also reduced the scale growth rate by gettering sulphur in the alloy. Both the addition of yttrium and the decrease of sulphur led to lower scale growth rates, however, in contrast to the addition of yttrium, lowering of sulphur content did not necessarily improve the scale adherence.

ACKNOWLEDGEMENTS

The authors wish to thank Mr. M. Beyss of KFA/IFF for his help in the preparation of the alloys, and their colleagues, E. Wallura, F. Els, H. Schulze and E. Gutzeit for scale analysis using SEM/EDAX, X-ray diffraction and optical microscopy. H. Jonas from KFA/IRE is gratefully acknowledged for his help in carrying out the acoustic emission experiments. One of the authors (ASK) would like to acknowledge with gratitude the Alexander von Humboldt-Foundation, Bonn, FRG, for supporting his stay in Germany.

REFERENCES

1. R.I. PRESSER and R. MCPHERSON, Scripta Met. 1977, 11, 745.
2. C.J. MIDDLETON, Met. Sci., 1981, 15, 154.
3. J G. SMEGGIL, A.W. FUNKENBUSCH and N.S. BORNSTEIN, Met.Trans. A, 1986, 17, 923.
4. J.G. SMEGGIL and G.G. Peterson, Oxid. of Met. 1988, 29, 103
5. J.L. SMIALEK, Met.Trans. A, 1987, 19, 164.
6. K.L. LUTHRA and C.L. BRIANT, Spring 1984 meeting of the Electrochemical Society, Cincinnati, OH, May 6-11, 1984.
7. K.L. LUTHRA and C.L. BRIANT, Oxid. of Metals, 1986, 26, 397.
8. W.J. QUADAKKERS, C. WASSERFUHR, A.S. KHANNA and H. NICKEL, Mater. Sci. and Technol., 1988, 4, 1119.
9. A. S. KHANNA, B. B. JHA and BALDEVRAJ, J. of Acoustic Emission, 1987, 6, 209.

THE ROLE OF CARBON ON THE GROWTH AND ADHESION OF OXIDE SCALES

H.J. Grabke

Max-Planck-Institut für Eisenforschung GmbH
P.O. Box 140260
D-4000 Düsseldorf, Fed. Rep. Germany

ABSTRACT

Oxidation in CO_2 or CO_2-CO-mixtures can be slow compared to oxidation in air, O_2 or H_2O-H_2 according to slow surface reaction kinetics. However, the presence of CO_2-CO within pores and holes of the scale can accelerate the oxidation of steels. As carbon does not dissolve in oxides such as Fe_3O_4, Cr_2O_3, Al_2O_3 etc., it is transported through scales only by diffusion of carbon-bearing molecules through cracks and pores. This inward transport can lead to internal carbide formation in the metal phase and to graphite deposition in the scale, both processes may deteriorate oxide scales. Carbides present at the metal surface, or formed upon initial attack in the transient stage, negatively affect the protectivity of scales, causing porosity. Carbides in the material, especially MC carbides (M = Ti,Zr,Nb,Ta...), cause scale fracture when reached by the oxidation front. Thus in general, the presence of carbon adversely affects scale growth and adhesion. Nitrogen and chlorine show similar effects as will be briefly discussed.

1. INTRODUCTION

Different 'active' elements can markedly affect the oxidation of high temperature alloys. Many alloying additions, such as Ce, Y, Hf, Ti etc., improve the oxidation behaviour and the adherence of oxide scales, whereas the nonmetal elements S, C, N and Cl generally have deleterious effects. In this paper the effects of carbon will be demonstrated and suggestions as to the possible influence of other nonmetal elements will be discussed briefly. Carbon-bearing gases - CO, CO_2, CH_4, hydrocarbons and organic compounds - may affect the oxidation kinetics and can transfer carbon to the alloy, generate carbides, or deposit graphite within the oxide scales. Carbides are present in most high temperature alloys and can detrimentally influence the oxide morphology. In most cases, the presence of carbon adversely affects the protectivity of oxide scales on high temperature alloys. The fundamentals of these phenomena are considered and corresponding effects of N and Cl are pointed out.

2. SURFACE REACTION KINETICS IN CO_2-CO MIXTURES

The rate of oxygen transfer from CO_2 to oxide surfaces

$$CO_2(g) \quad = \quad CO(g) + O \text{ (at the oxide surface)} \qquad (1)$$

is a relatively slow reaction compared to the oxygen transfer from O_2 or H_2O. Therefore, the oxidation of metals and alloys in CO_2-CO can be surface reaction controlled up to relatively large oxide scale thicknesses. The corresponding linear kinetics have been observed (Fig. 1) by several authors[1-5] for the oxidation of iron to wustite

$$CO_2 + Fe \quad = \quad CO + FeO \qquad (2)$$

For this reaction the rate of oxygen transfer is given by

$$v \quad = \quad k_1 \cdot a_O^{-m} \cdot p_{CO_2} \left(1 - \frac{a_O}{p_{CO_2}/p_{CO}}\right) \qquad (3)$$

In this expression, k_1 is the rate constant of the CO_2 decomposition, a_O is the oxygen activity at the surface of the growing oxide layer, p_{CO_2} and p_{CO} are the partial pressures of CO_2 and CO, respectively. The exponent $m = 0.5...1$ is dependent on temperature and varies for different oxides. The expression (3) was obtained from measuring the reaction (1) on oxide surfaces in equilibrium with a flowing CO_2-CO mixture;[5] this measurement is possible using isotope-exchange methods or relaxation measurements upon small deviations from equilibrium. Data are given in Fig. 2 for reaction (1) on wustite and magnetite.[5] Similar measurements are possible in H_2O-H_2 mixtures which yield rate coefficients k_2 which are about 1 to 2 orders of magnitude higher than k_1.[6,7] The rate coefficient for oxygen transfer from O_2 is considerably higher.[8] The oxygen activity a_O, which characterizes the state of the oxide surface concerning stoichiometry, ionic and electronic defect concentrations and concentration of adsorbed oxygen, can be defined as

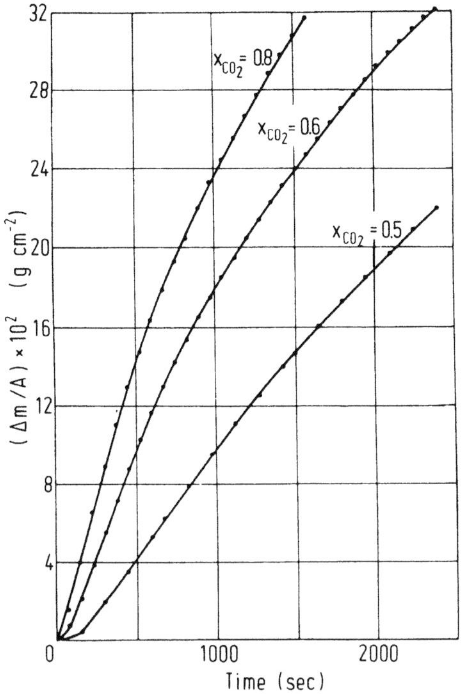

Fig. 1 a) Typical rate curves for the oxidation of iron in different CO_2-CO mixtures at 1100°C.[4]

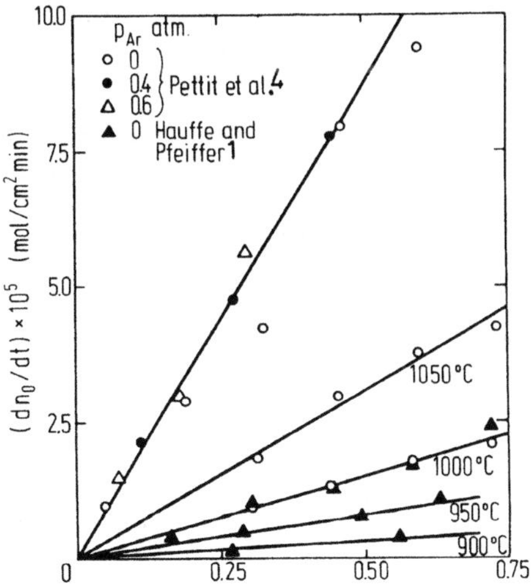

Fig. 1 b) Oxidation of iron in CO_2-CO mixtures, plotted according to equation (3).[1,2,4]

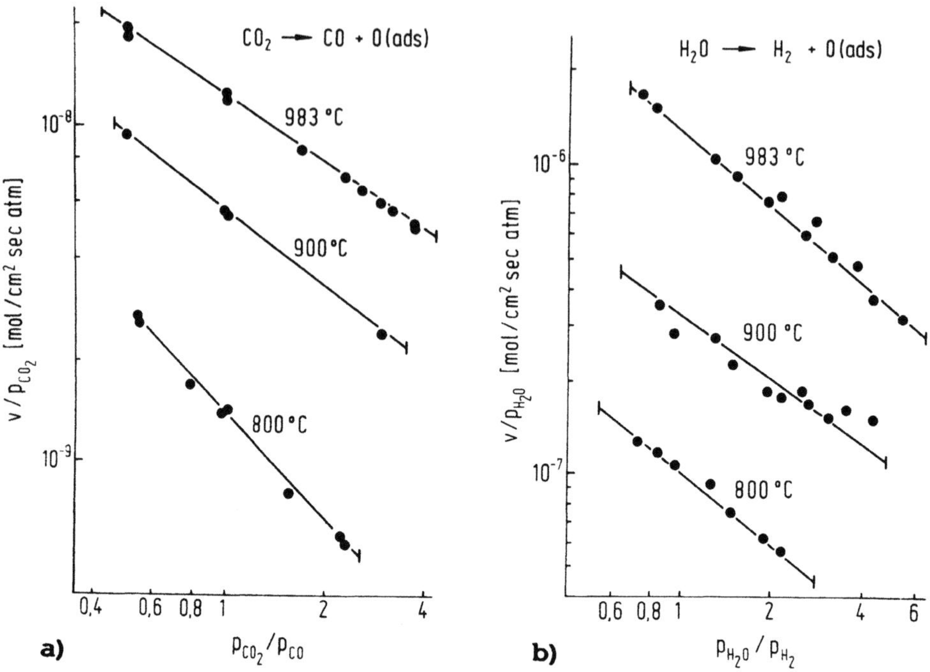

Fig. 2 Rate coefficients of oxygen transfer from CO_2 (a) and from H_2O (b) to the surface of wustite, determined by isotope exchange and resistance relaxation methods.[5,6,7]

$$a_O = (p_{CO_2}/p_{CO})_{eq} \tag{4}$$

i.e. the CO_2/CO ratio which would be in equilibrium with the actual state of the oxide surface. The dependence of the rate of oxygen transfer on the oxygen activity, given by the factor a_O^{-m}, arises from the participation of the electrons from the solid in the CO_2-decomposition, as was discussed in detail elsewhere.[9] Additionally, the factor a_O^{-m} can comprise a dependence on the degree of coverage with adsorbed oxygen (O(ad)).[9,10] This retards the oxygen transfer, such that the degree of coverage with O(ad) increases with increasing a_O and with decreasing temperature. Oxidation, i.e. oxygen transfer to the oxide, occurs if $a_O < p_{CO_2}/p_{CO}$. In the oxidation of iron, the surface of a thin wustite layer is virtually in equilibrium with the Fe beneath, and a_O is given by the value for the equilibrium of Fe,FeO,CO_2 and CO.

$$a_O = (p_{CO_2}/p_{CO})_{Fe,FeO} = K \tag{5}$$

Introducing this condition into equ. (3) and furthermore the condition $p_{CO_2} + p_{CO} = (p_{CO_2} + p_{CO})_{eq} = 1$ atm, the rate equation for oxidation of iron to wustite is given by,

$$v = k_1 K^{-m}(1+K)\ (p_{CO_2} - p_{CO_2}^{eq}) = k\ (p_{CO_2} - p_{CO_2}^{eq}) \qquad (6)$$

as was confirmed in the earlier investigations.[1,2]

However, with increasing thickness of the oxide, gradients of a_O and a_{Fe} arise in the oxide layer, and the value for a_O at the surface increases, see Fig. 3. This leads to a gradual decrease of the oxidation rate as can be seen from equ. (3) and corresponds to a gradual transition from linear to diffusion-controlled parabolic kinetics.[3,4] It should be emphasized that surface reaction controlled kinetics are never exactly linear, but a decrease of the 'linear constant' should be observed from the beginning - which can be mistaken for parabolic behaviour, see Fig. 1. Also, the partial pressure dependence is never exactly linear, as can be seen from equation (3).

Hence, the slow surface reaction, especially in the case of oxidation in CO_2-CO, may influence the growth kinetics up to large oxide thicknesses, for example in the growth of wustite up to > 100 µm.

On the other hand, it must be noted that CO_2-CO and/or H_2O-H_2, when trapped in voids or holes of oxide layers on iron and steels, can also accelerate the oxidation.[11] These gases enable a transfer of oxygen from the side of the hole near to the surface to the side near to the metal.

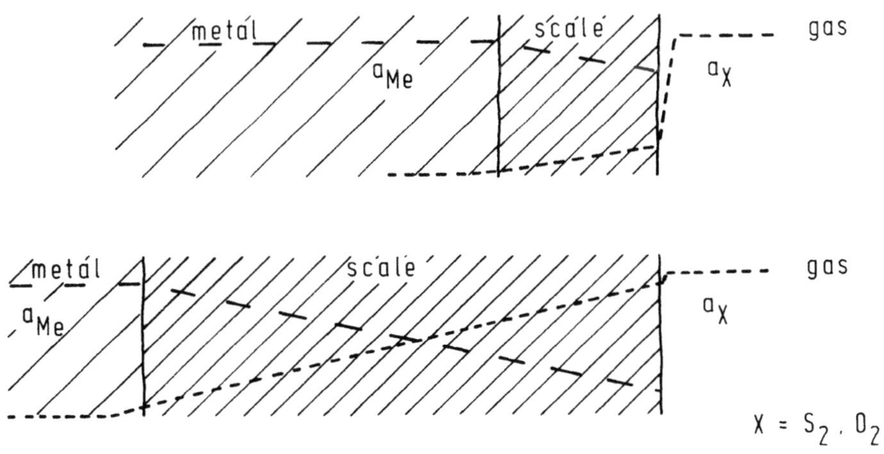

Fig. 3 Transition from linear to parabolic oxide growth, variation of the activities of nonmetal element (oxygen) and metal with increasing scale thickness.

3. CARBON IN OXIDES

Several authors[12-15] have discussed and assumed a certain solubility of atomic carbon in oxides, such as FeO, MnO, Fe_3O_4 and MgO. Effects of dissolved carbon on the disorder and diffusivities in FeO, MnO and Fe_3O_4,[12-13] have been advanced to explain the dependencies of iron and manganese oxidation in CO_2-CO on the total pressure (pCO_2 + pCO). High disorder in these oxides could lead to an uptake of carbon atoms in cation vacancies or interstitial sites or to the association of carbon with defect clusters. However, an earlier equilibrium study on the solubility of carbon in highly disordered FeO in CO_2-CO mixtures already indicated that the solubility of carbon was less than 1 wt.ppm.[16] Solubility of carbon had also been claimed for MgO[14,15] since carbon contents of some 100 ppm had been analyzed and a peculiar distribution and diffusion behaviour of the carbon was observed. However, those MgO crystals were grown by arc-fusion using graphite electrodes.

In a recent study[17] the solubility of carbon in various oxides - FeO, MnO, Fe_3O_4, Cr_2O_3, Al_2O_3 and MgO - was examined by equilibrating them at 1000°C in radioactive CO_2-CO mixtures. The ^{14}C which might have entered the oxide lattice could be detected with very high sensitivity, corresponding to < 0.01ppm C. In none of these oxides was any dissolved carbon detected. Radioactivity if present could be traced to macroscopic defects in the polycrystalline or single crystal oxides. It was demonstrated by autoradiography that carbon can enter only through pores or cracks into the oxides, transported by molecular diffusion - not by atomic diffusion in the lattice. Also, at grain boundaries no ingress of carbon was observed in any of the same oxides.[17] Thus carbon is not dissolved in oxide scales and cannot affect their disorder and growth kinetics - the observed dependency of growth kinetics on total pressure pCO_2 + pCO and carbon activity may rather be associated with effects of surface reaction kinetics (see chapter 2).

4. CARBON TRANSPORT THROUGH OXIDE SCALES

In the previous chapter it was stated that carbon does not dissolve in oxides, which form the protective scales on high temperature steels and alloys: Fe_3O_4, Cr_2O_3, Al_2O_3.[17] Carbon from the gas atmosphere therefore can only be transferred to the metal matrix by molecular transport, thus carbon-bearing molecules, such as CO, CO_2, CH_4 and hydrocarbons, must transport the carbon through pores or cracks in the scale. Carburization and internal carbide precipitation slowly occurs in materials on which the oxide scale becomes fissured due to stresses, creep, thermal cycling or fatigue.

The permeation of carbon through chromia scales on Fe-Cr-alloys has been investigated recently using ^{14}C as a tracer.[18,19,20] After preoxidation in a flowing H_2-H_2O mixture at 900°C for about 150h, the samples were exposed to the radioactive H_2-H_2O-CO-CO_2 at 900°C for 700h, and the carbon ingress was monitored with radioactive carbon. Carbon penetration profiles were determined in the scale and in the alloy, Fig. 4, and the lateral distribution was monitored by autoradiography. The carbon uptake decreased with increasing Cr-content achieving a minimum at 12.5 to 20%Cr, indicating good adherence and low porosity of the Cr_2O_3 scale.[18] An increase in carbon uptake, and hence in scale porosity was observed for Cr-contents between 25 and 50%Cr (Fig. 4). The reason for the increase in porosity for high chromium contents of the alloys is not clear as yet, but this effect may be due to chromium carbides in the surface of the alloy and thereby affecting the initial oxide growth (see chapter 7). The alloys contained about 50ppm C - with increasing Cr-content there will be more carbides which can have a disruptive effect on the scale growth. The effect of an increasing amount of chromium carbides at the surface on oxide growth was investigated with Fe-20%Cr alloys having different carbon contents (48, 93 and 130ppm C). As the amount of carbide precipitates at the interface oxide/metal increased, the extent of carbon ingress was enhanced, Fig. 5. Enhanced porosity also could be induced by distributing impurities such as SiC on the surface, obviously the chromia scale does not grow together and enclose such particles completely, but leaves some open pores, causing enhanced carbon uptake.[19]

Oxidation tests on the austenitic Alloy 800 demonstrated that cold working of the material before oxidation leads to a more protective, less porous chromia layer.[20] Cold work and other methods of surface deformation generate easy diffusion paths for fast Cr-diffusion to the surface promoting the formation of a protective oxide layer.

5. CARBURIZATION OF HIGH TEMPERATURE ALLOYS

Carburization of high temperature alloys is a phenomenon of considerable technical significance, since internal carbide formation in Cr-containing steels causes deleterious changes in materials properties.[21-25] As shown in the previous chapter, carbon can enter the material if the oxide scale has developed pores or cracks, or if the scale fails by conversion to carbides or by graphite deposition within the scale.[18] Conversion of a Cr_2O_3-scale to carbides occurs under special thermodynamic conditions: at high carbon activity, for example $a_C = 1$ under carbon deposits, and at high temperatures (> 1050°C). The case of graphite deposition in the scale will be discussed in the next

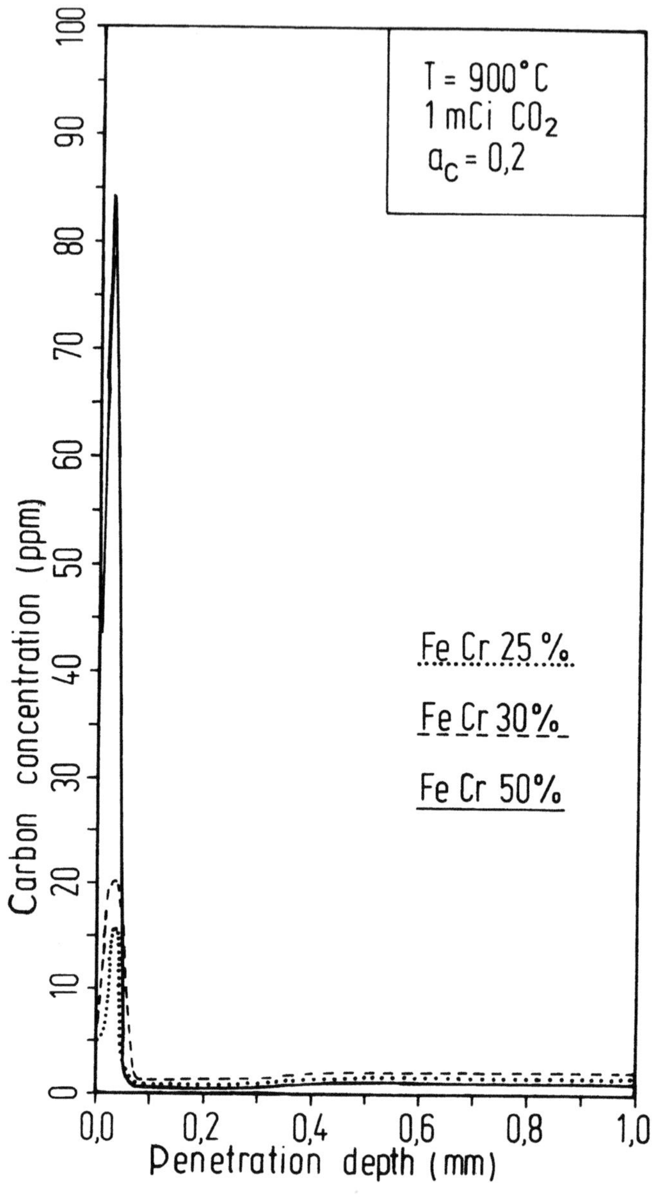

Fig. 4 Carbon penetration profiles, determined by measuring the radioactivity (^{14}C) upon stepwise polishing of specimens after exposure to H_2-H_2O-CO-CO_2 for 700h at 900°C, demonstrating the detrimental effect of an increasing content of chromium carbides: alloys with about equal C-content (50 ppm) and increasing Cr-concentrations[19].

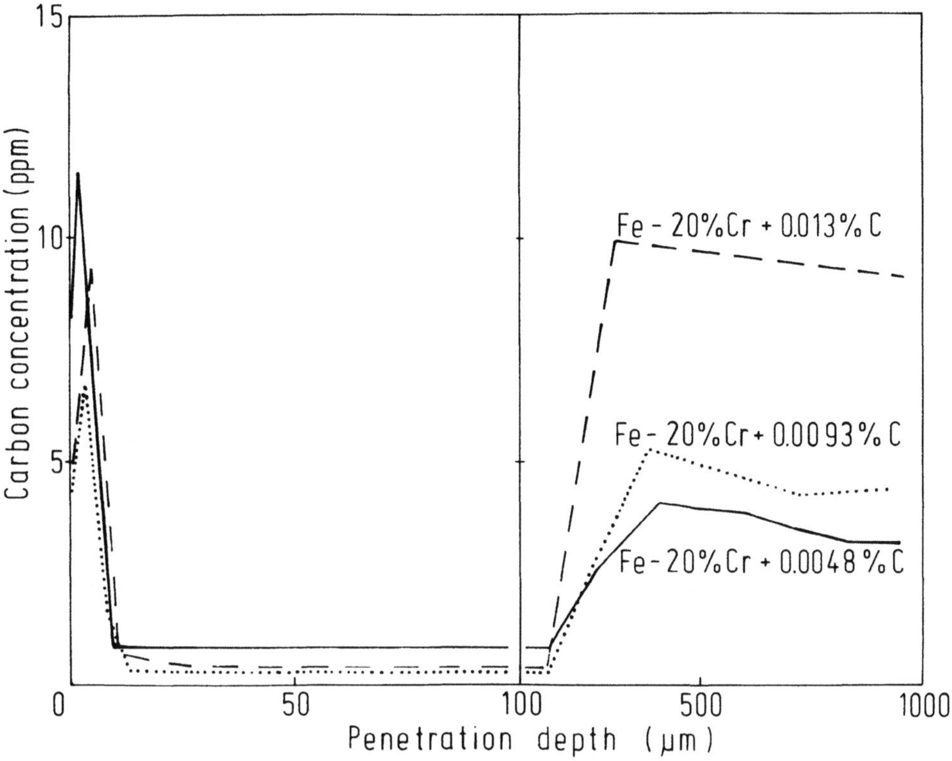

Fig. 5 Carbon penetration profiles, determined by measuring the radioactivity (^{14}C) upon stepwise polishing of specimens after exposure to H_2-H_2O-CO-CO_2 for 700h at 900°C, demonstrating the detrimental effect of an increasing content of chromium carbides: alloys with equal Cr-concentration and increasing C-content.

chapter. The kinetics and mechanisms of carburization will not be described here in detail, but it may be noted that precipitation of internal carbides leads to a volume increase and to internal stresses in the material.[23] These effects, in turn, can cause cracking of the oxide scales on high temperature alloys. In this way, carbon and carburization can negatively affect the protective oxides scales.

6. GRAPHITE DEPOSITION IN OXIDE SCALES

Graphite or coke deposition on top of an oxide scale is possible if the carbon activity in the gas atmosphere $a_C \geq 1$. Cr_2O_3 and Al_2O_3, however, are bad catalysts for graphite deposition and growth. It has been observed that these oxides can be exposed for long time in gas atmospheres at $a_C \geq 1$ without formation of deposits.[26] However, carbonaceous gases can easily deposit graphite at metal surfaces, if gas molecules can penetrate through cracks or pores in the scale to the oxide/metal interface. Even in gas atmospheres with $a_C < 1$ graphite deposition is possible at the oxide/metal interface, because the carbon activity in CO_2-CO mixtures rises within the scale according to the Boudouard-reaction:

$$2\,CO = CO_2 + C \tag{7}$$

$$a_C = K_B\, p_{CO}{}^2/p_{CO_2} = K_B\, p_{CO}/a_O \tag{8}$$

where the definition $a_O = (pCO_2/pCO)_{eq}$ is used (as in chapter 2, eq. (4)). Thus, assuming the equilibrium (7) is established within the scale or at least at the interface oxide/metal, the carbon activity a_C can rise to high values with decreasing a_O. The oxygen activity at the interface Cr_2O_3/alloy is low and carbon activities $a_C > 1$ can easily be achieved. Together with the catalytic effect of the metal phase, this situation leads to graphite nucleation and growth - causing destruction of the oxide scale.[27]

Even in an atmosphere of pure CO_2 (at high pressure), the failure mechanism described above plays an important rôle - causing breakaway oxidation of low-alloyed steels in CO_2-cooled nuclear reactors.[28-30] Below 600°C these steels form protective Fe_3O_4-scales, but after long-term oxidation in high pressure CO_2 carbon is deposited in the scale and at the oxide/metal interface. Upon growth of Fe_3O_4 by outward diffusion of cations, voids are formed at the interface oxide/metal, and CO_2 penetrating through the scale will equilibrate in these voids according to reaction (7), which may lead to $a_C > 1$ and carbon deposition. The graphite growth causes bursting and spalling of the scale, followed by 'breakaway oxidation'. These phenomena can be suppressed by alloying such steels with Si, since the SiO_2-subscale formed below a Fe_3O_4-layer does not catalyze graphite precipitation and growth.

7. EFFECTS OF CARBIDE PRECIPITATES IN THE ALLOY

In chapter 4 it has already been shown that the presence of chromium carbide particles at the surface of an alloy causes porosity in growing chromia scales. The detailed mechanism of this effect has not been elucidated but the effect was sufficiently confirmed by the 'carbon permeation technique'.[19] Chromium carbide particles probably are only effective in inducing scale porosity when present at the surface during oxidation. Chromium carbides beneath the surface, not directly attacked by oxidation, have been shown to dissolve before being reached by the oxidation front.[31]

In contrast, the more stable MC-type carbides (M = Ti, Nb, W, Ta) will not dissolve before being approached by the oxidation front. Therefore, the effect of these carbides can be much more deleterious, as has been shown for the oxidation of the Cr_2O_3-forming Ni-base alloy IN 939 and of the Al_2O_3-forming Ni-base alloy IN 738 LC.[32] Oxidation of the MC-carbides causes a relatively large increase in volume, as indicated by the ratios: $TiO_2/TiC = 1.28$, $Ta_2O_5/TaC = 2.23$, $Nb_2O_5/NbC = 2.22$. Oxides of these metals grow by inward diffusion of oxygen via oxygen ion vacancies. Most of these oxides are more stable than Cr_2O_3, so even underneath the chromia layer a carbide MC will become oxidized, and the outward growth of the oxides destroys the chromia layer, as shown in Fig. 6. Around the outgrowing oxides of Ti, Nb, Ta etc. shear stresses arise leading to scale cracking, thicker Cr_2O_3 scales, internal oxidation and nitridation.[32]

8. EFFECTS OF CARBIDES FORMED IN THE TRANSIENT STAGE

On an oxide-free 'virgin' alloy surface any compound can be formed upon attack of the environment for which the formation pressure (activity) is exceeded in that environment. Only after this transient stage during which transport in the gas phase or surface reactions are rate-controlling, equilibrium is gradually established between the environment and the surface of the scale, whereby the thermodynamically more stable phase should overgrow other phases.[33,34]

In the case of attack by an oxidizing and carburizing atmosphere, e.g. a CO_2-CO atmosphere, on a Cr-containing alloy, chromium carbides and -oxides can form in the transient stage.

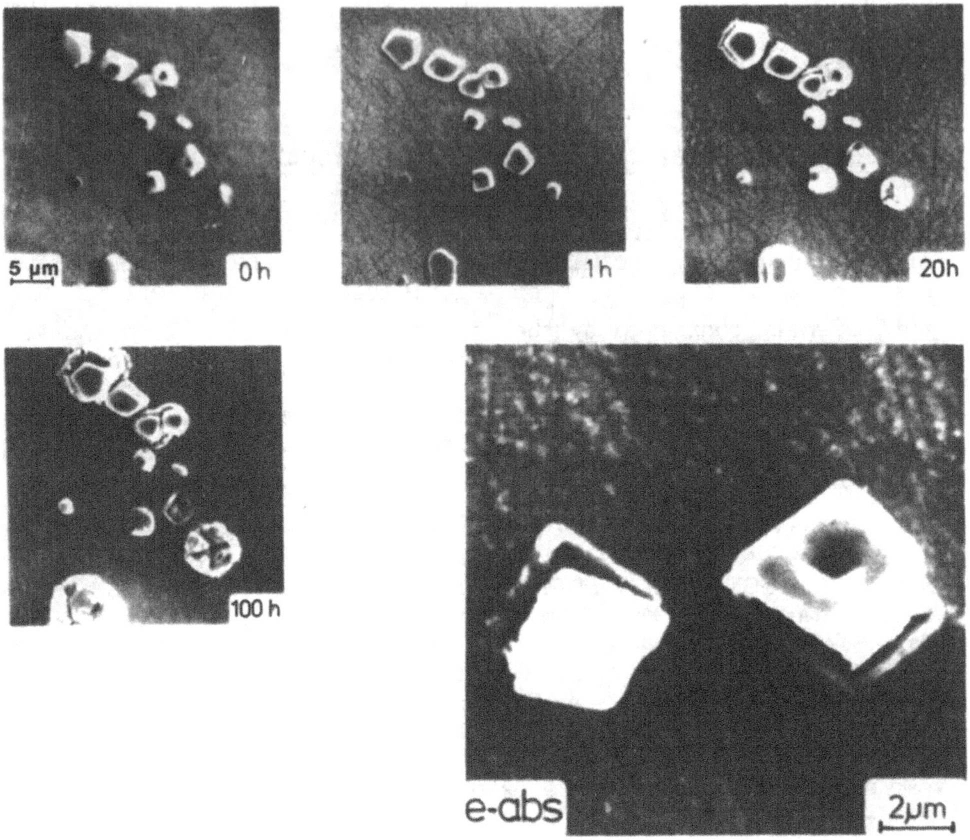

Fig. 6 Scanning electron micrographs, showing the outgrowth of TiO_2 through a chromia layer upon oxidation of IN 939 at 700°C in air.[32] The chromia layer is destroyed by TiO_2 formed at the sites of TiC-carbide particles.

The most stable phase generally will be Cr_2O_3, overgrowing the carbides which can dissolve into the alloy. However, a deleterious effect of this initial formation of chromium carbides on the morphology and protective properties of the later Cr_2O_3 scale may not be excluded, although no experiments appear to have been undertaken demonstrating this effect of carbon and carbides. However, chromium nitrides have a similar effect on the oxidation of Fe-20%Cr in H_2-H_2O compared with N_2-H_2-H_2O.[35]

Even more marked effects have to be expected in the exposure of carbide forming metals, such as Cr and Mn, to oxidizing and carburizing gas mixtures. Carbides and oxides will be formed on the surface of these metals during the transient stage, however, the carbides can be overgrown but they cannot dissolve into the metal phase. In this respect, the behaviour of pure metals is different from Fe- or Ni-base alloys where the carbides can dissolve after being overgrown.

9. SIMILARITIES IN THE EFFECTS OF CARBON, NITROGEN AND CHLORINE ON THE GROWTH AND ADHESION OF OXIDE SCALES

The effects demonstrated in the previous chapters for carbon will be discussed for nitrogen and chlorine in the following. Certainly in oxygen transfer (chapter 1), nitrogen and chlorine play no rôle except, maybe, retarding the oxygen transfer by adsorption on oxide scales.

Solubility of N or Cl in the oxides discussed (chapter 3) is highly improbable, in spite of the fact that solubility of Cl has been proposed by some authors[36] to explain the accelerated growth of Cr_2O_3 in complex environments.

The transport of the elements N and Cl from the gas phase to the metal probably occurs also by diffusion of N- or Cl-bearing molecules through cracks and pores, and as emphasized in chapter 3, most scales have some pores. In the same way as carbon, nitrogen can be dissolved in the metal phase at the oxide/metal interface, which leads to internal nitride formation[24] with consequences similar to those for internal carburization (chapter 4). Cl certainly cannot dissolve in the metal phase, but will react to form chlorides at the same interface due to the low oxygen activity at this interface. Depending on the temperature and chlorine pressure, the resulting chlorides can be solid, molten or volatile, some evaporation of chlorides will occur at any conditions. If condensed chlorides are formed in any stage of the corrosion process, it seems to be difficult that they are overgrown by the oxide layer. Hence, any chlorides formed in the initial stage of attack by a chlorine-containing gas atmosphere would probably detrimentally affect the protectiveness of oxide scales. The same is valid for formation of nitrides during the initial attack of ferritic Fe-Cr alloys,[35] but the effect is less pronounced for austenitic alloys. This phenomenon is related to the good epitaxial fit of the cubic CrN on ferritic alloys, which favours the nucleation of that nitride.[37] Also nitrides in high temperature alloys, such as TiN, ZrN or else, can have detrimental effects on the scale morphology, as reported for MC-type carbides (chapter 6).

So there are some similarities in the effects of carbon, which have been described in detail, and the effects of nitrogen and chlorine. However, the effects of chlorine are only shortly tackled here, - they are much more severe than the effects of carbon and nitrogen and less well understood.

REFERENCES

1. K. HAUFFE and H. PFEIFFER: Z. Metallkunde 1953, **44**, 27.

2. E.T. TURKDOGAN, W.M. MCKEWAN, and C. ZWELL: J. Phys. Chem. 1965, **69**, 327.

3. F.S. PETTIT, R. YINGER and J.B. WAGNER: Acta Met 1960, **8**, 617.

4. F.S. PETTIT and J.B. WAGNER: Acta Met. 1964, **12**, 35.

5. H.J. GRABKE: Ber. Bunsenges. physik. Chemie 1965, **69**, 48.

6. H.J. GRABKE, K.J. BEST and A. GALA: Werkstoffe u. Korrosion 1970, **21**, 911.

7. A. GALA and H.J. GRABKE: Archiv Eisenhüttenwes. 1972, **43**, 463.

8. S. STOTZ: Ber. Bunsenges. physik. Chemie 1966, **70**,769.

9. H.J. GRABKE: Annals New York Acad. Sciences 1973, **213**, 110.

10. H.J. GRABKE and H. VIEFHAUS: Ber. Bunsenges. physik. Chemie 1980, **84**, 152.

11. A. RAHMEL: Werkstoffe u. Korrosion 1965, **16**, 837.

12. G. SIMKOVICH: Werkstoffe u. Korrosion 1970, **11**, 973.

13. P. KOFSTAD: Oxid. Metals 1983, **19**, 129.

14. F. FREUND, G. DEBRAS and G. DEMORTIER: Journal of the Amer. Ceramic. Society 1973, **61**, 429.

15. F. FREUND: Physikalische Blätter 1981, **37**, 283.

16. H. MEURER and H. SCHMALZRIED: Archiv Eisenhüttenwes. 1971, **2**, 87.

17. I. WOLF and H.J. GRABKE: Solid State Communications 1985, **38**, 556.

18. H.J. GRABKE and I. WOLF: Mat. Sci. Eng. 1987, **87**, 23
T.N. RHYS-HONES and H.J. GRABKE: Mat. Sci. Techn. 1988, **4**,446

19. I. WOLF, H.J. GRABKE and H.P. SCHMIDT: Oxid. Metals 1988, **29**, 289.

20. S. LEISTIKOW, I. WOLF and H.J. GRABKE: Werkstoffe u. Korrosion 1987, **38**, 556.

21. H.J. GRABKE, U. GRAVENHORST and W. STEINKUSCH: Werkstoffe u. Korrosion 1976, **27**, 291.

22. A. SCHNAAS and H.J. GRABKE: Oxid. Metals 1978, **12**, 387.

23. A. SCHNAAS and H.J. GRABKE: Werkstoffe u. Korrosion 1978, **29**, 635.

24. H.J. GRABKE: in Proc. 8th Int. Congress on Metallic Corrosion, Vol. III, Sept. 1981, Ed. DECHEMA, p. 2143-2156.

25. J.M. HARRISON, J.F. NORTON, R.T. DERRICOTT and J.B. MARIOTT: Werkstoffe u. Korrosion 1979, **30**, 785.

26. H.J. GRABKE, P.C. PRASANNAN and E.M. MÜLLER: Metallurg. Trans 1986, **17A**, 915.

27. K. LEDJEFF, A. RAHMEL, M. SCHORR: Werkstoffe u. Korrosion 1980, **31**, 83.

28. N. BIRKS: Brit. Corros. J. 1968, **3**, 56.

29. A.M. PRITCHARD, J.E. ANTILL, K.R.J. COTTELL, K.A. PEAKALL and A.E. TRANSWELL: Oxid. Metals 1975, **9**, 181.

30. D.R. HOLMES, R.B. HILL and L.M. WYATT: Corrosion of Steels in CO_2. The British Nuclear Energy Society 1974.

31. L.A. MCDONALDO LOPEZ, H. GRÜBMEIER, A. NAOUMIDIS and H. NICKEL: Quantitative Mikrosondenuntersuchungen zur Oxidation von Hochtemperaturlegierungen bei niedrigen Sauerstoffpotentialen, Berichte der KFA Jülich Nr. 2120, IRW, März 1987.

32. J. LITZ, A. RAHMEL and M. SCHORR: Oxid. Metals 1988, **30**, 95.

33. B.H. KEAR, F.S. PETTIT, D.E. FORNWALT and L.P. LEMAIRE: Oxid. Metals 1971, **3**, 557.

34. H.J. GRABKE: in Proc. Seminar 'High Temperature Materials Corrosion in Coal Gasification Atmospheres', Petten Sept. 1981, Ed. J.F. Norton. Elsevier Appl. Sci. Publ. London 59-82.

35. D. BRAMHOFF, H.J. GRABKE and H.P. SCHMIDT: in Proc. Europ. Colloqu. 'The rôle of active elements in the oxidation behaviour of high-temperature metals and alloys', Petten, Dec. 1988.

36. M.K. HOUSSAIN and S.R.J. SAUNDERS: Oxid. Metals 1978, **12**, 1.

37. C. UEBING, H. VIEFHAUS and H.J. GRABKE: Appl. Surf. Sci. 1988, **32**, 363.

EFFECT OF CeO_2 COATINGS AND INTERFACIAL SEGREGATIONS OF S AND C ON THE OXIDATION BEHAVIOUR OF Fe-Cr ALLOYS IN O_2

P. Papaiacovou, R.J. Hussey, D.F. Mitchell and M.J. Graham

Division of Chemistry, National Research Council of Canada

Ottawa, Canada K1A 0R9

ABSTRACT

The influence of 4 nm thick CeO_2 coatings and interfacial segregation of S and C on the oxidation resistance of Fe-20Cr at 1173 K in 1 torr oxygen were examined using ^{18}O/SIMS, TEM and SEM/EDX for periods up to 19 hrs. The CeO_2 became incorporated into the oxide scale and promoted the formation and retention of an adherent layer displaying a 10 times lower oxidation rate. Tracer studies with $^{16}O/^{18}O$ have confirmed a change in the oxidation mechanism. During oxidation of both uncoated and coated alloys sulfur and carbon accumulate at the metal/oxide interface. The amount of segregated sulfur does not vary with time, whereas the level of segregated carbon increases. The total amount of segregants was found to be the same in both untreated and coated alloys.

PREOXIDATION FOR SULFIDATION PROTECTION OF FE-CR-NI-ALLOYS

R. Lobnig, P. Papaiacovou, H.P. Schmidt and H.J. Grabke

Max-Planck-Institut für Eisenforschung GmbH
P.O. Box 140260
D-4000 Düsseldorf 1, Fed. Rep. Germany

ABSTRACT

The effect of preoxidation upon sulfidation-oxidation behaviour of Fe-20Cr and Fe-20Cr-12Ni in H_2-H_2O-H_2S-mixtures has been studied at 973, 1073 and 1173 K, applying thermogravimetry and tracer(^{35}S)-studies. Preoxidation is most effective against sulfidation afterwards, under conditions were spinel formation is avoided and only chromia is formed. At sulfur pressures lower than necessary for FeS-formation only small amounts of sulfur can penetrate through a preformed chromia layer on alloys. The sulfur concentration in the oxide layer has a maximum at the boundary between the alloy and the oxide due to small amounts of chromium sulfides at this location. The chromia scales formed on Fe-20Cr-12Ni are more protective against internal sulfidation than on Fe-20Cr.

1. INTRODUCTION

In several technological processes, e.g. coal gasification, petrochemical or gas turbine applications, high temperature alloys are exposed to gas atmospheres caracterised by low oxygen potentials and high sulfur activities. In such atmospheres sulfides of iron and nickel are already stable phases while the oxygen pressure is too low to allow the formation of oxides of these metals except spinels with Cr_2O_3 and Al_2O_3.[1]
Under such conditions the accelerated attack by sulfidation appears to be the major problem.[2-4] Many studies of the behaviour of high temperature alloys have been made, mainly with austenitic FeCrNi-alloys with high Cr-contents.[3-12] Direct exposure of the

metallic alloy surface to multiple oxidant gases causes formation of a mixed scale of oxides, sulfides and carbides. Such a scale will not be protective and breakaway is observed after extended exposure even if the transient sulfide and carbide inclusions are overgrown by stable oxides. It has been shown that solid continuous layers of oxides are most effective in inhibiting the degradation of alloys, thus preoxidation under controlled conditions can lead to good corrosion resistance.[13]
In this study the effect of different oxygen pressures during preoxidation of FeCrNi-alloys was investigated, such that the protectiveness of pure chromia scales could be compared with that of spinel phases such as $FeCr_2O_4$ or Fe_3O_4.[10] Additionally penetration and interaction of sulfur with oxide layers was studied by radioactive tracer (^{35}S) techniques. The protectiveness of chromia scales grown on Fe-20Cr was compared with that of chromia scales grown on Fe-20Cr-12Ni and the effect of Cerium as alloy addition was investigated.

2. EXPERIMENTAL

The gas mixtures for the investigations were chosen by reference to a superimposed two-dimensional thermochemical stability diagram (Fig.1).[14-16] Although superposition of such diagrams, which have been calculated for pure metals, is not correct for alloys in the strictest thermodynamic sense, they are very useful to describe the possible corrosion products to a first approximation.

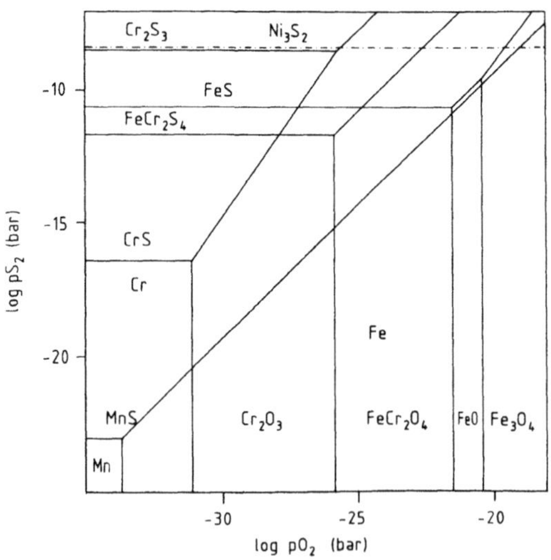

Fig.1: Thermochemical stability diagram for the system M-S-O with M = Fe,Cr,Ni,Mn at 973 K.

The kinetics of preoxidation and subsequent sulfidation and oxidation of the specimens were followed gravimetrically: The sample was suspended from a quartz rod in the middle of a vertical furnace whereby its mass increase during the experiment could be continuously recorded by an electronic microbalance (Fig. 2). The reaction tube was flushed with purified H_2 during heating to the reaction temperature of 973 K. Then the required S_2- and O_2-pressures were established by an appropriate mixture of H_2, H_2O and H_2S, which was passed through the furnace at a constant flow rate in order to avoid depletion of the gases.
The apparatus used for the ^{35}S-tracer measurements was a closed loop system in which, by means of a ventilator, the gases were circulated over the samples and through the H_2S- generating furnace and the oxalic acid saturator, constantly replenished with H_2S and H_2O (Fig.3). The oxalic acid saturator contains a mixture of oxalic acid dihydrate and oxalic acid anhydrate at a controlled temperature to establish the required H_2O-pressure. The H_2S-generating furnace contains a powdered mixture of Fe and FeS tagged with ^{35}S and Al_2O_3, here the temperature dependence of the reaction of H_2 with FeS to form H_2S is used to establish well-defined H_2S/H_2 ratios.

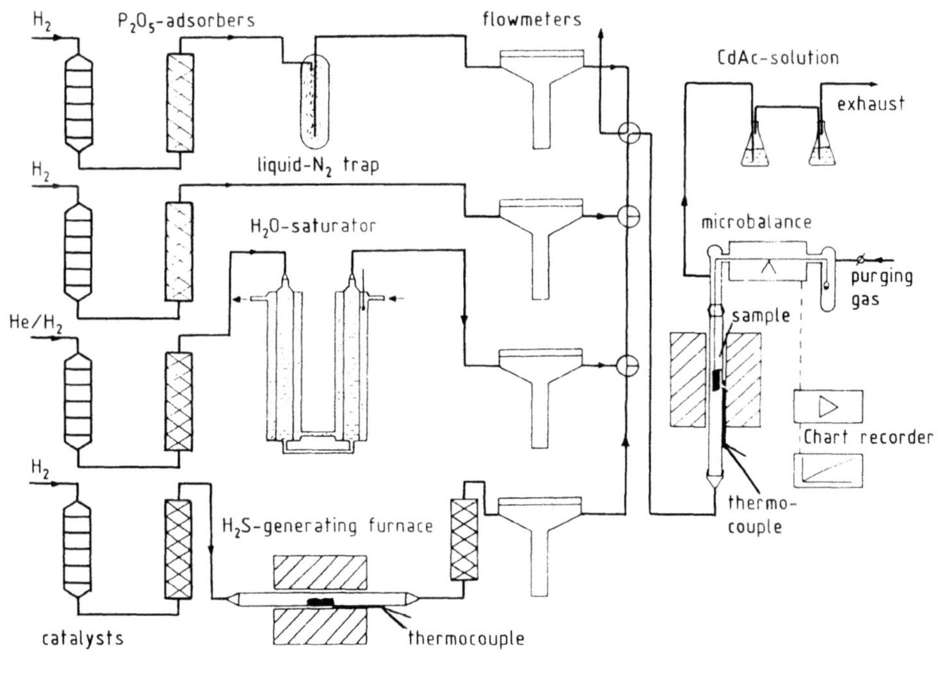

Fig.2: Schematic representation of the experimental apparatus for the gravimetric measurements.

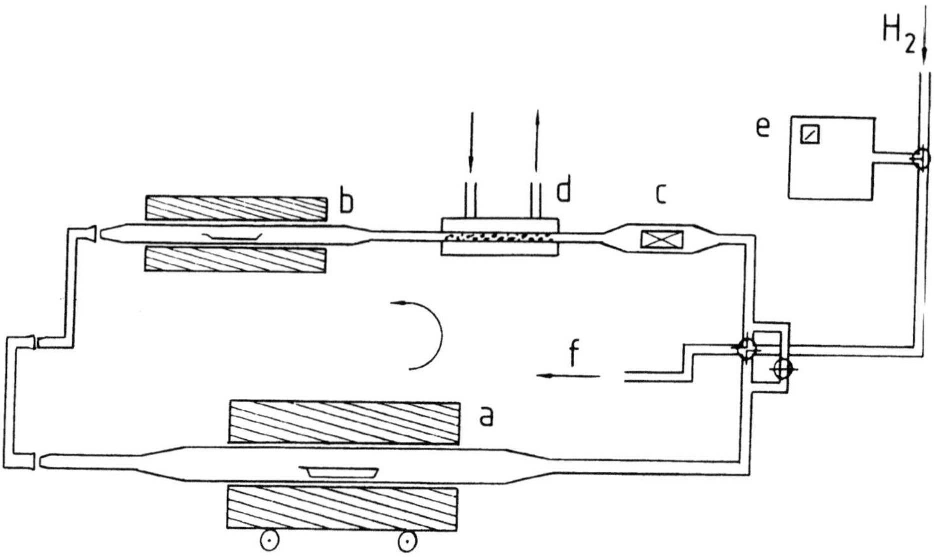

Fig.3: Schematic representation of the experimental apparatus used for the ^{35}S-experiments:
a) sample furnace, b) Fe/FeS-furnace, c) gas circulator
d) H_2O-saturator, e) pressure measuring device,
f) exhaust.

After the exposure experiments with ^{35}S, sulfur penetration profiles were calculated (detection limit 0.1 ppm S) from depth profiles of ^{35}S, determined by stepwise polishing and measurement of the residual activity, using a methane flow counting tube.
Both sets of samples were examined by scanning electron microscopy (SEM) combined with energy dispersive X-ray analysis (EDAX) and by X-ray diffraction (XRD).

3. RESULTS AND DISCUSSION

In a first experiment samples of Fe-20Cr-12Ni-0.5Ti were preoxidised in H_2/H_2O at 973 K for 21 hours, firstly, in the stability range of Cr_2O_3 only, i.e. at $pO_2 = 10^{-27}$ bar and secondly, in an atmosphere with a sufficiently high oxygen pressure allowing formation of $FeCr_2O_4$, i.e. $pO_2 = 10^{-23}$ bar. The subsequent sulfidation and oxidation was carried out at 973K in a highly sulfidizing $H_2/H_2O/H_2S$ gas mixture, where both CrS and FeS would be stable phases with $pO_2 = 10^{-23}$ bar, $pS_2 = 10^{-9.5}$ bar.
Fig.4 shows the increase in mass per unit area of the samples with time. It can be seen that the sulfidation rate is much higher after preoxidation in the spinel than in the chromia region of the thermochemical stability diagram.

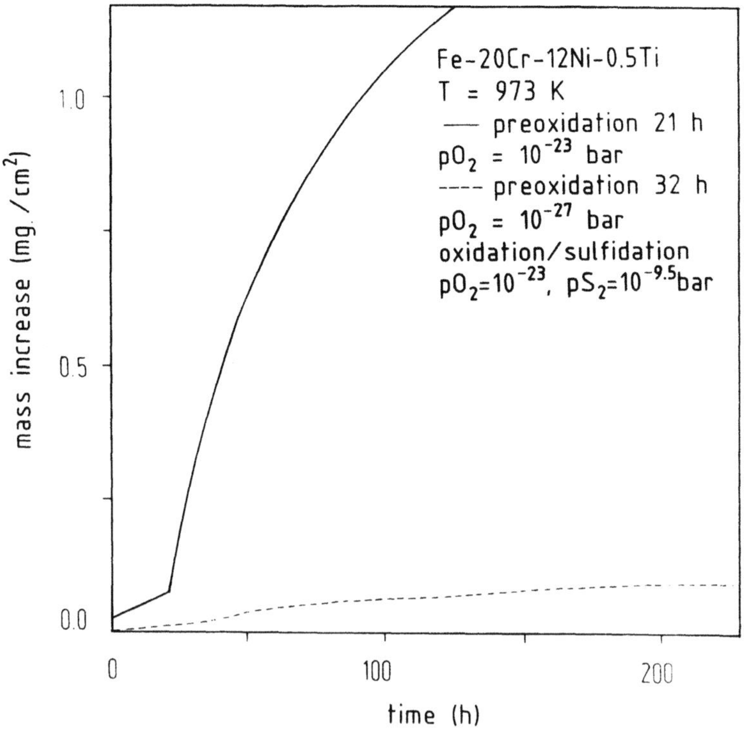

Fig.4: Mass increase of alloy Fe-20Cr-12Ni-0.5Ti as function of time at 973 K during preoxidation at different pO_2 and subsequent oxidation/sulfidation.

To obtain further information about the relation between the rate of sulfidation and the presence of spinel in the oxide scale, in a second series of experiments on Fe-20Cr-12Ni-1.5Mn samples the oxygen pressure of the preoxidation treatment was systematically increased from $pO_2 = 10^{-27}$ to 10^{-20} bar at 973 K.
Fig.5 shows the morphologies of the samples oxidized for 21 hours at $\log(pO_2) = -23$ with only islands of spinel in the chromia scale, at $\log(pO_2) = -21$, where the amount of spinel is already increased, and at $\log(pO_2) = -20$ where spinel completely covers the surface. XRD gave information on the scale composition: At the lowest $pO_2 = 10^{-27}$ bar the scales consisted of Cr_2O_3 only. From 10^{-23} to 10^{-22} bar the spinel had the composition $(Fe,Mn)Cr_2O_4$. At $pO_2 = 10^{-20}$ bar Fe^{2+}-ions can be incorporated and the spinel composition can be described as $(Fe,Mn,Cr)_3O_4$.

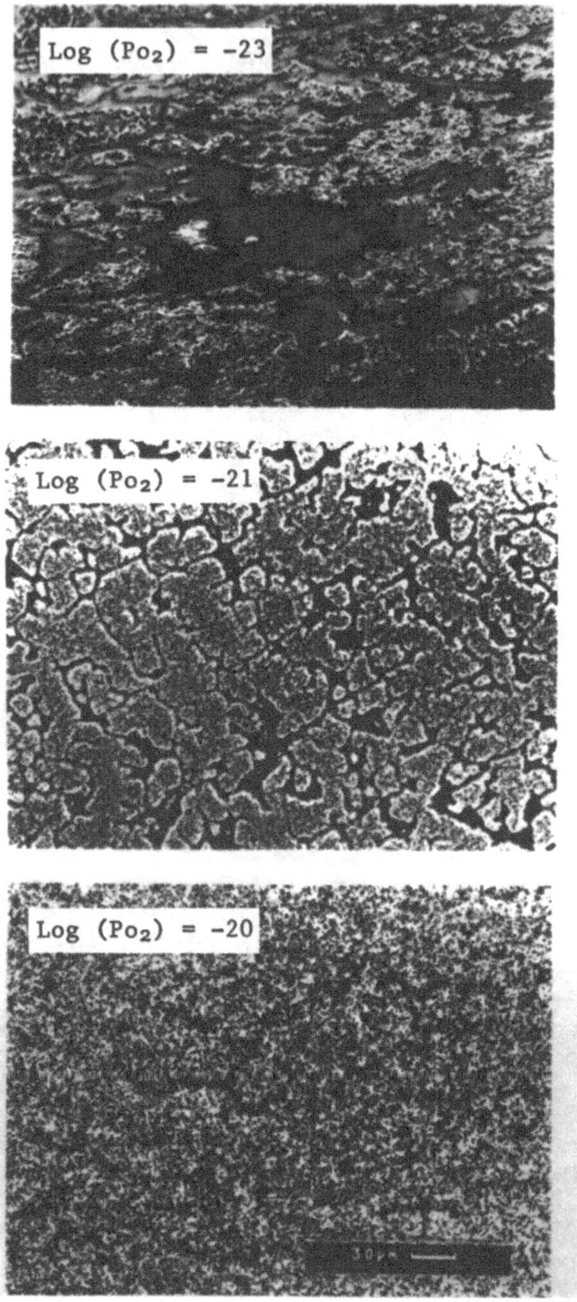

Fig.5: SEM-pictures of Fe-20Cr-12Ni-1.5Mn after preoxidation at different pO_2.

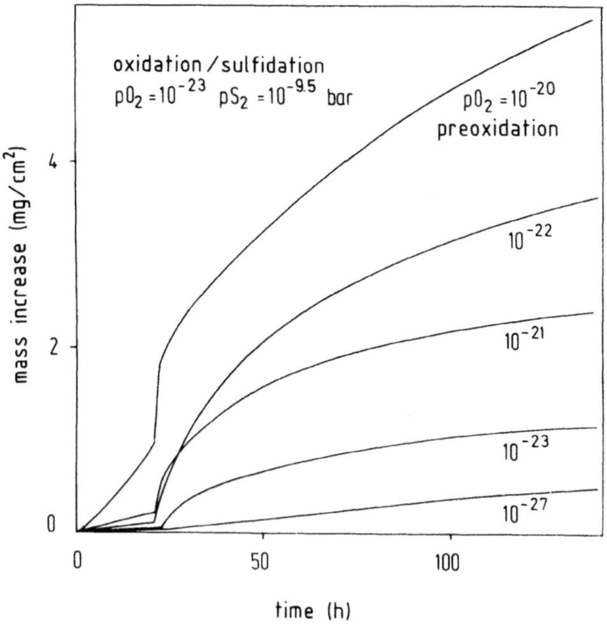

Fig.6: Mass increase of alloy Fe-20Cr-12Ni-1.5Mn as function of time at 973 K with preoxidation at different pO_2.

Fig.6 shows the kinetics of preoxidation and of sulfidation and oxidation of Fe-20Cr-12Ni-1.5Mn samples. It can be seen that also here the reaction rate considerably increased when the oxygen pressure during preoxidation was raised from 10^{-27} to 10^{-20} bar. After 150 h sulfidation and oxidation only small amounts of fine grained (Mn,Cr)S had been formed on the surface of the preformed Cr_2O_3-scale (Fig. 7a). On all other samples where spinel had been formed during preoxidation the oxide scales were overgrown by Fe-rich sulfides, which locally had a thickness of more than 100 µm (Fig.7b). In addition to the external growth of FeS internal sulfidation was observed, especially at grain boundaries.
Under the conditions of these experiments the deterioration of the materials was mainly caused by external growth of FeS. The situation is shown schematically in Fig. 8: Growth of external sulfides can occur by outward diffusion of double charged cations of Fe, Mn and Ni or of Cr^{3+} -cations through the chromia layer or the spinel islands. As the diffusion of Fe^{2+}, the most important cation under these conditions, is about 3 orders of magnitude faster in spinel than diffusion in chromia,[17-23] the corrosion

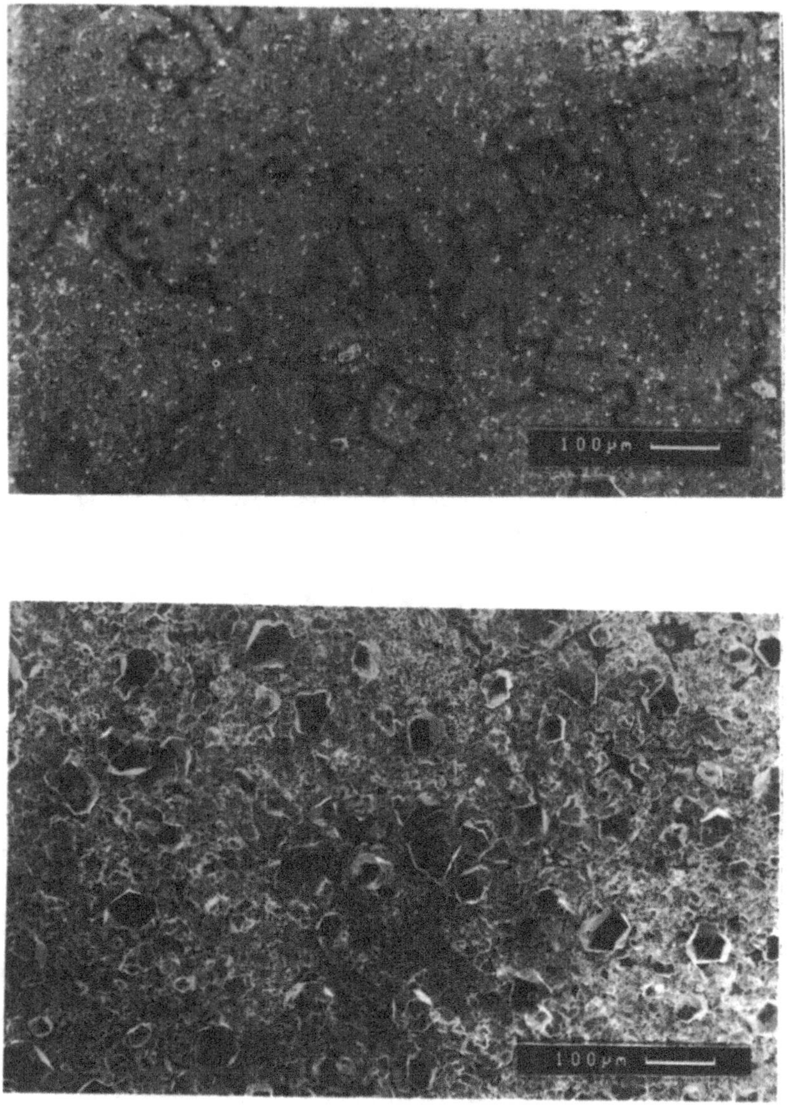

a)

b)

Fig.7: SEM-pictures of Fe-20Cr-1.5Mn-samples after preoxidation at a) $\log(pO_2) = -27$ b) $\log(pO_2) = -21$ at 973 K and subsequent sulfidation /oxidation at 973 K at $\log(pO_2) = -23$, $\log(pS_2) = -9.5$.

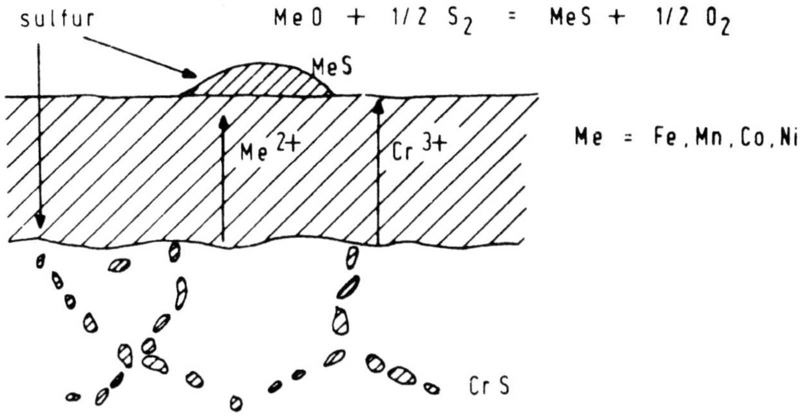

Fig.8: Schematic representation of external sulfide growth by outward diffusion of cations, a) through an intact chromia layer, b) by short-circuit diffusion at the sites of spinel crystallites in the layer.

rate must increase with the amount of spinel in the oxide scale, as found in the experiments. Additionally the transport of cations along grain boundaries and phase boundaries in the scale must be taken into account. With increasing pO_2 during preoxidation more islands of spinel are formed in the scale and the amount of spinel-chromia-interfaces increases, which may act as fast diffusion paths to further increase the sulfidation rates. The Fe-diffusion in the compact chromia scale is so slow that no FeS was formed during the 150 h exposure.

This shows that preoxidation is most effective under conditions where formation of spinel is avoided and only chromia is formed. However, after long term exposure or at higher temperatures even a chromia layer fails in the oxidizing- sulfidizing environment.[4,5] The breakdown is mainly caused by external sulfide formation on top of the preformed chromia layer but, additionally, internal sulfidation was observed.

Even at sulfur pressures lower than that necessary for FeS-formation uptake of sulfur into the chromia scale was detected by tracer measurements using ^{35}S. The exact experimental conditions are shown in Table 1. In order to obtain chromia scales thick enough to measure sulfur concentration profiles across the scale, preoxidation in H_2/H_2O was performed at 1173 K for all ^{35}S-tracer experiments. Subsequent exposure to $H_2/H_2O/H_2S$, tagged with ^{35}S, was performed at 973 K and 1173 K. Before sulfidation and oxidation at 973 K the preoxidized samples were annealed in H_2/H_2O for 3 days to establish the defect structure of the chromia.

The employed sulfur pressure was lower than that necessary for FeS-formation, the oxygen pressure was kept high enough for Cr_2O_3 to remain the stable phase compared to CrS.

Table 1: MEASUREMENT OF SULFUR PENETRATION THROUGH CHROMIA SCALES USING ^{35}S.

Treatment	Temp	Time	$\log(pO_2)$	$\log(pS_2)$
preoxidation	1173 K	150 h	-20.5	-
sulfidation/oxidation	1173 K	700 h	-20.0	-9.9
preoxidation	1173 K	150 h	-20.5	-
equilibration/annealing	973 K	75 h	-26.0	-
sulfidation/oxidation	973 K	700 h	-26.0	-11.5

Fig.9 shows a SEM-micrograph of a Fe-20Cr-12Ni sample after 150 h preoxidation, to form a chromia layer, and subsequent mixed sulfidation and oxidation for 700 h, both at 1173 K. Although the sulfidation time of 4 weeks was fairly long, the chromia scale is still dense and protective. No sulfides could be detected on the surface.

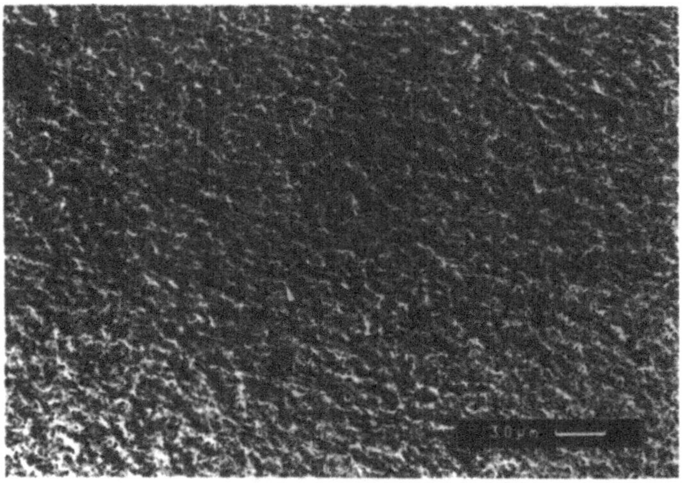

Fig.9. SEM-picture of a Fe-20Cr-12Ni-sample after 150 h preoxidation and 700 h mixed sulfidation and oxidation at 1173 K.

The values of the total sulfur uptake, from the surface to the centre of the cylindrical samples, were calculated from the successive radioactivity measurements at increasing depth from the surface and are shown in Table 2 for four different steels exposed to S-bearing gas mixtures at 973 and 1173 K. The table also shows the maximum penetration depth with sulfur concentrations above the detection limit.

Table 2: TOTAL SULFUR UPTAKE FROM THE SURFACE TO THE MIDDLE OF THE SAMPLES

ALLOY	TOTAL SULFUR UPTAKE (10^{-6}g)	PENETRATION DEPTH (10^{-6}m)	T (K)
Fe20Cr	3.5	57	973
Fe20Cr0.01Ce	2.2	54	973
Fe20Cr12Ni	1.2	55	973
Fe20Cr12Ni0.01Ce	1.1	62	973
Fe20Cr	18.0	352	1173
Fe20Cr0.01Ce	19.9	344	1173
Fe20Cr12Ni	11.0	350	1173
Fe20cr12Ni0.01Ce	5.8	338	1173

At 1173 K the total sulfur uptake is about 5 to 10 times higher than at 973 K. Comparison of the sulfur uptake by the austenitic samples and the ferritic alloys shows that the chromia layer on the austenitic samples was much more protective, keeping the sulfur content 2 to 3 times lower than that of the corresponding ferritic alloys with comparable Cr content. A small alloy addition of Ce appears to further improve the protectiveness of the Cr_2O_3 layer.

Fig.10 shows the sulfur penetration depth profiles at 1173 K. The profiles of the samples which had been annealed at 973 K are similar. The sulfur content at the chromia scale/alloy interface is markedly increased and at greater depths it quickly decreases to very low levels. The high radioactivity at the oxide/metal interface can be attributed to small amounts of chromium sulfides. Fig.11 shows a cross section of a Fe–20Cr-sample: Between the chromia scale and the FeCr-alloy a CrS-particle was detected by EDAX. This shows that the solubility of sulfur in the alloy was exceeded.

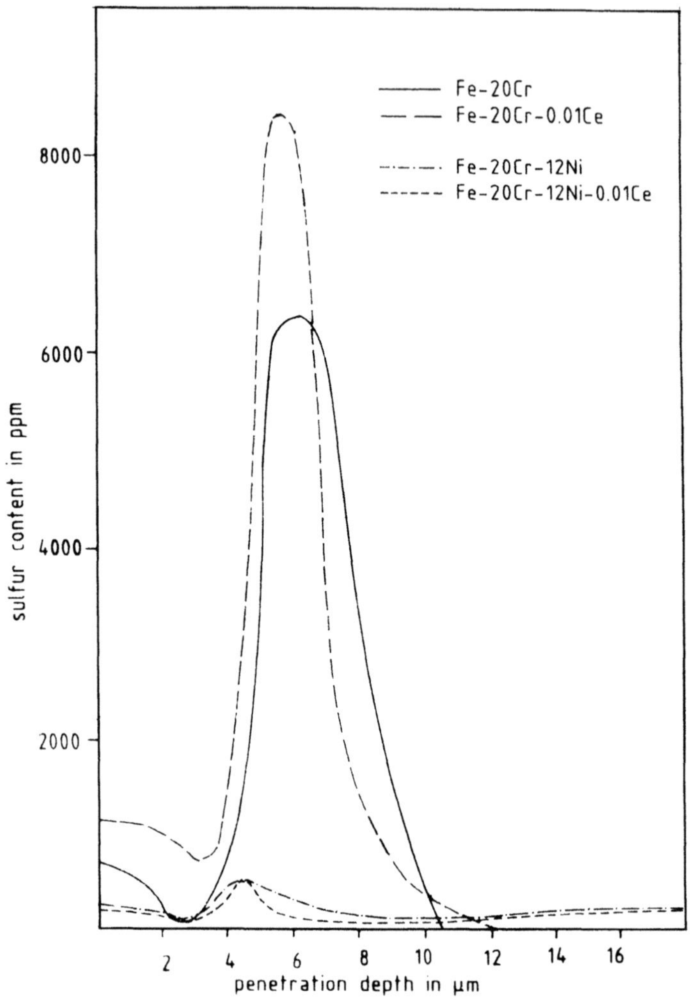

Fig.10: Sulfur penetration profiles after 150 h preoxidation and 700 h mixed sulfidation and oxidation at 1173 K.

Ce appears to affect the S-distribution across the sample. Although the total S-uptake in Ce-containing samples is smaller they show a higher S-concentration at the chromia/metal interface than Ce-free alloys. This may be associated with the very high S-affinity of Ce, which can act as a sulfur getter below the scale.

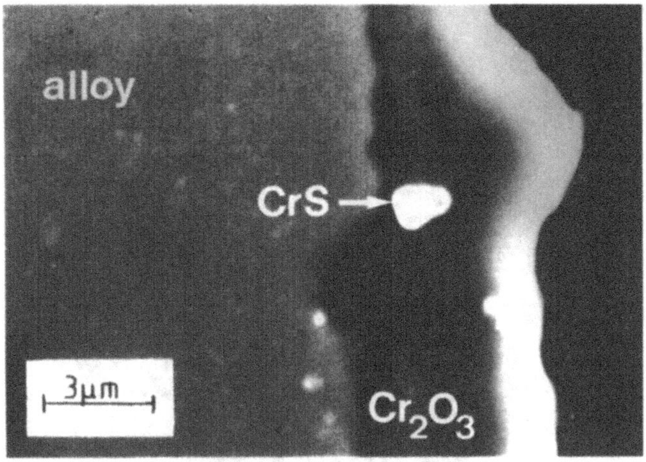

Fig.11: Metallographic cross-section of the oxide scale formed on Fe-20Cr after 150 h preoxidation and 700 h sulfidation/oxidation at 1173 K, showing CrS at the scale/alloy interface.

The drastic drop in sulfur concentration immediately below the chromia scale can also be shown by autoradiographs. Fig.12 shows an autoradiograph of a FeCrCe-sample on which the chromia layer had been removed from the upper part. Radiation of ^{35}S has darkened the film only in regions where the chromia scale had remained on the surface.

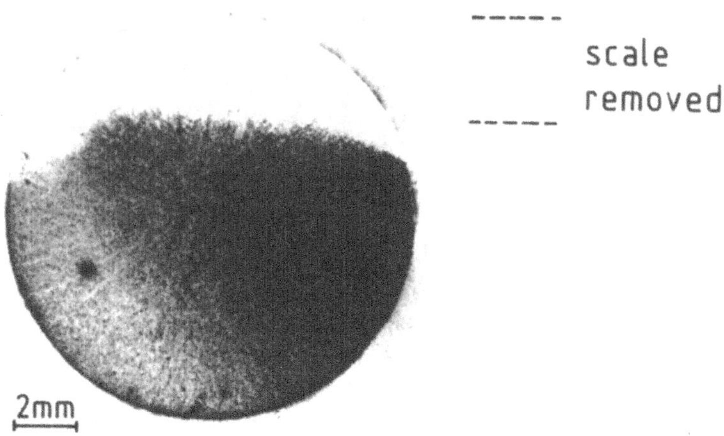

Fig.12: Autoradiograph of an Fe-20Cr-0.01Ce-sample after 150 h preoxidation and 700 h sulfidation/oxidation at 1173 K.

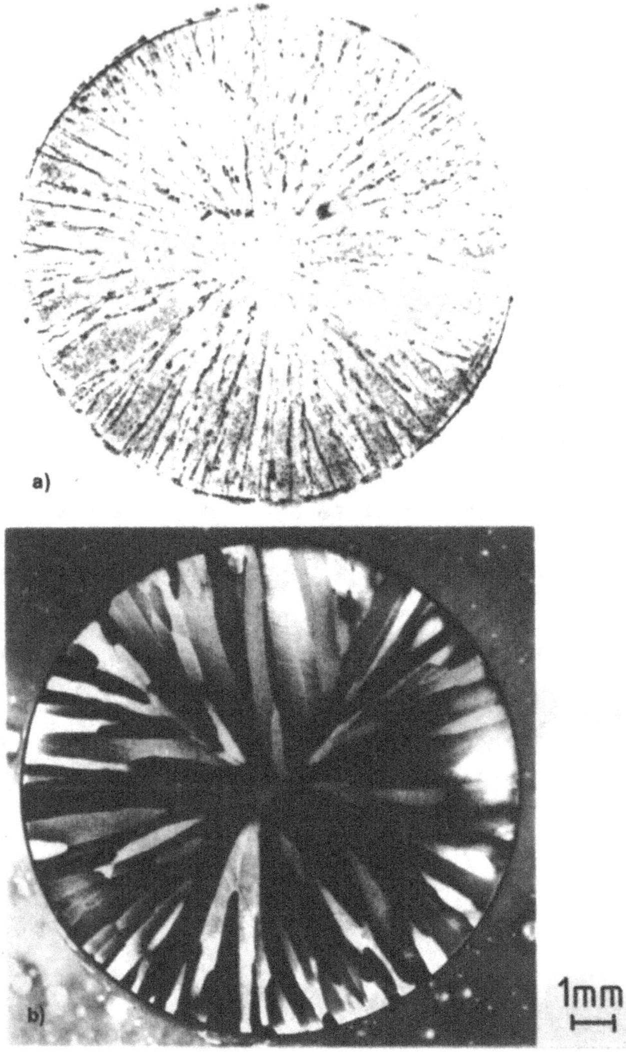

Fig.13: a) Autoradiograph of an Fe-20Cr-12Ni-sample,
b) Macroscopic cross section of the bulk sample.

On some samples the sulfur enrichment below the scale is enhanced especially at grain boundaries of the alloy. Fig.13 shows an autoradiograph taken on an exposed FeCrNi-sample after having removed half the scale thickness, together with a photo of the macrostructure of the bulk sample. The sulfur enrichment at the grain boundaries is clearly evident. The reason for this particular behavior is the slow diffusion of chromium in austenitic steels, which causes chromium depletion in the alloy region adjacent to the chromia scale during preoxidation. Both factors, low Cr-concentration and small Cr-diffusion coefficient, lead to inner sulfidation ($c_S D_S > c_{Cr} D_{Cr}$). The higher chromium activity at grain boundaries causes the enhanced CrS-formation there.

4. CONCLUSIONS

High oxygen partial pressure during preoxidation enhances spinel formation in the scale creating fast diffusion paths for the outward diffusion of cations, especially of Fe^{2+}. The presence of spinel in the oxide scale therefore promotes sulfide growth on the oxide surface.
Preoxidation, under conditions where spinel formation is avoided and only chromia is formed is most effective against sulfidation.
Chromia scales formed, on Fe-20Cr-12Ni alloys are more protective against sulfur penetration than those on Fe-20Cr alloys. Small additions of cerium further improve the protectiveness of chromia scales.
At the chromia/alloy interface Cr and Ce act as sulfur getters, leading to a marked increase of sulfur concentration at this interface, especially at alloy grain boundaries.
The amount of sulfur penetrating through an intact chromia scale is extremely low even after 700 h exposure to S-bearing gases.

Acknowledgement

This study was initially supported by the Bundesminister für Forschung und Technology (BMFT) in the COST 501 programme and later by the Deutsche Forschungsgemeinschaft.

REFERENCES

1. J.F.LANCASTER: Int. Met. Rev., 1978,**3**, 101.

2. H.J.GRABKE: in 'Aufbau von Oxidschichten auf Hochtemperaturwerkstoffen und ihre technische Bedeutung',(ed. A.Rahmel), 89-120, 1983, Oberursel, Deutsche Gesellschaft für Metallkunde.

3. D.S.WILLIAMS, R.MÖLLER and H.J.GRABKE: Oxid. Met., 1981,**16**, 253.

4. R.A PERKINS: in 'Environmental degradation of high temperature materials', March 1980, The Institution of Metallurgists, London, 5/1.

5. R.A.PERKINS: in Proc. Petten Int. Conf. on Alloy 800, 1978, eds. W.Betteridge et al., North Holland Publ.Company, 213.

6. K.NATESAN: Corrosion, 1985, **41**, 646.

7. M.DANIELEWSKI and K.NATESAN: Oxid. Met., 1978,**12**, 227.

8. C.S.GIGGINS and F.S.PETTIT: Oxid. Met., 1980,**14**, 363.

9. M.A.LABRANCHE, A.GARRETT-REED and G.J.YUREK: J. Electrochem. Soc., 1983, **130**, 2415-2413.

10. H.J.GRABKE, P.PAPAIACOVOU and P.SCHMIDT: Berichte zu COST 501, Projekt D 5, 'High temperature corrosion of FeNiCr alloys in oxidizing and sulfidizing gases', Jahresberichte 1983,1984, 1985

11. R.HUSSEY, P.PAPAIACOVOU and H.J.GRABKE: in Proc. 8e. Congres europeen de Corrosion, Nice, Nov.1985, Vol.1, 51-1-9.

12. H.J.GRABKE and P.PAPAIACOVOU: in Proc. 8e Congres europeen de Corrosion, Nice, Nov. 1985, Vol.1, 52-2-6.

13. F.H.STOTT, F.M.F.CHONG: in 'Corrosion resistant materials for coal conversion systems', (eds. D.B.MEADOWCRAFT and M.J.MANNING), 491-516, 1983, London, Applied Science Publishers.

14. J.D.TRETJAKOW and H.SCHMALZRIED: Berichte der Bunsengesellschaft für physikalische Chemie, 1865, **69**, 396-402.

15. D.J.YOUNG, W.W.SMELTZER and J.S.KIRKALDY: J. Electrochem. Soc., 1973, **9**, 1221.

16. I.BARIN and O.KNACKE: 'Thermochemical properties of inorganic substances',1977, Berlin/Heidelberg, Springer.

17. J.W.HALLORAN and H.K.BOWEN: J.Am.Ceram.Soc., 1980, **63**, 58.

18. R.DIECKMANN and H.SCHMALZRIED: Ber.Bunsenges., 1977, **81**, 344, 1977, **81**, 414.

19. R.DIECKMANN, T.O.MASON, J.D.HODGE and H.SCHMALZRIED: Ber.Bunsenges., 1978, **82**, 778.

20. J.D.HODGE: J.Electrochem.Soc., 1978, **125**, 55C.

21. O.KUBASCHEWSKI: Verein Deutscher Eisenhüttenleute, Ber.Nr. 0.015, Düsseldorf, 1983.

22. G.V.SAMSONOV: in 'The oxide handbook', IFI/Plenum, 1973, New York.

23. R.D.KINLOCH: Ph.D.Thesis, University of Delaware, 1970, University Microfilms, Ann Arbor, Michigan.

EFFECTS OF HYDROGEN CHLORIDE AND OF NITROGEN IN THE OXIDATION OF FE-20CR

D. Bramhoff, H.J. Grabke, H.P. Schmidt

Max-Planck-Institut für Eisenforschung GmbH
P.O. Box 140260
D-4000 Düsseldorf, Fed. Rep. Germany

ABSTRACT

The oxidation behavior of alloy Fe-20Cr in He-H_2-H_2O-HCl and N_2-H_2-H_2O-HCl was investigated at 1173 K by thermogravimetry. The rate of mass increase in He-H_2-H_2O-HCl decreases with increasing p_{HCl}; the kinetics can be described by growth of Cr_2O_3 according to a parabolic rate law and simultaneous evaporation of $CrCl_2$ following a linear rate law. The oxidation in He-H_2-H_2O-HCl causes no enhanced porosity of the oxide scale, compared to oxidation in He-H_2-H_2O. The protectiveness of the scales is adversely affected only if condensed chlorides are formed. During exposure in N_2-H_2-H_2O-HCl the oxidation is faster and the evaporation is enhanced in comparison to He-H_2-H_2O-HCl. The different behaviour is attributed to chromium nitride formation in the initial stage of corrosion attack.

1. INTRODUCTION

In a number of important industrial processes metallic materials are exposed to high temperature environments containing chlorine. These processes include the incineration of municipal waste, the combustion of chlorine contaminated coal and heat recuperation from industrial waste combustion. The behaviour of metals and alloys in chlorinating environments is not well known because it is complicated by the volatility of many metal chloride species.[1] Metals or alloys which form protective scales in oxidizing environments often do not so in chlorinating environments. The alloy performance is deteriorated by chloride vaporisation and adhesion loss of the oxide layer. In mixed

oxidizing-chlorinating environments, the kinetics of the reactions are very complex and not easily to interpret, because the oxidation reaction leads to a mass increase while the volatilisation reaction leads to a decrease in the mass of the specimen.[2-6]
The aim of this study was, firstly to determine the effect of small HCl(g)-addditions and of nitrogen on the scaling characteristics of a Fe-20Cr alloy, and secondly to investigate and clarify the vaporisation mechanism by means of kinetic measurements.

2. EXPERIMENTAL DETAILS

In order to obtain fundamental information on the reaction mechanisms in complex environments, the experiments must be carried out under well defined conditions, where the number of the possible reactions and involved species is minimized.
Therefore two sets of experiments were carried out in this study by performing thermogravimetric measurements in H_2-H_2O-HCl-atmospheres with He and with N_2 as carrier gas, respectively. In all cases the reaction temperature was 1173 K, the oxygen pressure was 1×10^{-20} bar and the concentration of HCl(g) was varied between 0....2000 vppm. Under these conditions Cr_2O_3 is the only condensed reaction product which is thermodynamically stable and chromium chloride is formed only as a vapour species.[7] In order to avoid the formation of condensed chlorides during heating, the experiments generally were started after heating the samples to reaction temperature in flowing H_2, which had been purified by passage over an active copper catalyst and through P_2O_5. At reaction temperature, the employed HCl pressure was below that necessary for the formation of condensed chlorides.
The thermogravimetric experiments were carried out in an apparatus consisting of two parts: a glass part for purification and mixing of the reactive gases and a vertical quartz reaction tube with a microbalance by which the mass gain of the sample was continuously recorded (Fig. 1).
In order to avoid depletion of the reactive gases at the surface of the sample, the measurements were carried out in flowing gas mixtures. The flow rate was maintained at 2.5 ml/sec in all cases. Helium or nitrogen were used as carrier gases to which the reactive gases (H_2, H_2O, HCl) were added in suitable amounts. The HCl partial pressure was established by passing H_2 through and over an aqueous solution of hydrogen chloride surrounded by a thermostated water bath. By appropriate choice of the solution concentration and the bath temperature, the required HCl(g) partial pressure was obtained. The water vapor pressure was established by passing H_2 through a temperature-controlled mixture of oxalic acid dihydrate and anhydrous oxalic acid.

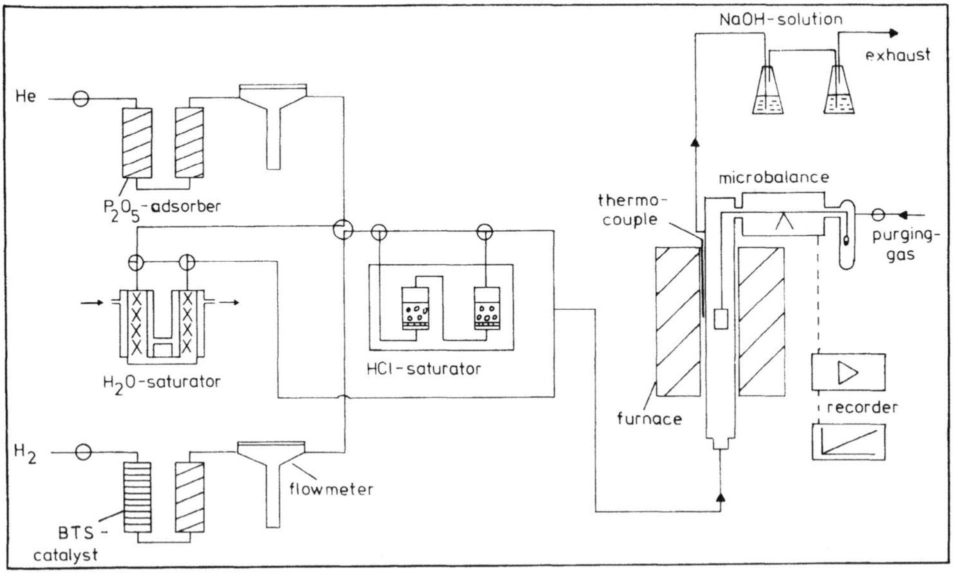

Fig. 1 Schematic representation of the experimental apparatus for thermogravimetric measurements.

The experiments were initiated after heating the sample to reaction temperature in flowing H_2. When the temperature was reached, the gas mixture H_2-H_2O-HCl was introduced into the reaction tube. During the test the exhaust gas was passed through a sodium hydroxide solution which removed any unreacted HCl and at the end of the test this solution was analysed for chloride content.

3. RESULTS AND DISCUSSION

3.1. He-H_2-H_2O-HCl-Atmospheres

Fig. 2 shows a series of mass gain curves for the oxidation of alloy Fe-20Cr in He-H_2-H_2O-HCl gas mixtures with different amounts of HCl(g). In the environment without HCl(g), Fe-20Cr shows an oxidation rate which is very close to parabolic indicating a diffusion-controlled process. Energy dispersive X-ray analysis (EDX) and X-ray diffraction (XRD) showed that the surface scale consisted only of chromia. The influence of HCl(g)-additions is also illustrated in Fig. 2. The mass increase in HCl(g)-containing gases is less than in gases without HCl(g), and decreases with increasing HCl pressure.

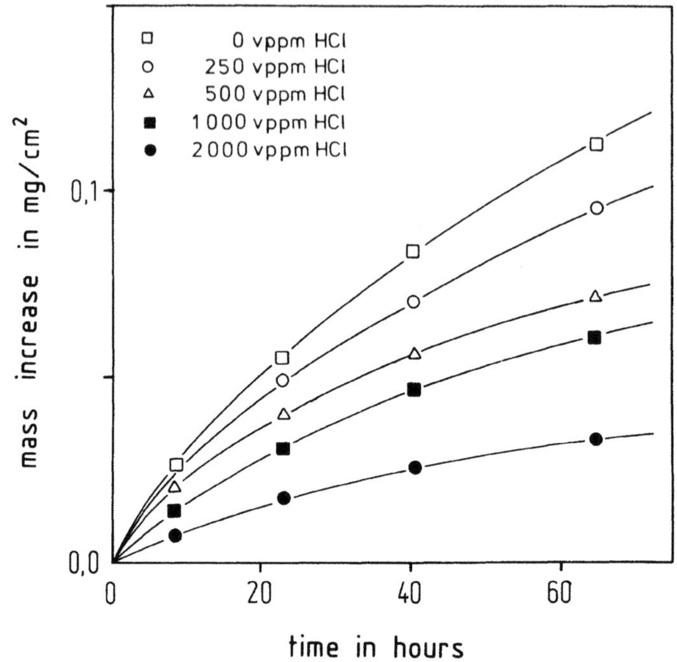

Fig. 2 Mass increase vs time of alloy Fe-20Cr oxidized at 1173 K in different $He-H_2-H_2O-HCl$ gas mixtures.

This behavior is consistent with the paralinear rate equation which is used to describe the oxidation of chromium and chromium containing alloys at high temperatures.[8] This equation describes the corrosion of a material by two simultaneous reactions, one which leads to an increase in mass of the specimen and obeys a parabolic rate equation, and the other one leading to a decrease in mass following a linear rate equation:

$$d(m/A)/dt = k_p/(m/A) - k_v\, f[p(HCl),\, p(H_2),\, p(H_2O)] \qquad (1)$$

where m is the mass of the scale, k_p is the parabolic rate constant ($g^2\ cm^{-4}\ s^{-1}$), and k_v is the linear constant rate ($g\ cm^{-2}\ s^{-1}$).

The formation of Cr_2O_3, being the only stable condensed phase, leads to an increase in mass of the samples. The consumption of Cr_2O_3 by reaction with HCl(g) can be described by the following equation:

$$Cr_2O_3(s) + 4\ HCl(g) + H_2(g) = 2\ CrCl_2(g) + 3\ H_2O(g) \qquad (2)$$

Due to its volatility, $CrCl_2$ does not stay in or on the scale but evaporates and condenses on the cooler parts of the outlet tube.

The partial pressure of $CrCl_2$ at the gas/solid interface in equilibrium with Cr_2O_3 depends on the HCl(g) partial pressure and is in the range $10^{-8}...10^{-9}$ bar. Using these values, calculations indicate that the rate-determining step of the evaporation is the diffusion of $CrCl_2$(g) through a laminar diffusion boundary layer of gas into the main gas stream.
Fig. 3 shows the overall mass gain of Fe-20Cr alloy in atmospheres with various HCl(g) contents, due to growth of Cr_2O_3 and simultaneous evaporation of $CrCl_2$, and the calculated mass loss by evaporation of $CrCl_2$(g).

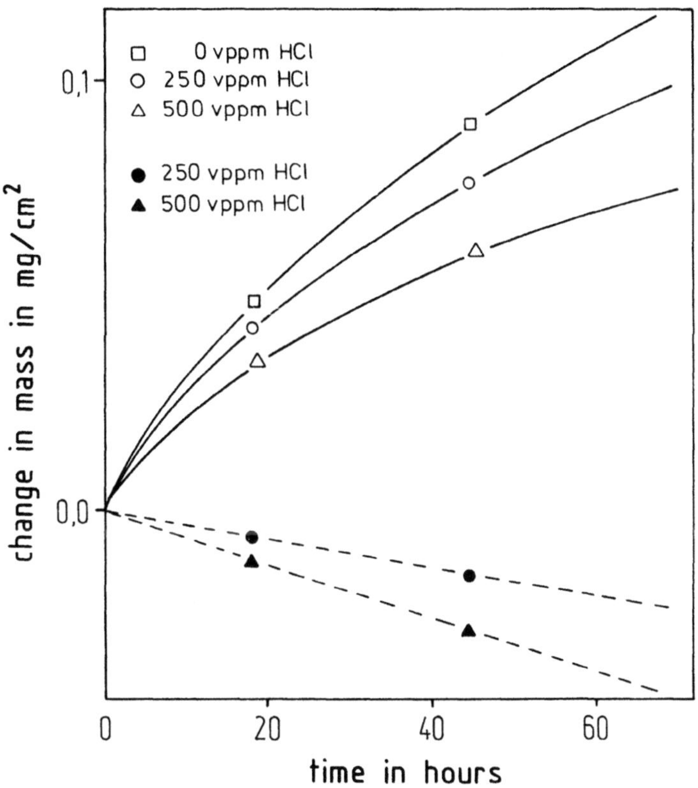

Fig. 3 Plot of thermogravimetric measurements: solid lines = overall mass gain; dashed lines = calculated mass loss by evaporation of $CrCl_2$.

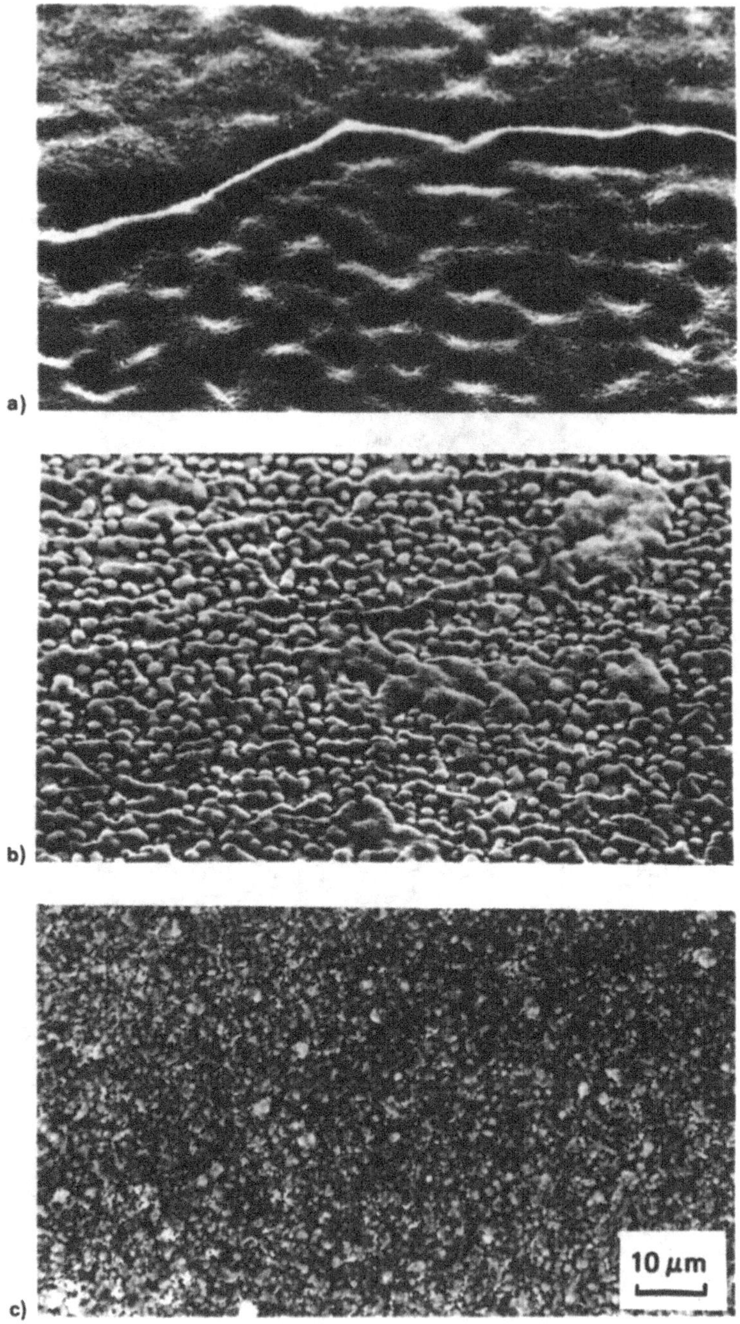

Fig. 4 Influence of HCl(g) additions on the scale morphology of Fe–20Cr alloy: a) 0 vppm, b) 500 vppm, c) 2000 vppm

The presence of HCl(g) in the gas mixture caused marked differences in the morphology of the Cr_2O_3-scales formed on Fe-20Cr. This is shown in Fig. 4a-c for environments with 0, 500 and 2000 vppm HCl(g). In all cases an adherent dense Cr_2O_3-layer is formed during the exposure but the roughness of the scales increases with increasing HCl(g) pressure. In particular, the scale formed in the environment without HCl(g) is very dense, approximately 1.5 μm thick and shows no evidence of HCl attack on the surface. The scale formed in the environment with 500 vppm HCl(g) is approximately 1.0 μm thick and shows a rough oxide scale, indicating that the scale had been attacked by HCl(g). The scale formed in the environment with 2000 vppm HCl(g) is much thinner (ca. 0.5 μm) than the previous ones and the surface roughness of the oxide is further increased. The observed decrease in thickness of the oxide scales with increasing HCl(g) pressure is in agreement with the paralinear corrosion model because the rate constant for the volatilization process depends on the partial pressure of HCl(g).

3.2 Tracer Studies

In order to get further and more detailed information on the actual porosity of oxide scales formed in HCl(g)-containing environments, 'carbon permeation tests' with ^{14}C-tracer were performed.[9,10] For these studies a separate glass apparatus in which the gases can be circulated over the samples, was utilized (Fig. 5). The samples, 10mm in diameter and 6mm in thickness, were inserted into a quartz tube located in a movable furnace and preoxidized for 140 hours in a H_2-H_2O or H_2-H_2O-HCl gas mixture at 1173 K. After preoxidation radioactive CO-CO_2 gas, tagged with ^{14}C, was added, and the samples were exposed to the radioactive gas mixture for 700 hours. The required p_{HCl} and p_{H_2O} in the system were established by HCl- and H_2O-saturators as described above.

After the experiments, the carbon uptake of the samples was determined by stepwise polishing and subsequent measurement of the surface radioactivity using a methane counting tube. From these measurements carbon concentration profiles were calculated. The penetration profiles, ppm C plotted versus penetration depth, are shown in Fig. 6. The total carbon uptake after both preoxidation treatments is very low which shows that the preformed Cr_2O_3-scales are dense and almost impermeable for carbon-bearing molecules. Furthermore, the profiles show no significant differences in the carbon penetration depth of the samples with different preoxidation treatments.

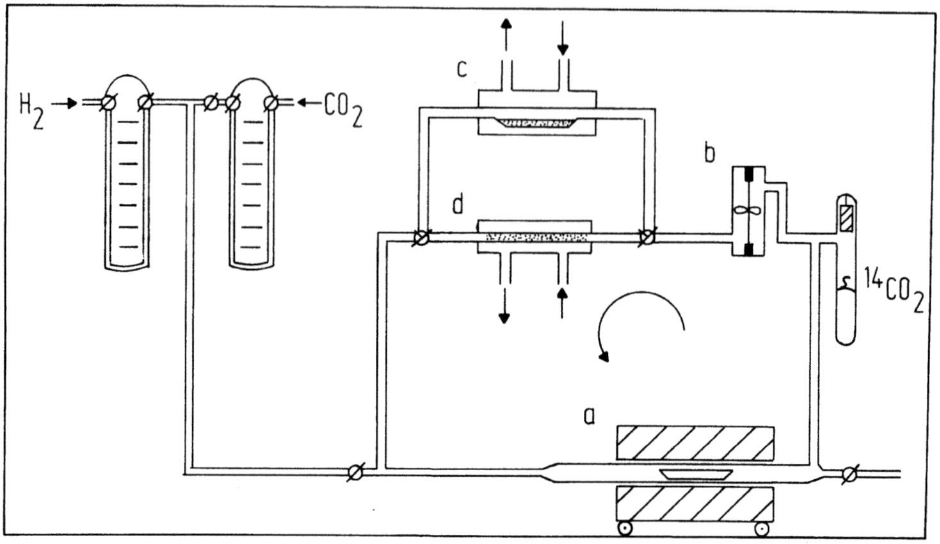

Fig. 5 Schematic representation of the gas circuit for preoxidation and exposure of alloys to radioactive H_2-H_2O-CO-CO_2-HCl gas mixtures: a) sample furnace, b) ventilator, c) HCl-saturator, d) H_2O-saturator.

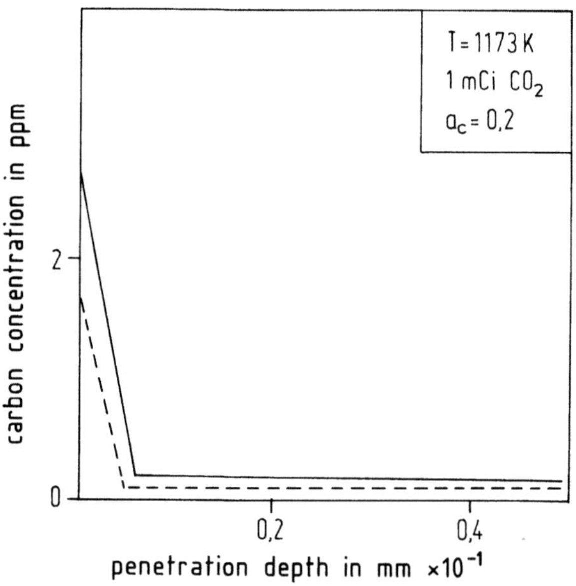

Fig. 6 Concentration profiles of carbon in Fe-20Cr after preoxidation in H_2-H_2O (---) and H_2-H_2O-HCl (—) and subsequent exposure to radioactive H_2-H_2O-CO-CO_2-HCl gas mixtures.

This result indicates that the presence of HCl(g) during the oxidation did not cause any enhanced porosity of the oxide scales compared to oxidation in H_2-H_2O. However the structure, adherence and protectiveness of the scale is detrimentally affected when condensed chlorides are formed, for example during heating of the samples in H_2-H_2O-HCl gas mixtures.
This could be shown in short-term experiments where samples were heated in H_2-H_2O-HCl to 1273 K in 15 seconds and subsequently exposed to the same gas mixture for 1 minute. Scanning electron microscopy (SEM) showed (Fig. 7) that during this test large crystals were formed at the surface of Fe-20Cr samples which can be seen in the top-view of the surface. X-ray diffraction analysis indicated the presence of iron chlorides and possibly of Cr_2O_3 and $FeCr_2O_4$. The formation of condensed chlorides during heating of the samples can be explained by the temperature dependence of the stability of the condensed chlorides. At temperatures lower than the usual reaction temperature of 1173 K, the HCl(g) partial pressure of the used gas mixture is above the pressure required for the formation of condensed chlorides.

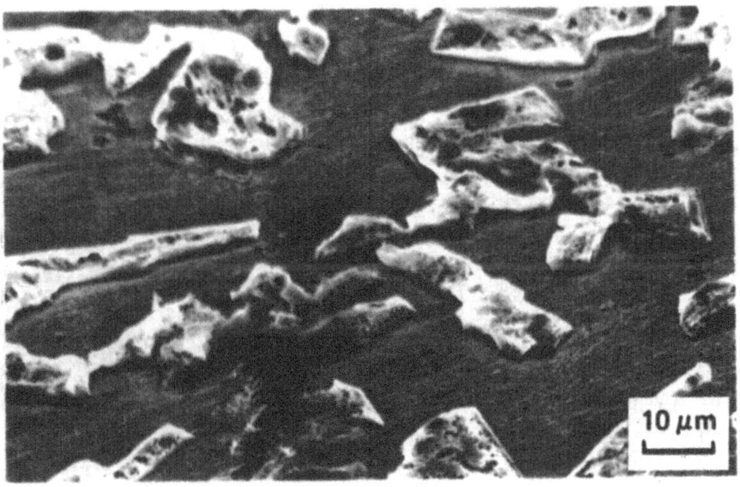

Fig. 7 Scanning electron micrograph of the chlorides formed on Fe-20Cr after heating in H_2-H_2O-HCl and subsequent exposure for 1 min at 1273 K.

The initial formation of condensed chlorides during heating adversely affects the structure and adherence of the growing oxide scale also in long-term testing, as can be seen in Fig. 8.

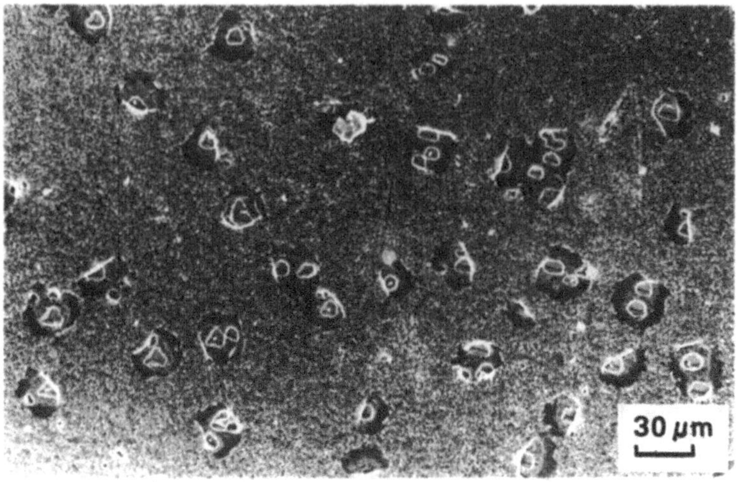

Fig. 8 Scanning electron micrograph of the oxide scale formed on Fe–20Cr after heating in H_2-H_2O-HCl and subsequent exposure for 50h at 1173 K.

After heating-up in H_2-H_2O-HCl and subsequent exposure for 50 h at 1173 K, the sample was covered with a dense, wrinkled oxide scale and shows locally spalled areas. The absence of any scale healing suggests that the oxide spalling had occured upon cooling. X-ray diffraction measurements indicated only the presence of Cr_2O_3 with no chloride compounds being detected. The spallation of the oxide scale may be explained by the presence of voids at the scale/metal interface resulting from the evaporation of condensed chlorides which had been formed in the initial stage of oxidation attack. The presence of voids reduces the adherence of the oxide scale and induces the severe spallation upon cooling of the samples after long-term exposure.
Heating of the samples in dry purified hydrogen or preoxidation in H_2-H_2O eliminated these adverse effects of hydrogen chloride.

3.3 N_2-H_2-H_2O-HCl-Atmospheres

The mass gain rates of Fe–20Cr samples in four N_2-H_2-H_2O-HCl gas mixtures at 1173 K are shown in Fig. 9. The general behavior is similar to that obtained in He-H_2-H_2O-HCl-atmospheres. The mass gain rate decreases with increasing HCl(g) pressure from 0...2000 vppm. Thus the kinetics in nitrogen atmospheres can also be described with the paralinear corrosion model: Growth of Cr_2O_3 following a parabolic rate law and simultaneous evaporation of $CrCl_2$(g) following a linear rate law.
Comparison of the thermogravimetric results obtained in helium

and in nitrogen atmospheres show however marked differences (Fig. 10). The increase in mass of samples oxidized in N_2-H_2-H_2O was at least twice that obtained in He-H_2-H_2O. Note also that in nitrogen atmospheres containing HCl(g) the evaporation rate is enhanced compared to helium atmospheres with HCl(g).

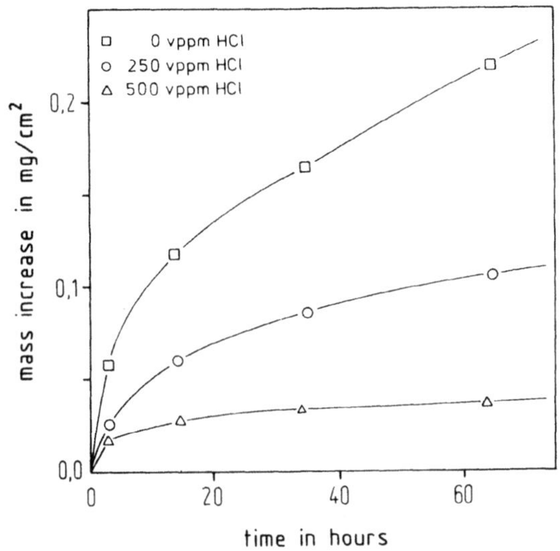

Fig. 9 Mass increase vs time of alloy Fe-20Cr oxidized at 1173 K in different N_2-H_2-H_2O-HCl gas mixtures.

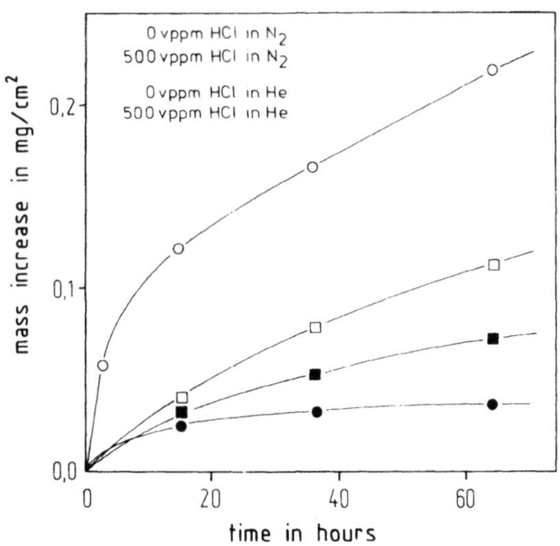

Fig. 10 Comparison of the thermogravimetric results obtained in He-H_2-H_2O-HCl and N_2-H_2-H_2O-HCl gas mixtures.

This difference can be explained by chromium nitride formation during the initial attack (transient stage). The initial formation of nitride was proved by taking Auger electron depth profiles and Auger electron spectra, after 1 min and 2 min exposure of Fe-20Cr samples in N_2-H_2-H_2O-atmospheres at 1173 K. The depth profiles give the relative elemental concentrations as a function of depth expressed as sputter time in minutes (sputter rate 2nm/min).

The results are shown in Fig. 11 and Fig. 12. After 1 min exposure Cr, N and C are enriched at the surface of Fe-20Cr. After 2 min exposure no enrichment of N and C could be anymore observed. Only chromium and oxygen are enriched, indicating that the surface scale consists predominantly of Cr_2O_3.

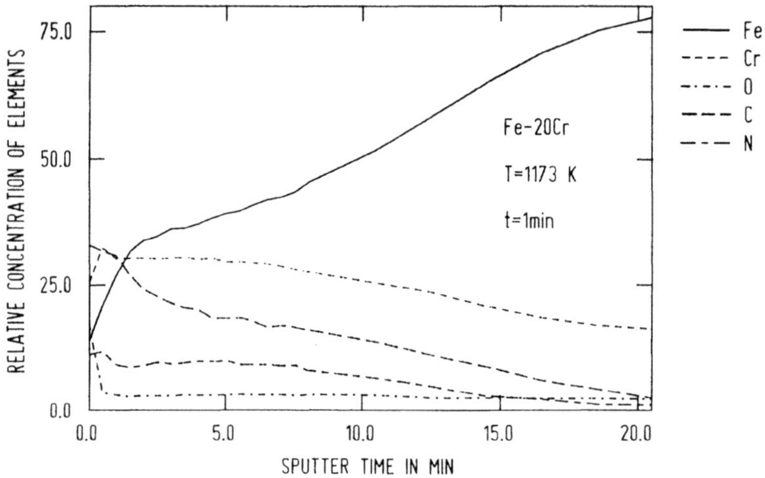

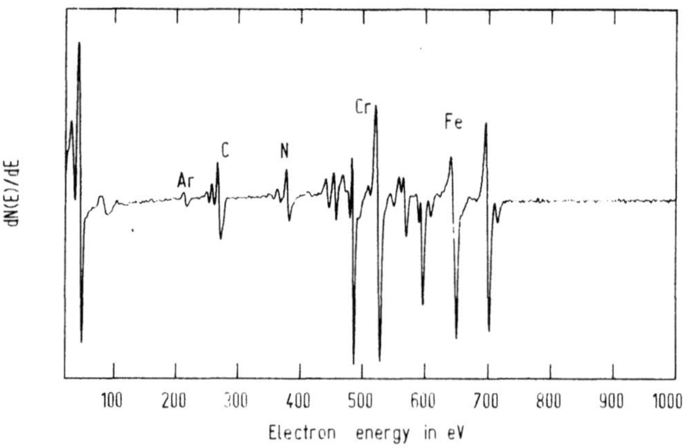

Fig. 11 Auger depth profile and Auger spectrum of alloy Fe-20Cr after 1 min oxidation in N_2-H_2-H_2O at 1173 K.

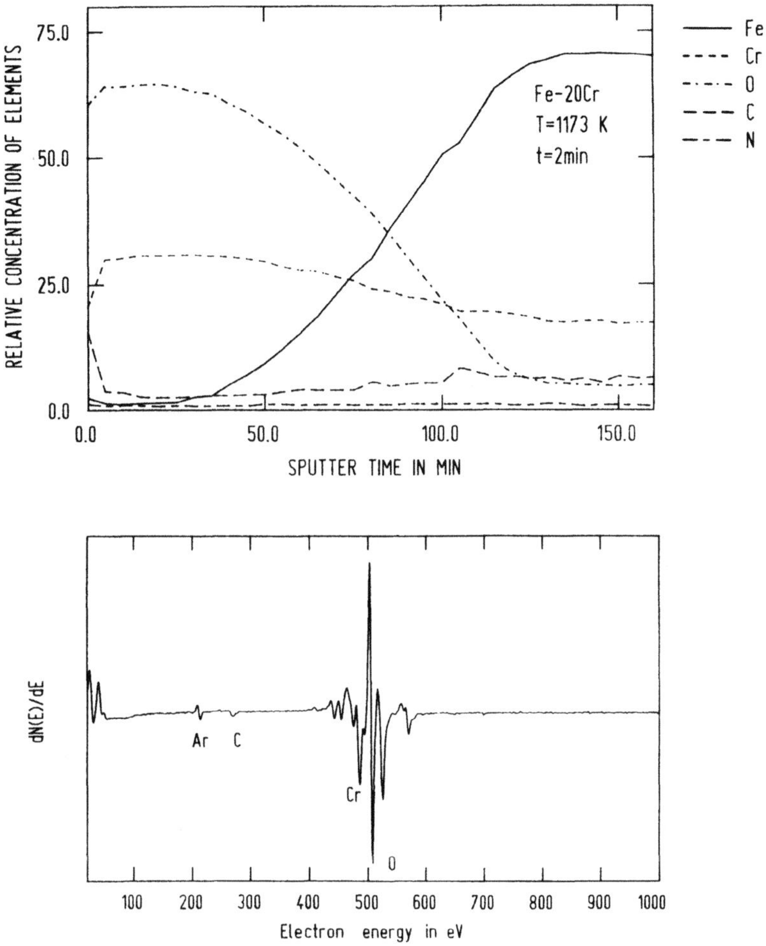

Fig. 12 Auger depth profile and Auger spectrum of alloy Fe-20Cr after 2 min oxidation in N_2-H_2-H_2O at 1173 K

These observations suggested that CrN, which had been formed during 1 min exposure, will react very rapidly with oxygen to the thermodynamically more stable Cr_2O_3, and already after 2 min all nitrogen is dissolved in the interior of the alloy.
However, somehow the initial formation of nitrides adversely affects the structure and protectiveness of the overgrowing oxide scale, as already observed for other surface impurities.[11] The growth of Cr_2O_3 on the nitrides prevents the Cr_2O_3-contact with the metallic matrix and the formation of a dense, pore-free and intact chromia layer. The porosity of the oxide scale enables the

transport of H_2O and HCl(g) through the scale, most likely by gaseous diffusion along the pores to the base metal. The H_2O and HCl(g) can react with the metallic matrix to form additional oxide or volatile chlorides. These reactions lead to the increase of the observed oxidation and evaporation rates in the presence of nitrogen in the reaction gas.

4. CONCLUSIONS

The influence of HCl(g) and nitrogen on the high temperature corrosion of alloy Fe-20Cr in H_2-H_2O-HCl atmospheres containing low oxygen partial pressures was studied.
With increasing p(HCl) in the gas atmosphere, the evaporation of Cr_2O_3 through formation of $CrCl_2$ is enhanced leading to paralinear kinetics. Under conditions where no condensed chlorides are formed, the porosity of the chromia scale is not enhanced, compared to oxidation in HCl(g)-free atmospheres.
In N_2-H_2-H_2O-HCl atmospheres, the oxidation rate is faster and also the evaporation is enhanced in comparison to He-H_2-H_2O-HCl. This is attributed to chromium nitride formation in the initial (transient) stage of attack.

Acknowledgement

We gratefully acknowledge the financial support of this study by the Bundesministerium für Forschung und Technologie.

R E F E R E N C E S

1. P.L. DANIEL and R.A. RAPP: 'Halogen Corrosion of Metals', in: 'Advances in Corrosion Science and Technologie', (ed. M.G. Fontana and R.W. Staehle), Vol.5., 55-172, 1976, New York, Plenum Press.

2. M.J. MC NALLEN, W.W. LIANG, J.M. OH and E.T. KANGI Oxid. Met., 1982, **167**, 371.

3. N.S. JACOBSEN, M.J. MC NALLEN and Y.Y. LEE: Metall. Trans. A, 1986, **17A**, 1223.

4. M.J. MALONEY and M.J. MC NALLEN: Metall. Trans. B, 1985, **16B**, 751.

5. J. HALFDANARSON and K. HAUFFE: Werkst. Korros., 1973, **24**, 8.

6. K. REINHOLD and K. HAUFFE: J. Electrochem. Soc., 1977, **124**, 875.

7. O. KUBASCHEWSKI and C.B. ALCOCK: 'Metallurgical Thermochemistry', 5th ed., 1979, Oxford, Pergamon Press.

8. C.S. TEDMON: J. Electrochem. Soc., 1966, **113**, 766.

9. I. WOLF and H.J. GRABKE: Solid State Commun., 1985, **54**, 5.

10. I. WOLF and H.J. GRABKE: in Proc. of the Eight Congress European Corrosion, 1985, Nice, p. 48.

11. I. WOLF, H.J. GRABKE and H.P. SCHMIDT: Oxid. Met., 1988, **29**, 289.

SESSION 5

PRACTICAL IMPLICATIONS AND OUTLOOK

THE APPLICATION OF ACTIVE ELEMENTS IN INDUSTRIAL PRACTICE

J.H. Davidson

Imphy S.A.

58160 Imphy - France

ABSTRACT

The beneficial influence of active elements on the oxidation behaviour of high temperature alloys usually reaches a maximum for concentrations of a few tens, or at most a few hundreds of p.p.m., and higher contents can have catastrophic consequences on workability. Residual levels must therefore be controlled within a fairly narrow range. This is extremely difficult to achieve, due to the high affinity of the active elements for oxygen and their tendency to react with furnace refractories. More widespread use of active elements will require a better understanding of their constitutional effects in heat-resisting alloy matrices, in order to rationalize their choice and concentration, together with improved addition techniques to reduce reaction times and to give more reliable yields.

1. INTRODUCTION

In view of the spectacular improvements in high temperature oxidation and corrosion resistance which can be obtained with small additions of active elements, it is natural to ask why their use is not more widespread in heat-resisting steels and alloys. It is the purpose of this paper to attempt to reply to this question and to outline some of the problems involved.

2. METHODS OF ANALYSIS

An early problem encountered with the use of active elements was the lack of an accurate chemical analysis technique, capable of measuring concentrations from a few p.p.m. to several thousand p.p.m.

The method initially developed at IMPHY involved concentrating the active element by a double separation. The specimen is first of all dissolved and the active elements are co-precipitated with calcium fluoride to separate the bulk of the matrix. The residue is then taken back into solution and the active elements are reprecipitated with calcium oxalate in a second enrichment and purification stage. The filtered precipitate is analysed by X-ray fluorescence spectroscopy, calibrated using synthetic solutions.

The detection threshold of this technique is a few p.p.m., with an accuracy of about 5 to 8 p.p.m., but it can also be used for high concentrations (e.g. for the analysis of mischmetal). The major drawback is the lengthy procedure involved, a total of nearly 3 days being required.

This method has now been largely replaced at IMPHY by inductively coupled plasma emission spectroscopy, which can be carried out in only a few hours, since no separation is necessary. Calibration is performed with synthetic solutions. The detection threshold is in the range 10 to 30 p.p.m., while the accuracy is $\leqslant$ 3 p.p.m. The method can also be applied to high concentrations (e.g. mischmetal).

3. TYPE OF ADDITIVE AND COST

In early work at IMPHY, the active elements were added in elemental form, or as Ni- or Co- based binary alloys with compositions adjusted to match the melting range of the material to which they were to be added.

However, experience showed that this was not a major problem, whereas considerable difficulty was experienced in obtaining supplies of conventiently sized pieces. In effect, most elemental active elements are too ductile to be easily broken down, and tend to spontaneously ignite during cutting operations. Yttrium is now most commonly used in the form of a crushable Al-Y master alloy, while mischmetal can be obtained as chopped lengths of wire-rod.

While many active elements are much more abundant in nature than the term "rare earths" might suggest,[1] they tend to be widely disseminated and their ores are highly stable. The different elements often occur together and are particularly difficult to separate from one another, due to their similar chemical affinities. Costs can vary by at least two orders of magnitude, depending on demand, on the nature of the element and on the purity required. The cheapest form is mischmetal, in which the active elements, Ce, La, Nd and several others are mixed in the propor-

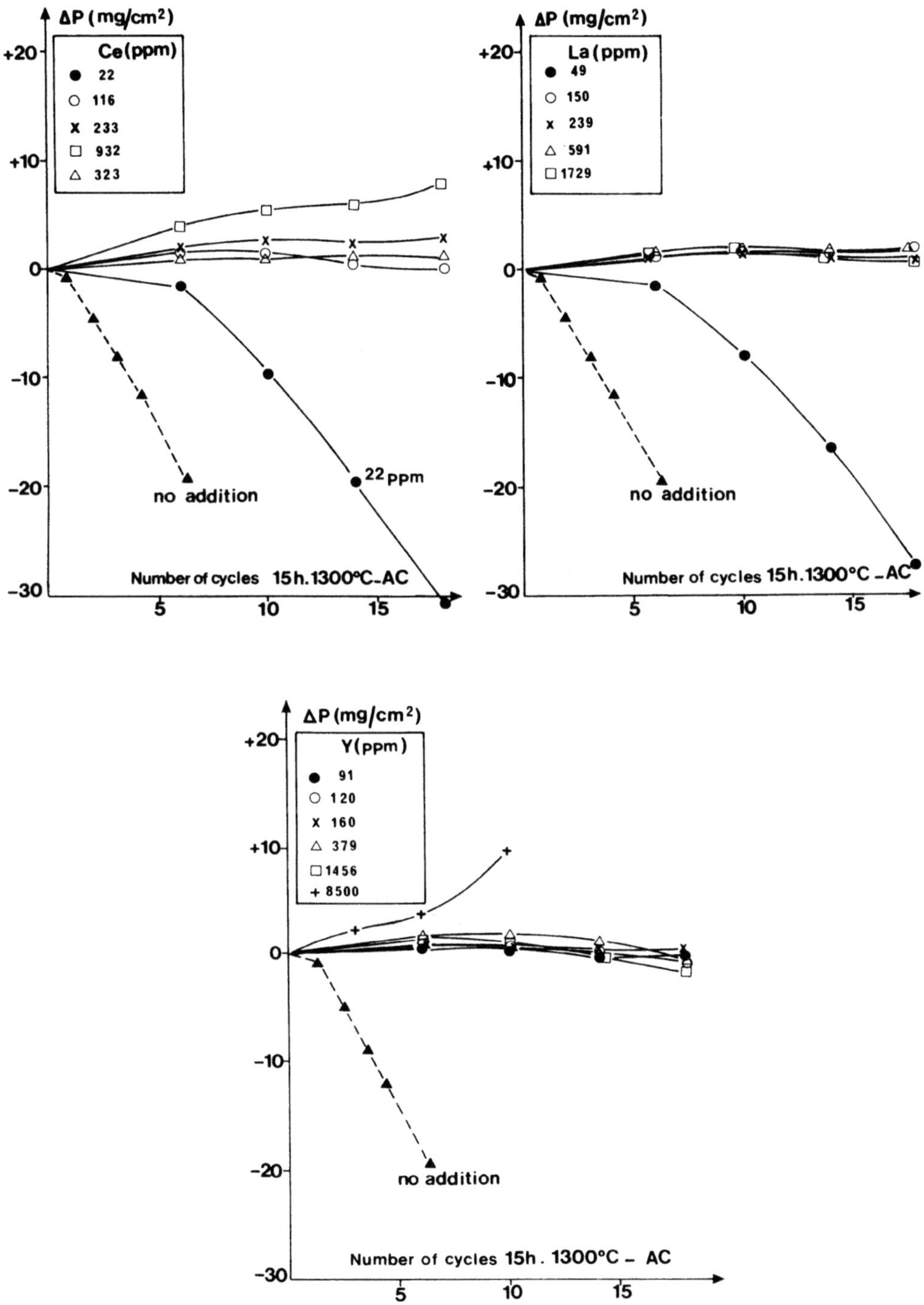

Fig. 1: Effect of active element content on the cyclic oxidation resistance of Fe-45Ni-25Cr-5Al alloys.[3]

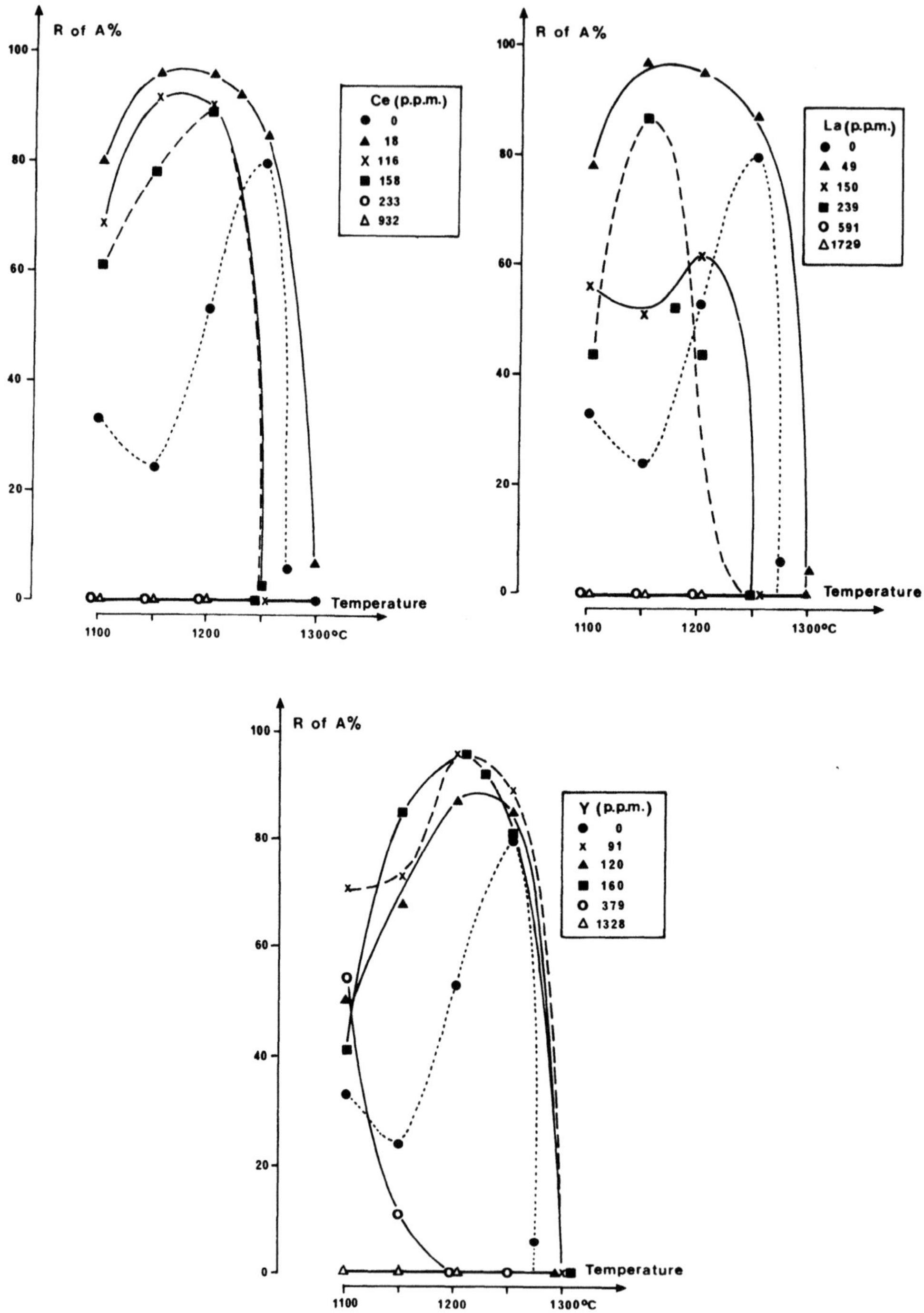

Fig. 2: Effect of active element content on the hot tensile ductility of as-cast Fe-45Ni-25Cr-5Al alloys.[3]

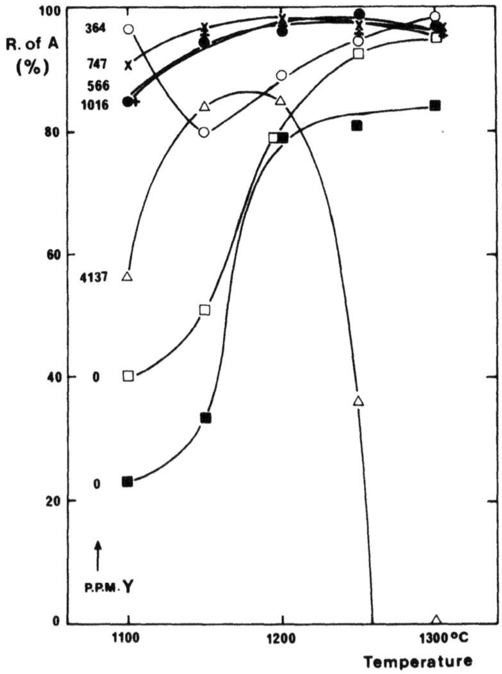

Fig. 3: Effect of Y content on the hot tensile ductility of as-cast Ni-20Cr alloys. The points for zero addition are for 2 different melts.

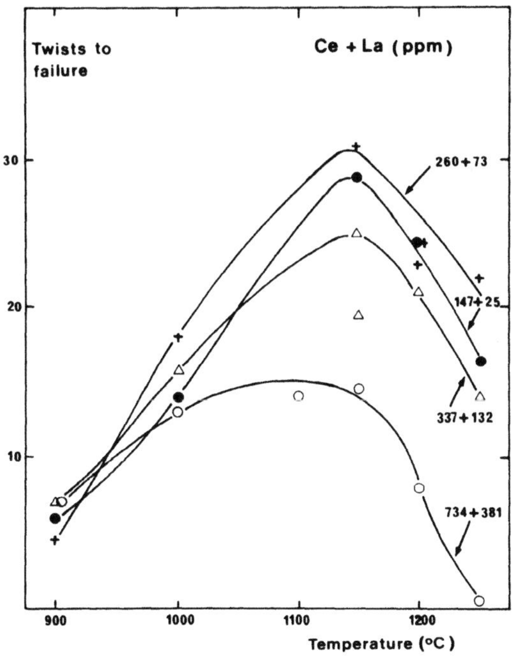

Fig. 4: Effect of mischmetal additions on the torsional hot ductility of industrial Ni-20Cr alloy billet. Note that such heating element grades always contain active elements.

tions in which they occur in the ores.[2]

At the time of writing, typical prices for mischmetal were about \$ 17/kg, whereas those for 95 % pure Y were around \$ 300/kg (95 % pure Y contains 95 % Y and a maximum of 1 % "impurities, the remainder being other active elements).

4. EFFECT OF CONCENTRATION

In general, the beneficial effect of active elements on oxidation behaviour requires a few tens or at most a few hundreds of parts per million, beyond which no marked further improvement is observed. This is illustrated for an Fe - 45 Ni - 25 Cr - 5 Al matrix in Fig. 1.[3] Higher levels are used in Fe - Cr - Al alloys chiefly to control grain size through intermetallic precipitation.[4]

However, the active elements generally have a marked influence on workability, the latter being extremely sensitive to concentration. This is illustrated in Figs. 2 to 5. While these figures show certain differences, which will be discussed in more detail in the paragraph on the choice of active elements, there are a number of common features. Small additions improve hot ductility significantly, and this is often true even when the residual level is below the detection limit, Fig. 5. It can be reasonably assumed that this effect is due to scavenging of detrimental elements such as sulphur and oxygen. Beyond a certain concentration, workability is reduced, often catastrophically. This can invariably be associated with the formation of intermetallic phases and local incipient melting. Fig. 6 shows an example for an Fe - 36 Ni - 18 Cr alloy with 850 p.p.m. Ce and 350 p.p.m. La (i.e. 1600-1700 p.p.m. total rare-earths, resulting from the addition of mischmetal). In this case, two rare-earth-containing intermetallic phases were observed, one of them containing phosphorus (alloy phosphorus content = 120 p.p.m.).

5. CONTROL OF YIELD

Because of the often fairly narrow range of residual concentrations necessary to avoid hot workability problems, it is important to closely control the yield of active element additions. Unfortunately, this is extremely difficult due chiefly to their high reactivity towards oxygen and refractory furnace linings although high vapour pressure could be an additional factor, particularly in the case of yttrium. Fig. 7 shows the kinetics of Y loss in a 35 kg bath of Fe - 20 Ni - 20 Cr - 5 Al alloy melted in air in an alumina-magnesite spinel crucible. Figs. 8 and 9 illustrate the yield of active elements, respectively for 5 kg and 50 kg ingots melted in argon, the additions being made 2 minutes before teeming. It can be seen that higher and more reproducible

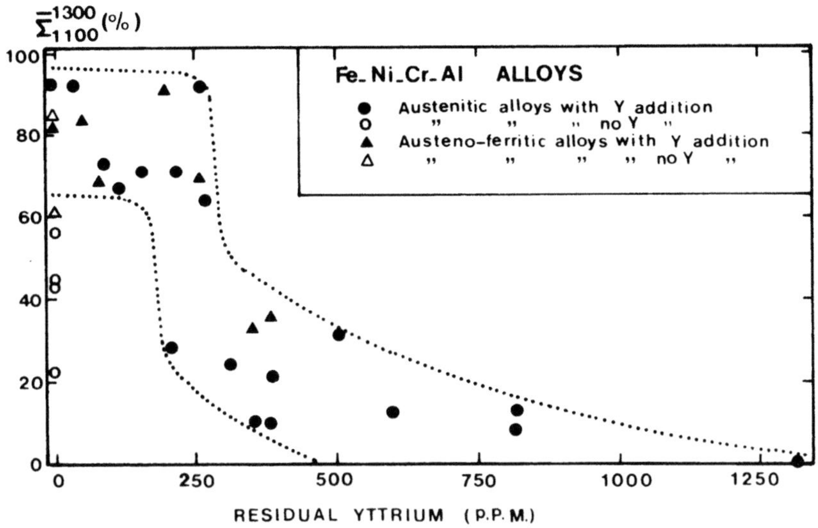

Fig. 5: Effect of Y additions on the tensile hot ductility of as-cast Fe-Ni-Cr-Al alloys (mean reduction of area between 1100 and 1300°C).[3]

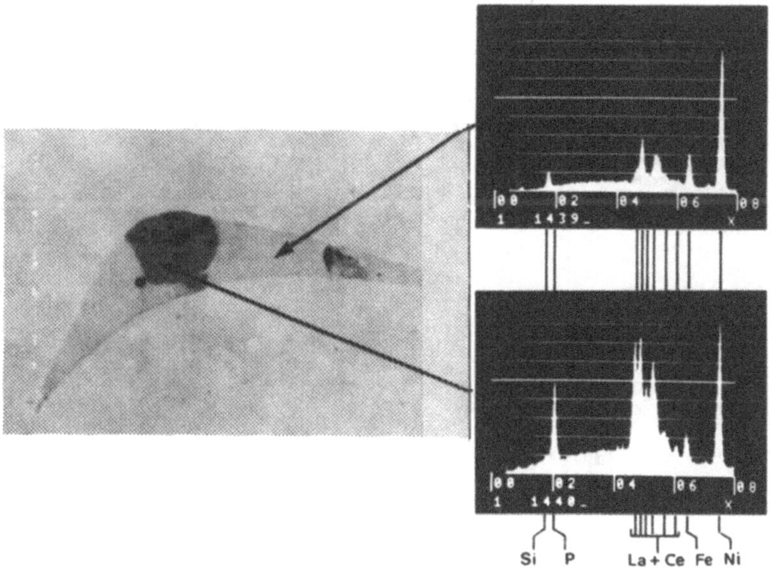

Fig. 6: Active element-rich phases in an industrial Fe-36Ni-18Cr billet which cracked during rolling (~ 1600-1700 ppm total rare earths).
The lenticular morphology is evidence for incipient melting, and the cracks (not shown) followed the grain boundary constituents.

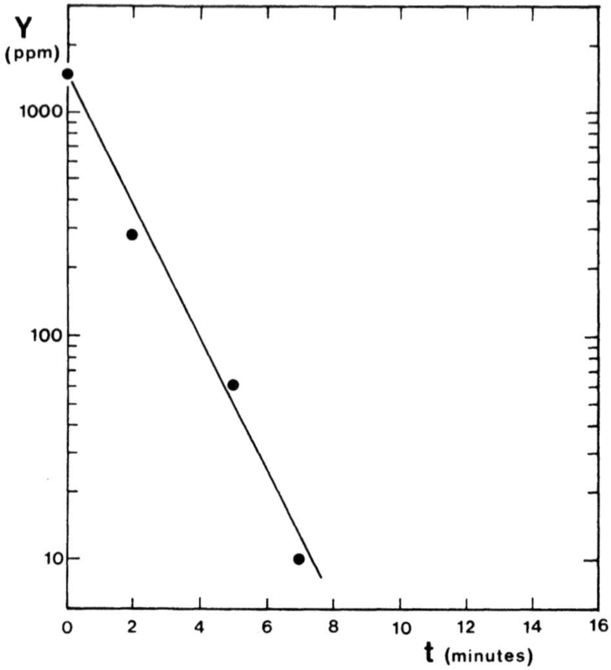

Fig. 7: Kinetics of Y loss in a 35 kg bath of Fe-20Ni-20Cr-5Al alloy exposed to air.

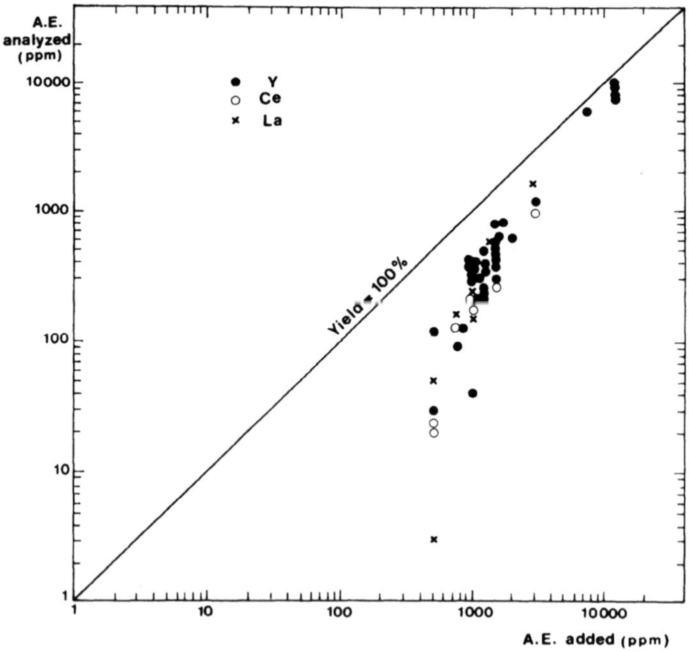

Fig. 8: Yield of active elements in 5 kg ingots of Fe-Ni-Cr-Al alloys melted in argon. Active element addition 2 minutes prior to teeming.

yields are obtained for greater volume/surface ratios and for larger initial additions. Fig. 10 shows the variation in yield from one melt to the next for a given alloy, and indicates a certain saturation, the rate of reaction with the furnace lining gradually decreasing.

Although the detailed mechanisms involved remain poorly understood, it is evident from these results that reaction times must be reduced as much as possible, and even though losses are proportionally not as high in industrial scale melts, the time necessary for making additions, homogenizing and pouring is significantly greater than for laboratory heats. Considerable development work is therefore necessary for each different furnace/alloy/active element combination, and this costly procedure, together with the scatter in results, represents the greatest obstacle to more widespread use of active element additions.

6. CHOICE OF ACTIVE ELEMENTS

The effects of the different active elements on oxidation behaviour are often fairly similar (cf Fig. 1). In contrast, the solubilities, and the influence on workability when they are exceeded, can vary widely.

Little data is available on the solubility of active elements. Figs. 3 and 4, and general industrial experience, suggest that, for a Ni - 20 Cr base, the values for Y lie somewhere between 1000 and 4000 p.p.m., whereas the combined solubilities of Ce, La etc... (mischemetal) are of the order of 1000 p.p.m. For Fe - Cr - Al alloys, the levels are probably of the order of 1000 p.p.m. for Y and 300 to 400 p.p.m. for mischmetal. In austenitic Fe - Ni - Cr - Al alloys, the solubility of Y is about 300 p.p.m., while Fig. 2 indicates considerably lower levels for Ce and La.

The consequences of exceeding the solubility limits depend both on the active element concerned and on the matrix. Thus, whereas workability remains acceptable up to at least 4000-5000 p.p.m. Y in Fe - Cr - Al alloys, ductility drops sharply for effective mischmetal additions greater than about 400 to 500 p.p.m. While complex phase diagrams are generally not available, the examination of simple binaries suggests the cause of these differences. In effect, the binary diagrams involving Fe, Ni or Co on one hand and Ce, La or Y on the other are basically of two types,[5-7] illustrated in Figs. 11 and 12 by the Co - Y and Co - La systems, respectively.[5]

In one group, the first intermetallic which occurs forms via a eutectic reaction, at a temperature which varies from one system to another, but which generally remains fairly high (Fig.

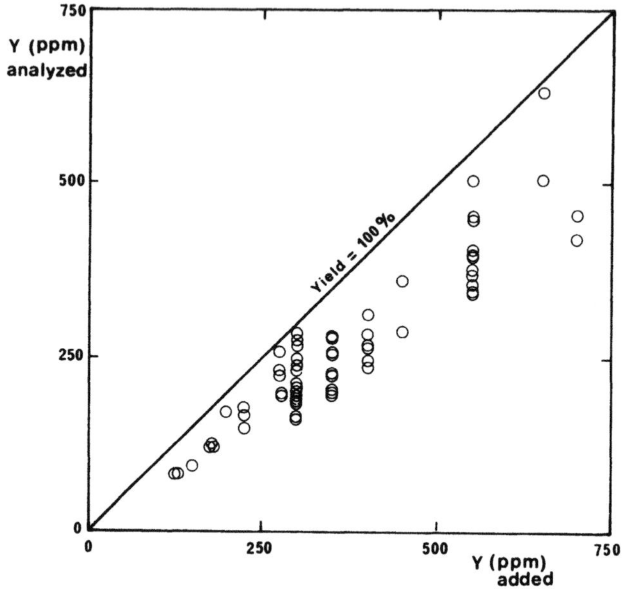

Fig. 9: Yield of Y in 50 kg ingots of Fe-Ni-Cr-Al alloys melted in argon. Active element addition 2 minutes prior to teeming.

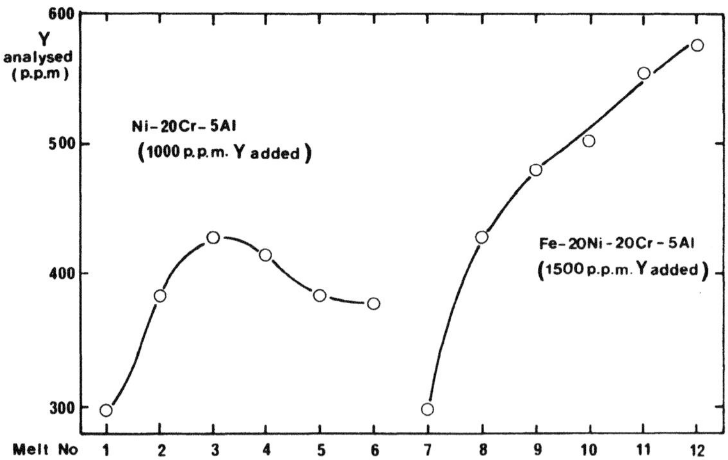

Fig. 10: Variation in Y-yield from one 5 kg melt to the next (2 series of 6 melts). Y-additions in argon 2 minutes prior to teeming.

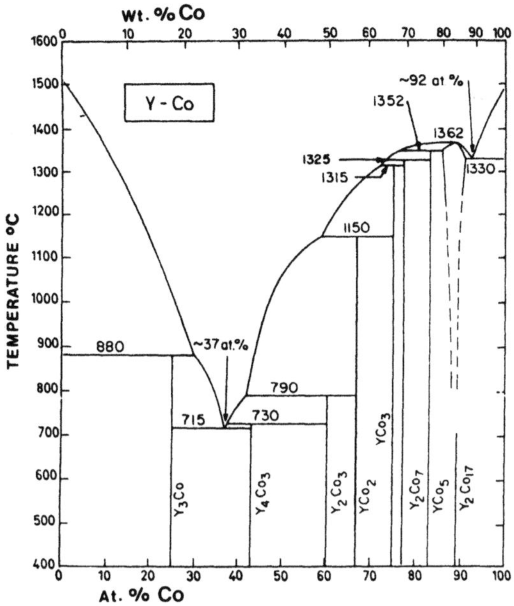

Fig. 11: Y-Co binary system.[5]

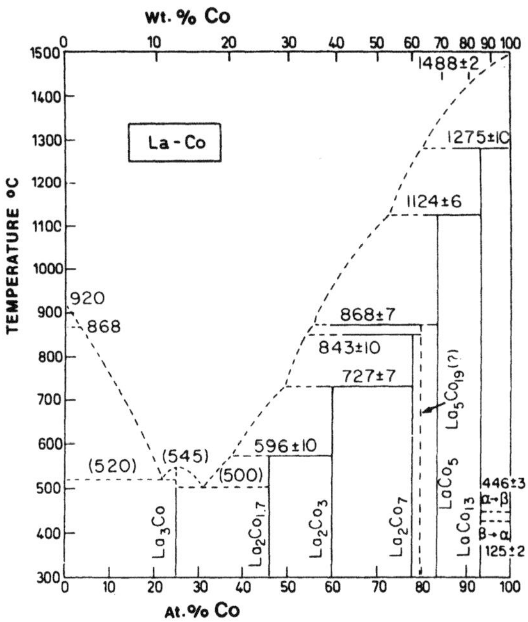

Fig. 12: La-Co binary system.[5]

11). The eutectic represents a limiting composition for areas of local segregation. In contrast, in the other group of systems, there is a peritectic "cascade", so that pronounced segregation can lead to very low liquidus temperatures (Fig. 12). The characteristics of the 9 selected binaries are given in Table 1. The behaviour of yttrium can be seen to be always of the "eutectic" type, whereas cerium and lanthanum fall in this category only for the nickel-based systems, the eutectic temperatures being lower than for yttrium.

Table 1: CHARACTERISTICS OF SELECTED ACTIVE-ELEMENT BINARY SYSTEMS

SYSTEM	PERITECTIC "CASCADE" DOWN TO	HIGH TEMPERATURE EUTECTIC AT
Fe-Ce	640°C	
Ni-Ce		1210°C
Co-Ce	443°C	
Fe-La	785°C	
Ni-La		1245°C
Co-La	500°C	
Fe-Y		1350°C
Ni-Y		1280°C
Co-Y		1330°C

Thus, while yttrium is much more expensive than mischmetal, it is frequently easier to use, with greater solubility levels and higher incipient melting temperatures when these limits are accidentally exceeded.

7. FUTURE REQUIREMENTS

Ever-increasing demands on materials performance will inevitably lead to more widespread use of active elements. However, in order to accelerate this trend, more data is required on their constitutional effects in heat-resisting matrices, enabling rationalisation of their choice and concentration.

Considerable progress is also necessary to adapt melting practice and methods of addition, in order to control residual levels within a fairly narrow range. This is particularly important in view of the continuing tendency for many advanced alloys to be produced by processes such as powder metallurgy, continuous casting or remelting, all of which involve long potential reaction times.

8. REFERENCES

1. B.T. KILBOURN, J. Metals, 1988, 40, (5), 22-25.

2. ANON, Informations Chimie, 1977, 168, 107-117.

3. J.H. DAVIDSON et al., in Proc. Int. Conf. "Behaviour of High Temperature Alloys in Aggressive Environments", (ed. I. Kirman et al.), 209-224, 1980, London, The Metals Society.

4. S.R. KEOWN, in "Stainless Steels '87", 345-356, 1988, London, Institute of Metals.

5. A.E. RAY, Le Cobalt, 1974, 1, 13-20.

6. M. HANSEN, "Constitution of Binary Alloys", Mc Graw-Hill, N. York, 1958, 451-2, 457-8, 887-8.

7. R.P. Elliott, "Constitution of Binary Alloys, First Supplement", Mc Graw-Hill, N. York, 1965, 417-8, 442-3, 677-8.

APPRAISAL KEYNOTE: FUTURE TRENDS AND DEVELOPMENTS

Per Kofstad

Department of Chemistry, University of Oslo,

P.B. 1033, Blindern, 0315 Oslo 3, Norway

ABSTRACT

The paper briefly summarizes important effects of active elements on the oxidation behaviour of metals and alloys - particularly using chromia scales as examples. In the author's evaluation available results suggest that the effects may generally be interpreted on the basis of the following mechanistic model: i) scale growth mainly takes place through grain boundary diffusion and ii) the active elements exert their effects through enrichment/segregation at grain boundaries and thereby on the diffusional transport through the scales, grain growth, high temperature creep, etc. of the scales.

With regard to future trends and developments the author is convinced that the key to understanding effects of active elements - and also many other aspects of protective oxidation - lies in improved understanding of interfaces and grain boundaries (structures, enrichment/segregation phenomena, transport properties) and the resultant effects on growth and properties of protective scales. Such knowledge can only be obtained through the characterization coupled with carefully controlled corrosion studies. It is proposed that a coordinated, cooperative research programme should be initiated in order to secure faster progress in the field and to obtain necessary basic data for developing high temperature materials.

1. INTRODUCTION

My task is to predict future trends and developments. Unfortunately, crystal balls are not always crystal clear, and images they provide are often diffuse and hazy. An alternative is to look back and view the gradual developments over time. With this as a basis we can make some reasonable extrapolations into the future. I shall attempt this approach.

The presence of small amounts of so-called active elements (which now include yttrium and the lanthanides, zirconium, hafnium, thorium a.o.) may greatly improve protective properties of oxide scales in high temperature oxidation of metals and alloys.[1] The effects of active elements have been known for more than 50 years. At that time, and for many years after, the effects were just empirical facts. We had no clear-cut theories or experimental evidence on which to base interpretations. Today, and as we also have learned at this colloquium, we are in a better position. But let me first go back a few years.

During the 1950's and 60's we had one important tool available to interpret the effects of impurities or small alloying additions on properties of oxide scales: that was the semiconductor valence approach to alloy oxidation - sometimes also called the Wagner-Hauffe approach.[1] This is basically correct and can be applied provided all the necessary conditions for its application are fulfilled. However, as we have learned over the years, these conditions are not met in numerous cases of oxidation of metals. And specifically, the approach has not had any success in explaining the effects of active elements.

During the 50's and 60's we also generally assumed that dense or continuous scales grow by lattice diffusion of the ions of the reactants through the scales. Lattice diffusion is important for a few systems where the oxide scales contain relatively large concentrations of point defects. But for highly protective scales, which oxides have low concentrations of point defects and correspondingly low lattice diffusion rates, other transport processes become important. In my view a most important development over the more recent years is the realization - based on solid experimental evidence - that numerous high temperature corrosion processes are to a large extent controlled by diffusion along grain boundaries and other easy diffusion paths. This particularly applies to slow-growing oxide scales consisting, for instance, of chromia and alumina. As we all know, they provide corrosion protection to high temperature superalloys, and their properties are in focus at this colloquium.

While the valence approach is important in interpreting lattice diffusion controlled oxidation, other questions become important in elucidating the effects of impurities, dopants and small alloying additions when grain boundary diffusion controls the oxidation. Do the impurities or dopants become enriched or segregate at grain boundaries? If so, what effects do they have on "defect structures" at grain boundaries, on grain boundary transport, creep phenomena governed by grain boundary diffusion, grain boundary sliding, grain growth, etc. These are all properties and processes of importance in growth and behaviour of scales in high temperature corrosion of metals. In my evaluation these are also pertinent questions in understanding the rôle of active elements in high temperature corrosion.

2. DIFFUSIONAL TRANSPORT IN CHROMIA, OXIDE GROWTH

Knowledge of defects and diffusional transport in oxides provides the basis for understanding high temperature corrosion phenomena. Despite the practical importance of chromia and alumina as protective scales, their defect structures and defect-dependent properties are not well enough understood. One reason for this is that these oxides exhibit very small deviations from stoichiometry and correspondingly the point defect concentrations in these oxides are difficult to measure. Furthermore, lattice diffusion is extremely slow and long time periods are therefore needed to achieve equilibria and to measure equilibrium properties.[1]

Even so, new data have become available during recent years, particularly for chromia. An important example is chromium self-diffusion in chromia. Tracer experiments have shown that chromium diffusion in single crystalline chromia is several orders of magnitude slower than in polycrystalline material.[1 3] Furthermore, values of the parabolic rate constants for growth of chromia estimated form the chromium lattice self-diffusion coefficient are generally orders of magnitude smaller than the experimentally measured values. These results, coupled with studies of growth behaviour and properties of chromia scales, lead to the conclusion that chromia scales grow by grain boundary diffusion. In all probability this also applies to chromia scales on chromia-forming alloys, although in this case the detailed effects of other alloying elements, e.g. Ni, Fe a.o. on diffusional properties of chromia are not known.

In the case of alumina even less is known about the details of the defect structure, but available results on diffusion in alumina, growth behaviour of alumina scales, etc. also strongly indicate that growth of alumina scales predominantly takes place by grain boundary diffusion.[1,4,5]

At this colloquium many other aspects of the growth behaviour of chromia and alumina scales, their adhesion to metal substrates etc. have been summarized.[5] These also serve as an important basis for considering the detailed effects of active elements, but will not be repeated here.

3. EFFECTS OF ACTIVE ELEMENTS

The detailed effects of active elements differ somewhat for chromia and alumina scales. As an illustration let us consider their effects on chromia growth in more detail.

The following effects of active elements on growth of chromia scales have been observed:[1,6,7]

- the rate of growth of chromia is reduced,
- less chromium is needed in the alloys to achieve continuous and protective chromia scales,
- grain growth is impeded in chromia scales,
- spallation and adherence of chromia scales to alloy substrates are improved particularly during thermal cycling,
- the diffusional transport mechanism through chromia scales is changes from predominant outward chromium diffusion to inward oxygen diffusion (in the presence of active elements).

In early studies the active elements were generally added as small alloying additions. (The amount of yttrium as an active elements is of the order of 0.1%. For larger additions yttrium forms intermetallic compounds at grain boundaries, e.g. YNi_9, and these enhance internal oxidation along grain boundaries.[1]) In more recent years numerous studies have shown that the same effects are obtained by ion implantation of the active elements.[7] Similar effects are also achieved by adding the elements as dispersed oxide particles in alloys.[6] Finally, the effects are also observed by coating alloy surfaces with compounds of the active elements, which on heating decompose to the oxides of the oxygen active elements.[8]

Over the years various hypotheses and models have been advanced to explain the effects. The improved adherence of the scales, for instance, has in some models been proposed to be due to:[1,6]

- oxide keying or pegging due to intergranular oxidation at grain boundaries at the alloy surface,
- improved mechanical properties of scale/substrate interface,
- elimination of voids at or near the oxide/alloy substrate interface due to the presence of dispersed oxide particles serving as vacancy sinks.

Although these and other models are used in numerous discussions in papers dealing with the rôle of active elements, none of these have been generally accepted.

In order to obtain a more basic understanding of the effects of active elements, it is necessary to know exactly where they are located on an atomic scale. Are they dissolved in the oxide lattice in the scales or are they located at the grain boundaries and/or the interfaces? Where and in what way do they exert their effects?

It is important to note that it can be experimentally difficult to pinpoint their exact location. If, for instance, they segregate or become enriched at grain boundaries, they may well be confined to a cross-sectional area which may be less than 50Å wide. This in turn, means that conventional x-ray diffraction and scanning electron microscopy with energy dispersive analysis will generally fail to detect their presence or their exact distribution and location. Rather it is necessary to use advanced, high resolution surface analytical techniques. And in this respect the use of TEM with associated energy dispersive analysis and selective area electron diffraction, SIMS, etc. have in recent papers provided detailed and important information.[7,9]

A general result of such studies is that yttrium is found to be present as particles (e.g. Y_2O_3, $YCrO_3$) in the scales, and even more importantly, that yttrium segregates or becomes enriched at the oxide grain boundaries.[7,9] It is interesting to correlate these results with the fact that oxide growth takes place by grain boundary diffusion and that the diffusional transport is changed from predominant chromium to predominant oxygen diffusion (with active elements). From this it is reasonable to conclude that the enrichment/segregation of yttrium (and other active elements) is responsible for the change in the diffusional transport along grain boundaries. In this respect it is also interesting to note that the oxides of the active elements (Y_2O_3, La_2O_3, CeO_2, ZrO_2, HfO_2) are all oxides in which oxygen diffusion is much faster than metal diffusion. The same applies to rare earth chromites. An overall effect of this change in diffusional transport may also be that the overall rate of oxide growth is reduced.

Segregation of yttrium (and other active elements) may correspondingly explain the reduced rate of grain growth in the scales. The small grain size may, in turn, have a beneficial effect on the mechanical properties (resistance to cracking) of the scales.

Why do active elements improve adhesion (decrease spallation) of the scales? This is a complex question that probably involves several aspects at or near the scale/metal interface. But it is not unreasonable to think that this in gross terms may be related to the change in diffusional growth mechanism. When outward chromium diffusion predominates, voids and pores are eventually formed at or near the scale/metal interface due to the shrinking of the metal phase and the inability of the scale to deform sufficiently to maintain adherence of the scale to the metal substrate. However, when the growth takes place by inward oxygen diffusion, oxide formation takes place at the metal/scale interface, and in principle voids should then not be formed. This may constitute a general explanation.[1]

In considering spallation of the scales we commonly use the term loss of adhesion of the scales to the metal substrate. But in most cases scales do not fail exactly at the metal/scale interface, but rather in the scale near the interface. Thus spallation is often a result of cohesive rather than an adhesive failure. This overall problem is in need of improved understanding.

So far no mention has been made of the sulfur effect. The basis for this is that sulfur, present as an impurity in many alloys or included in the scales during exposure to sulfur-containing atmospheres, accumulates and segregates at the metal/scale interface.[10,11] It is argued that this in turn, embrittles the interface region and makes the scale more prone to decohesion/cracking/spallation. It is proposed that the beneficial effect of yttrium is that it has a gettering effect on sulfur or reacts with the sulfur to yttrium-containing sulfides and thereby removes the embrittling effects. But it also appears that sulfur may have additional effects in that the rate of oxide growth may also be increased when a sufficiently high amount of sulfur is present in the alloy or in the scale.[12]

4. FUTURE TRENDS AND DEVELOPMENTS

The above examples illustrate developments that have taken place in the field of active elements. What extrapolations can we make?

In my evaluation the key to understanding the rôle of active elements lies in the atomistic domain. Available data strongly suggest that grain boundaries and interfaces, their structures, compositions and properties play a decisive rôle in many part-processes in the overall oxidation behaviour. These part-processes include grain boundary diffusion, high temperature creep and grain boundary sliding, grain growth a.o. The understanding of these properties requires knowledge of the distribution and location of the active elements on an atomic scale and that we are able to identify enrichment/segregation phenomena at grain boundaries and interfaces. Experimentally this requires the use of high resolution surface analytical techniques. It also appears necessary to couple such experimental studies with atomistic modelling of the processes involved.

Although these comments apply to the rôle of active elements, the same approach is necessary to understand the rôles of other elements. An example in case is sulfur. It is established that sulfur tends to accumulate at the metal/oxide interface. But it has also been shown that sulfur affects the growth rates of scales. Does excess sulfur become enriched at grain boundaries in the scales and thereby affects the grain boundary transport? These aspects should be further studied and elucidated.

Another important aspect of the oxidation behaviour of chromia- and alumina-forming high temperature alloys is to get a more detailed understanding of the oxidation behaviour in the absence of active elements. The scales of such alloys are often described as consisting of chromia and alumina. But detailed studies show that additional phases may also be present. Furthermore, it is also well known that different alloys may oxidize at considerably different rates and exhibit differences in morphology and growth behaviour of scales. A guestion that has not been sufficiently studied is what effects other alloying elements have on growth properties of the scales. By way of example, do "chromia" scales on Fe-Cr alloys have the same properties as "chromia" scales on Ni-Cr alloys? It is not unreasonable to speculate that nickel or iron will become enriched at grain boundaries in the chromia and in this way affect the scale properties. More basic information is needed.

These are important problems to elucidate and also represent suggestions for future work. To me these developments in the field of high temperature oxidation also reflect a general trend in materials research in general: the gap between basic research and applied research and development is gradually becoming smaller. In the same terms, the further development of advanced engineering materials is based on an atomistic understanding of structures, compositions, and transport properties of materials.

Further improvements of high temperature resistant alloys and coatings require - as emphasized above - the use of high resolution methods for materials characterization in studies of oxidation behaviour. Relatively few laboratories have as yet available equipment - and particularly well trained personnel -

for such research. In order to make good use of available facilities and achieve further developments, efforts should be made to establish a cooperative European programme in the field. A tentative aim and title for this may be: "Improved high temperature materials through microstructural and atomistic design and engineering".

REFERENCES

1. KOFSTAD, P., "High Temperature Corrosion", Elsevier Applied Science, London/New York, 1988.

2. ATKINSON, A., this conference.

3. KOFSTAD, P. and LILLERUD, K.P., Oxid. of Met., 1982, 17, 127,177.

4. WOOD, G.S. and STOTT, F.H., in "High Temperature Corrosion", R.A. Rapp, Ed., NACE, Houston TX, 1983.

5. STOTT, F.H., this conference.

6. WHITTLE, D.P. and STRINGER, J., Philos. Trans. R. Soc. London, Ser. A, 1980, 295, 309.

7. BENNET, M. and MOON, D.P., this conference.

8. ECER, G.M., SINGH, R.B. and MEIER, G.H., Oxid. of Met., 1982, 18, 55.

9. YUREK, G.T., PRZYBYLSKI, K. and GARRAT-REED, A.J., J. Electrochem Soc. 1987, 134, 2643.

10. SMEGGIL, J.G., this conference.

11. SMEGGIL, J.G. and PETERSON, G.G., Oxid. of Meta., 1988, 28, 103.

12. KHANNA, A.S., QUADAKKERS, W.J. and WASSERFUHR, C., this conference.

GPSR Compliance

The European Union's (EU) General Product Safety Regulation (GPSR) is a set of rules that requires consumer products to be safe and our obligations to ensure this.

If you have any concerns about our products, you can contact us on ProductSafety@springernature.com

In case Publisher is established outside the EU, the EU authorized representative is:

Springer Nature Customer Service Center GmbH
Europaplatz 3
69115 Heidelberg, Germany

Batch number: 10607330

Printed by Printforce, the Netherlands